MOLECULAR
POPULATION GENETICS

Not for Profit. All for Education.

Oxford University Press USA is a not-for-profit publisher dedicated to offering the highest quality textbooks at the best possible prices. We believe that it is important to provide everyone with access to superior textbooks at affordable prices. Oxford University Press textbooks are 30%–70% less expensive than comparable books from commercial publishers.

The press is a department of the University of Oxford, and our publishing proudly serves the university's mission: promoting excellence in research, scholarship, and education around the globe. We do not publish in order to generate revenue: we generate revenue in order to publish and also to fund scholarships, provide start-up grants to early-stage researchers, and refurbish libraries.

What does this mean to you?
It means that Oxford University Press USA published this book to best support your studies while also being mindful of your wallet.

Not for Profit. *All* for Education.

As a not-for-profit publisher, Oxford University Press USA is uniquely situated to offer the highest quality scholarship at the best possible prices.

OXFORD
UNIVERSITY PRESS

MOLECULAR POPULATION GENETICS

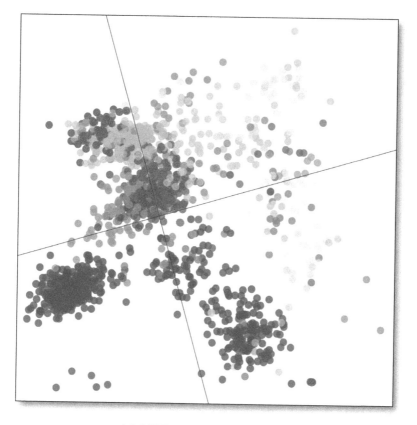

MATTHEW W. HAHN
Indiana University

 SINAUER ASSOCIATES

NEW YORK   OXFORD
OXFORD UNIVERSITY PRESS

Address editorial correspondence to:
Sinauer Associates
23 Plumtree Road
Sunderland, MA 01375 U.S.A.
publish@sinauer.com

Address orders, sales, license, permissions, and translation inquiries to:
Oxford University Press U.S.A.
2001 Evans Road
Cary, NC 27513 U.S.A.
Orders: 1-800-445-9714

Library of Congress Cataloging-in-Publication Data
Names: Hahn, Matthew William, 1975- author.
Title: Molecular population genetics / Matthew W. Hahn.
Description: New York : Oxford University Press ; Sunderland, MA : Sinauer
    Associates, [2018] | Includes bibliographical references and index.
Identifiers: LCCN 2017053388 | ISBN 9780878939657 (paperbound)
Subjects: | MESH: Genetics, Population | Molecular Biology | Genetic Phenomena
Classification: LCC QH352 | NLM QU 450 | DDC 577.8/8—dc23
LC record available at https://lccn.loc.gov/2017053388

Pod.

For Leonie, Tristan, and Asher,
my favorite people

CONTENTS

CHAPTER 5
POPULATION STRUCTURE   79

CHAPTER 8
LINKED SELECTION 165

CHAPTER 9
DEMOGRAPHIC HISTORY   203

CHAPTER 10
POPULATION GENOMICS   249

PREFACE

Molecular population genetics is one of the most active and exciting fields in biology, combining advances in molecular biology and genomics with mathematical and empirical findings from population genetics. Its results are regularly presented in top journals, and there are multiple scientific meetings that feature work in this field every year. Yet the bar for entry into molecular population genetics is almost unbelievably high. For instance, there are at least five different meanings attached to statistics represented by the letter "D," there are almost as many meanings of the commonly used term "allele," and researchers regularly and without explanation rename each other's terms to fit their own system of notation. There are widely known, uncorrected errors in the literature, and practitioners can sometimes appear to be speaking a secret language known only to other molecular population geneticists. Despite the many excellent books covering the field of population genetics as a whole, none has brought together the methods and results of molecular population genetics in a single volume. While this book is not intended to codify molecular population genetics, hopefully it will serve as a record of the field as it stands now.

This book is my attempt at turning a one-semester graduate course into a practical text that can be used as a reference by novices and experts alike. It came about because I was forced to cobble together pieces of many different books, reviews, and primary research articles into a coherent users' guide to the field of molecular population genetics. Without trying to teach such a course I don't think I would have remembered just how hard it was for me to pick up many of the pieces by myself earlier in my career. But once the idea of a book took hold, I couldn't stop noticing and noting all of the complex ideas, methods, and terminology being used without sufficient explanation. And so, nine years later, here we are.

Molecular population genetics is in many ways distinct from the other two main branches of the field: classical population genetics and quantitative genetics. Since the time of R. A. Fisher, Sewall Wright, and J. B. S. Haldane, classical and quantitative population genetics have been largely concerned with predicting the course of evolutionary change over time. Whether this change is examined in terms of allele frequencies (Δp) or phenotype frequencies (Δz), these two disciplines consider the effects of evolutionary forces such as natural selection, genetic drift, mutation, and migration on the future trajectory of populations. Molecular population genetics, on the other hand, is explicitly a historical science, concerned with inferring the actions of evolutionary forces that have operated in the past. Knowledge of the larger field of population genetics is necessary for any researcher—after all, we would not be able to understand the effects of natural selection on molecular data without understanding its effects on individual mutations or phenotypes—but molecular population genetics is largely devoted to analyzing data that tell us about the past rather than to predicting the future. We are, in Norman Johnson's words, "Darwinian detectives."

Researchers use molecular population genetics techniques to address three main types of problems. First, for those lucky enough to have identified the gene or mutation underlying a phenotypic trait of interest, molecular variation can be used to understand the forces that have acted on the phenotype. Few methods can tell us so directly about how evolution has acted in the past on a phenotype. Second, molecular population genetics is used to examine the relative effects of evolutionary forces across the genome. While this may be an indirect indicator of selection on phenotypes, the questions asked in such studies are generally concerned with the forces maintaining genetic variation, whether or not there is a direct connection to phenotypic variation. Finally, many studies utilize molecular variation to infer the history of populations. Molecules can tell us about the movement of individuals and the timing of population demographic shifts and can therefore inform our understanding of recent evolutionary history. As sequencing technologies have become cheaper and faster—and more and more loci can be sequenced as a result—these three areas have quickly been converging: in order to understand selection on individual genes we must understand the general effects of selection across the genome; in order to understand both of these we must also understand the historical context in which selection acts and the effects of this history on molecular variation. My hope is that this guide can be used by researchers asking about any one of these three areas, and that the techniques and methods used by one area can easily be picked up for use in another.

As I require the students in my molecular population genetics class to write programs that implement many of the methods covered, I have also written this book with the intention of providing enough detail so that anybody can do the same. While I do not think the casual user will necessarily need this level of detail (there are certainly enough software packages available to carry out these analyses, e.g., Librado and Rozas 2009; Excoffier and Lischer 2010), I also believe that good scientists must understand much more than how to use a pull-down menu. This being said, I am fairly superficial in my theoretical treatment of many areas—there are people more qualified than I who have written excellent mathematically oriented texts on all of the subjects covered here. Hopefully the many references to these papers and books given in each chapter will point those looking for more detail in the right direction.

I have assumed that readers of this book have a basic understanding of classical population genetics and evolutionary thinking. Without such knowledge, it will be very difficult to understand the basis for many of the methods presented here. With that knowledge, the book can either be read straight through or used as a reference for specific subjects. There are of course a huge number of methods in molecular population genetics, many more than could possibly be covered in a coherent manner. I have tried to be as inclusive as possible, but I am sure there are important papers and important areas of research that I have failed to acknowledge. I apologize in advance to all of the authors of these studies and hope there is a chance to include them in future editions of this book.

ACKNOWLEDGMENTS

There were many people who made this book possible. I have been lucky enough to have had some wonderful advisors along the way, including Rick Harrison, Mark Rausher, Chuck Langley, and John Gillespie. I have also had some equally wonderful unofficial mentors, including Sara Via, Nelson Hairston, Jr., Greg Wray, Sergey Nuzhdin, and Dave Begun. John Willis, Marcy Uyenoyama, and Cliff Cunningham all also contributed to my knowledge in this area at an early stage in my career. I am continually amused at the fact that I can now call these people my colleagues. Andy Kern, Dick Hudson, Dan Schrider, Matt Rockman, and Leonie Moyle all read through the entire book and offered excellent advice, opinions, and corrections that immeasurably improved the content. This was a big task, and I really appreciate all the work they put into it. A number of people also read individual chapters, including Mike Whitlock, Martin Lascoux, Justin Fay, John Wakeley, Bret Payseur, John Pool, Casey Bergman, Yaniv Brandvain (and his lab), John Novembre, and Beth Shapiro. David Rand gave helpful feedback on early versions of all the chapters. Many people answered my scientific and/or historical questions, including Debbie Nickerson, Joe Felsenstein, Justin Fay, Magnus Nordborg, Walter Bodmer, John Wakeley, Guillaume Achaz, Peter Tiffin, Brandon Gaut, Michael Nachman, Bruce Weir, Chuck Langley, Joel McGlothlin, Andy Kern, Kevin Thornton, Benjamin Peter, Dave Begun, Chip Aquadro, John Kelly, and Noah Rosenberg. Needless to say, all of these people are absolved of responsibility for any remaining errors. Grad students and postdocs in my lab read through various versions of the text, tracked down results and references for me, made the occasional figure for me, and have simply been great to have around; I am especially grateful to Dan Schrider, Mira Han, Melissa Toups, James Pease, Tami Cruickshank, Fábio Mendes, Jeff Adrion, Gregg Thomas, Geoffrey House, and Rafael Guerrero for this help. The faculty at Indiana University have always been supportive, and I have all of them to thank for the amazing environment in which I work. Emily Bernhardt gave me an especially useful piece of advice about writing a book, which I have tried to follow. At Sinauer Associates, first Azelie Fortier and more recently Rachel Meyers have been wonderful editors, and both they and Andy Sinauer have always expressed complete confidence that I would eventually finish this project, even when there appeared to be little basis for this belief. They are a great group of people to have written this with.

Finally, as with much of what I write, Leonie Moyle has provided feedback, advice, criticism, and lots of encouragement. She has had a big influence on my approach to research and teaching and is a wonderful person to share life with outside of work. I doubt I would have finished this book without her.

Bloomington, Indiana
August 2017

MODELS OF EVOLUTION

Evolution begins as one mutation, on one chromosome, in one individual. Molecular population genetics is the study of such mutations as they rise and fall in frequency in a population. Various evolutionary forces can accelerate or impede the movement of a mutation through the population, and the action of these forces can be inferred from patterns of molecular variation among individuals.

Although the use of genetic markers goes back to the discovery of the ABO blood groups in 1900, the beginning of true "molecular" population genetics dates to the groundbreaking studies of Harris (1966) and Lewontin and Hubby (1966). These researchers showed that there was much more variation among individuals at the molecular level than had previously been anticipated from studies of morphological phenotypes. However, these studies used allozymes (a portmanteau of "allele" and "enzyme"; Prakash, Lewontin, and Hubby 1969) to uncover molecular variation and were therefore still only observing a fraction of all variation—those mutations that caused proteins to run at different speeds through a gel because of a change in electrical charge. It was not until 1983 that the first studies of molecular variation at the nucleotide level appeared (Aquadro and Greenberg 1983; Kreitman 1983). By sequencing every nucleotide, these studies allowed us to fully observe the genetic variation segregating in natural populations.

Molecular population genetic studies ask a wide range of questions about the effects of evolutionary processes on natural populations. In order to do this, they generally use DNA sequences from a small sample of individuals to understand the forces that have acted on the whole population. The pattern of genetic variation obtained from even a single locus can be used to make inferences about the forces of mutation, recombination, and natural selection, as well as details of the demographic history of a population—for instance, its relative size or history of migration. These inferences can be made because a large body of population genetics theory has been developed in the past 100 years that tells us what we should expect to observe when each of these forces is acting. Early theoretical research in population genetics was not done with molecular data in mind, but the rise of molecular methods has inspired a growing body of work explicitly concerned with modeling the processes of molecular evolution.

Molecular population genetics theory is essential in making inferences from DNA sequence data, and it is therefore important to at least review the predominant models and their assumptions here. The brief introduction provided in this chapter is not intended to cover the fundamentals of population genetics, and it assumes that the reader is familiar with many basic concepts. Instead, this chapter attempts to distill the most relevant theory and models that are applied to molecular population genetic data. These models are those used in the field to make inferences from sequence data, so understanding their structure is key to understanding how these inferences are being made. Additionally, this chapter attempts to clarify the often-confusing way in which common terms are used in this field and defines the way in which they will be used throughout the book. It closes with a discussion of the neutral theory of molecular evolution, attempting to explain the meaning of this theory and some of its common misconceptions.

BASIC SEQUENCE TERMINOLOGY

The DNA sequence data collected in molecular population genetic studies will resemble the alignment shown in **FIGURE 1.1**. This alignment shows four DNA sequences, each with 15 nucleotides, from the same *locus* (location) on a chromosome. I will refer to the four homologous DNA strands as *sequences* or *chromosomes*, because the data come from four unique homologous chromosomes, whether or not the sequences themselves are unique. This terminology will be used throughout the book, but it should be noted that there are many terms used in the literature for the four DNA sequences, including *genes*, *alleles*, *samples*, *cistrons*, and *allele copies*. Use of the word *genes* to imply multiple sequences sampled at a single locus is not as common as it was 20 years ago, especially now that individual researchers regularly collect multiple sequences from multiple genes within a species. But many studies still use the term *allele* to refer to each sampled chromosome, in effect using the "different by origin" definition of allele (Gillespie 2004, pp. 6–8). I will use the word *allele* only to refer to the individual nucleotides (or amino acids) when they differ at a single position in the alignment, as in the A or C alleles. This usage is referred to as the "different by state" definition of allele. So we might say that there are $n = 4$ chromosomes in the alignment in Figure 1.1. Note that this terminology does not depend on whether the four sequences come from two random diploid individuals, four haploid individuals, or four separate inbred (isogenic) diploid lines. In all cases, we have still sampled four chromosomes from nature.

Within the alignment we can see that there are differences among sequences at several positions, with one nucleotide or another present in different individual chromosomes. We will mainly focus on biallelic sites because they are the most common type of variation observed, although there can be more than two alleles

```
1  2  3  4  5  6  7  8  9  10 11 12 13 14 15
T  T  A  C  A  A  T  C  C  G  A  T  C  G  T

T  T  A  C  G  A  T  G  C  G  C  T  C  G  T

T  C  A  C  A  A  T  G  C  G  A  T  G  G  A

T  T  A  C  G  A  T  G  C  G  C  T  C  G  T
```

FIGURE 1.1 An alignment of four sequences. In this example the sample size is $n = 4$, with each sequence having a length of $L = 15$. In total there are six sites with nucleotide differences among the sequences (at positions 2, 5, 8, 11, 13, and 15), so $S = 6$.

at any position in the alignment. There are many different terms used to describe these DNA differences: we can say that there are six *polymorphisms* or *segregating sites* or *mutations* or *single nucleotide polymorphisms (SNPs)* in our sample (at positions 2, 5, 8, 11, 13, and 15). The terms *polymorphism* and *segregating site* have been the most common terms historically, though *SNP* (pronounced "snip") has become much more common recently. (The earliest use of the abbreviation SNP dates to Nikiforov et al. [1994].) The set of alleles found on a single sequence is referred to as a *haplotype*.

Different fields treat the term *mutation* quite differently. It can be used to mean the process by which a change in DNA takes place or the new allele generated by this process. Sometimes *mutation* is used as a synonym for any polymorphism, or refers only to rare polymorphisms (those that occur <1–5% of the time or are present in only a single sequence) in more medically oriented population genetics (Cotton 2002). Because all polymorphisms must arise as mutations initially, and because I will be discussing the evolutionary origin of variation in this book, I attempt to confine my use of *mutation* to mean the variation-generating process and the new mutations this process generates. Finally, I will reserve the term *substitution* to mean only those DNA differences observed between species, as distinct from variation within species.

We generally do not consider insertion/deletion (indel) polymorphisms as segregating sites, though indels 1 base pair (bp) in length are sometimes included in this classification. The reason for this is that it is difficult to count the actual number of differences between two sequences with multinucleotide indels—is a 2-bp indel counted as one polymorphism or two? The answer depends on whether we think that a single 2-bp mutation or two separate 1-bp mutations have occurred. Alignment columns with indels or missing data of any kind are often not considered in analyses, and therefore the value of the sample size (n) may differ across sites (see Chapter 3).

MODELS OF EVOLUTIONARY PROCESSES

Models of populations

In all populations, for all polymorphisms, *genetic drift* acts to change allele frequencies. Drift is simply the stochastic change in allele frequencies that is due to the finite nature of all populations and occurs because in each new generation some chromosomes leave more descendants than others. Drift differs from natural selection (a deterministic force) because of the fact that there is no consistent difference among alleles or genotypes in their numbers of offspring, and therefore each individual allele does not consistently rise or fall in frequency as a result of drift alone.

Models of drift are necessarily models of how individuals within populations replace themselves from generation to generation. One of the most commonly used models is the *Wright-Fisher model* (Fisher 1930b; Wright 1931). This model imagines a population of constant size with N diploid hermaphrodites; we require them to be hermaphrodites (i.e., monoecious) so that all individuals can mate with one another, but the model can be extended to populations with different sexes. Since individuals are diploid, there are $2N$ chromosomes in the population every generation for autosomal loci. If our model included sex chromosomes and equal numbers of individuals from two separate sexes, there would be $1.5N$ X or Z chromosomes, $0.5N$ Y or W chromosomes, and $0.5N$ mitochondrial or chloroplast genomes, depending on the biological system being studied. In order to form the next generation of individuals, we will assume that individuals mate at random and that chromosomes are sampled uniformly with replacement to leave descendants. No individuals survive into the next generation—instead, the entire population is replaced by its descendants. This model is most applicable to species with non-overlapping generations, such as annual plants or insects that appear once a year (annual vertebrate species are rare but do exist; see, e.g., Karsten et al. 2008).

To see the effect of drift on changes in allele frequencies in the Wright-Fisher model, consider a single nucleotide position with two alleles, A_1 and A_2. In generation t there are i chromosomes carrying allele A_1, which is at frequency $p_t = i/2N$. This implies that there are $2N - i$ chromosomes carrying allele A_2, which is at frequency $q_t = 1 - p_t$. The sampling of chromosomes for the next generation is equivalent to sampling from a binomial distribution with parameters $2N$ and $i/2N$. Therefore, the mean and variance of p in the next generation for the Wright-Fisher model are:

$$E(p_{t+1}) = p_t \tag{1.1}$$

$$Var(p_{t+1}) = p_t q_t / 2N \tag{1.2}$$

where $E(\bullet)$ represents the expectation (mean) of a random variable and $Var(\bullet)$ represents the variance. These equations say that when only drift is acting (no mutation and no selection), mean allele frequencies are expected to stay the same over time. Because the expected change in allele frequency is 0, no predictions can be made about the rise or fall of any particular allele. On the other hand, the variance in the process is directly related to the population size, such that there will be predictably larger changes in allele frequency in smaller populations and with intermediate allele frequencies. Importantly, even though no mean change is expected, independent populations starting at the same allele frequency will inevitably begin to differ in their average allele frequencies, leading to evolutionary divergence. Alleles will drift toward frequencies of 0 or 1, at which point the allele that is at frequency 1 will be said to be "fixed." Once fixation of one or the other allele occurs, no further change is possible because one of the two alleles has been lost from the population.

A related consequence of these changes is that the level of genetic variation in a population is expected to decline when drift is the only force acting. If we define *heterozygosity* as the probability that two chromosomes chosen at random will have different alleles, then the expected amount of heterozygosity for a biallelic locus in a randomly mating population is $2pq$ (this idea is covered in greater depth in Chapter 3). As one allele becomes more common than the other, heterozygosity will decline (because it is maximized at $p = q = 0.5$); in the Wright-Fisher model the expected heterozygosity declines as a function of $1/2N$ per generation. Although the decline in heterozygosity is not a direct measure of change in allele frequency, the above result provides insight into the very slow rate at which allele frequencies change when drift is the only force acting.

A second population model—more realistic than the Wright-Fisher in some respects and more mathematically tractable for some purposes—is the *Moran model* (Moran 1958). In the Moran model individuals of different ages can co-exist, and we do not wholesale replace the population with new individuals each generation. Strictly speaking, the Moran model only applies to haploid populations, but for ease of comparison with the Wright-Fisher model we will consider a constant-size population of 2N haploid individuals. At a given time point a single individual is chosen at random to reproduce, and a second individual (not necessarily different from the first) is chosen at random to die. At the next time point the new offspring individual as well as all surviving individuals are available to reproduce, and again one individual is chosen to reproduce and one is chosen to die. If we were to repeat this birth-and-death step for 2N time points, we would have the equivalent of a single generation in the Wright-Fisher model. This is because, on average, each individual will have been replaced; however, some individuals will live less than a "generation" and some individuals will live longer.

Changes in allele frequency under the Moran model occur when an individual carrying one allele is chosen to reproduce and an individual carrying the alternative allele is chosen to die. It can of course also be the case that the individual chosen to reproduce and the one chosen to die carry the same allele, and in this situation there will be no allele frequency change. After 2N time points repeating the birth-and-death operations, we can ask what the mean and variance of allele frequencies are in the next generation. Again considering a biallelic locus of the same type described for the Wright-Fisher model, the mean and variance of the allele frequency p in the next generation for the Moran model are:

$$E(p_{t+1}) = p_t \qquad (1.3)$$

$$Var(p_{t+1}) = 2p_t q_t / 2N \qquad (1.4)$$

As with the Wright-Fisher model, no change in the mean allele frequency is expected. However, the variance in allele frequencies in a Moran population is twice as large as that in a Wright-Fisher population. This is due to the fact that the variance in offspring number per individual is twice as large in the Moran model. An intuitive explanation for this increased variance in offspring number relative to the Wright-Fisher model is the variance added by individuals who reproduce at variable numbers of time points (i.e., who are "alive" for different amounts of time). A consequence of this difference is that there is twice as much drift in the Moran model, and therefore heterozygosity is lost at twice the rate $(1/N)$. Evolution by drift is still very slow in the Moran model, but it is twice as fast as in the Wright-Fisher model.

Neither of the two population models described here are realistic for most species, and for some applications there are other more realistic models used in population genetics—notably, the *Cannings model* (Cannings 1974). The Cannings model can have any arbitrary variance in offspring number and is a generalization of the Wright-Fisher model. However, the Wright-Fisher model both is immediately intuitive and allows for the derivation of many important evolutionary results. As we will see in the next section, it also acts as a touchstone for results derived under other population models, serving as a model to which all others can be compared.

The effective size of a population

A central concept in population genetics is the *effective population size*, usually denoted N_e (Wright 1931). In contrast to the census population size (which is simply a count of the number of individuals at a given time), the effective population size is an abstract value that allows real populations to be modeled as Wright-Fisher populations with the equivalent amount of genetic drift. Because there are many forces that increase the variance in offspring number in natural populations beyond the value expected in a Wright-Fisher model, the effective population size is usually smaller than the census population size.

The effective size of a population gives us a way to compare different populations and species to each other in a single frame of reference—namely, the amount of drift expected in an idealized Wright-Fisher population. In this way, we have a single value that helps to quantify the role of genetic drift in determining the effectiveness of mutation, selection, recombination, and migration. We can even conceive of different effective population sizes for different regions of the genome, each with a history equivalent to that which would be seen in a Wright-Fisher population with different levels of genetic drift. These differences can then be manifested as differences in many evolutionary processes across the genome (see Charlesworth 2009 for a review).

However, effective population size can also be a vague and widely misunderstood concept as a result of both its definition and its application. One problem is that there are at least four ways in which the effects of drift can be characterized, and therefore four different ways in which any given population could be equivalent to some aspect of drift in a Wright-Fisher population. This leads to multiple definitions of effective population size: the variance effective size, the inbreeding effective size, the eigenvalue effective size, and the coalescent effective size. In equilibrium populations that meet the assumptions of the Wright-Fisher model (and in many nonequilibrium populations), these measures of the effective population size will be equal. But in some populations with nonequilibrium histories these effective sizes can differ substantially from one another and can even be undefined—that is, there is no equivalent Wright-Fisher population in these cases (Ewens 2004). For molecular sequence data the most applicable effective size is likely the coalescent effective population size, although it is still undefined in some circumstances (Sjödin et al. 2005; Wakeley and Sargsyan 2009). Under this definition we are equating the amount of genetic drift to the rate of coalescence (see Chapter 6).

The concept of an effective population size is also commonly misused, or at least its meaning and application are broadened unnecessarily. As stressed by Ewens (2004, p. 38), "it would be more indicative of the meaning of the concept if the adjective 'effective' were replaced by 'in some given respect Wright-Fisher model equivalent.'" The essential connection of N_e to an idealized Wright-Fisher population model (and specifically to drift in this model) is often missed or assumed to be more important than it really is. As discussed in the previous section, the Wright-Fisher model is fundamental to much of theoretical population genetics, acting as a central comparator in understanding the behavior of many different populations of interest. But the size of this Wright-Fisher population is not equivalent to the effective number of breeding individuals, the number of individuals contributing to the next generation, or any "real" population size at all. We could just as easily have defined the effective size as that equivalent to a Moran population with the same amount of genetic drift, though in this case the numerical value of N_e would be half as large. This all means that while the numerical value of N_e is not extremely useful, the rank-order values among regions of the genome or among organisms may still tell us about the relative expected strength of drift and selection (see Chapter 3).

Models of mutation

There are many models of the process of DNA mutation. Multiple models are needed both because there are many different types of mutations and because different molecular biology techniques have provided varying amounts of information about the underlying variation. Some models are used simply because they are more mathematically tractable. It is important to realize that these are not specific molecular models of how mutations arise in the germline (e.g., how the DNA polymerase and associated proofreading enzymes incorporate incorrect bases). Instead, these models are intended to explain in general quantitative ways how the polymorphisms detected in current-day samples arose many generations ago and how different rates of mutation can result in different numbers of observed polymorphisms. Population genetic models of mutation can also be simpler than many phylogenetic models of this process, largely because we expect to find many fewer differences between sequences being compared.

Despite the differences among population genetic models of mutation, there are also many similarities. All assume that mutation is random, although what is meant by *random* may differ from general usage. Mutations in nature are highly nonrandom in many ways. For instance, all mutations are not equally likely. It is well known that nucleotide transition mutations (those between purines or between pyrimidines) are more common than nucleotide transversions (mutations from a purine to a pyrimidine, or vice versa). All loci are also not equally likely to mutate—for instance, there are large regional differences in mutation rate along chromosomes (Hodgkinson and Eyre-Walker 2011), and mutation rates may be 10-fold higher at CpG sites (those where a G follows a C along the DNA strand) relative to nearby sites (Bird 1980). However, *random* in the evolutionary sense only means that mutations that are advantageous or deleterious in one environment are not relatively more or less likely to arise in another environment, though the overall mutation rate may change. Regardless of the many complexities of the mutational process that could be taken into account, these mutation models generally ignore the details of the nucleotide changes involved, considering only whether or not a mutation has occurred. Furthermore, in the models of mutation considered here, generally only mutations with no effect on fitness (neutral mutations) are considered.

The randomness our mutation models require relates to their origin in time and space, which both can be described by a Poisson process. Neutral mutations are assumed to accumulate independently among sites and along lineages at a constant rate μ (expressed per generation), such that the number of mutations observed after t generations will be Poisson-distributed with mean μt. The Poisson distribution is appropriate because mutation rates are very low, on the order of 10^{-8} to 10^{-9} per generation per site in eukaryotes (Lynch 2010). The mutation rate may be expressed per locus or per site, and in the latter case the number of mutations observed in a single generation across L nucleotide sites will be Poisson-distributed with mean μL. In models that consider the arrangement of nucleotide changes along a sequence, we assume that each site has an independent probability of mutating. Given newer methods that consider the spacing of polymorphisms along a chromosome (e.g., Li and Durbin 2011), assumptions about the independence of mutations and constant rates of mutation among sites may be very important. Unfortunately, neither assumption is always true (e.g., Schrider, Hourmozdi, and Hahn 2011; Harris and Nielsen 2014), though the effect of these violations on most methods may not be large.

The simplest mutation model involves two alleles (**FIGURE 1.2A**). In classical population genetics these are often denoted as A and a or as A_1 and A_2, and could represent two different nucleotides, two different amino acids, or two different haplotypes. (I will generally use A_1 and A_2 in this book,

FIGURE 1.2 Models of mutation. Four different models are demonstrated in (A): The two-allele model refers to single sites with two alleles, though these sites are also consistent with other models. The infinite sites model refers to sequences in which at most one mutation occurs per site. The finite sites model refers to sequences in which multiple mutations can occur at the same site. The infinite alleles model treats each entire sequence as an allele, rather than alternate states at a single site (such that there are three alleles shown here). (B) The stepwise mutation model, under which mutation can cause changes only between neighboring alleles, generally differing in their number of repeats.

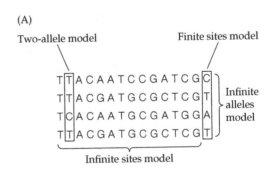

(A)

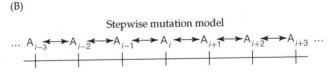

(B)

except in Chapters 5 and 9, where I use other denotations to avoid confusion with alleles in populations 1 and 2.) This *two-allele model* is the most common mutation model used when considering data from single sites. One of the two alleles can be assigned to be the ancestral state if such assignments are also known in the data being modeled, though these assignments are not necessary. An important feature of the model that must be specified in each usage is the number and type of mutations that are allowed. Often only $A_1 \to A_2$ (or $A_2 \to A_1$) mutations are allowed, in which case we say that there is no possibility of a back-mutation. This is not the same as allowing only a single mutation from $A_1 \to A_2$ to occur in the history of a sample, as sometimes this condition can be relaxed without relaxing the restriction on back-mutations. But often only a single origin of each mutation, with no back-mutation, is modeled.

A straightforward extension of the two-allele model to larger DNA sequences is known as the *infinite sites model* (Kimura 1969). In this model we imagine that we have DNA sequences of sufficient length such that there are multiple segregating sites in our sample. Because the mutation rate is low, the infinite sites model assumes that every mutation in the history of the sample occurred only once, that there are no back-mutations, and that each mutation to a new allele occurred at a new site in the sequence (Figure 1.2A). These assumptions ensure that individuals will be similar at the DNA level because of a shared history rather than convergent mutations. The assumptions also mean that each segregating site will be at most biallelic, with only two alleles (e.g., A_1 and A_2) present in the sample. The mutations are assumed to occur independently according to a Poisson process.

In some cases, multiple mutations may have occurred at a site in the history of the sampled chromosomes. Because we no longer meet the infinite sites assumptions, we must instead use a *finite sites model*. Finite sites models are most commonly encountered in comparisons of DNA sequences among species, for which the long time-periods involved mean that sites may have changed multiple times. Several methods, such as the Jukes-Cantor correction (Jukes and Cantor 1969; Chapter 7), are available to accurately estimate distances between sequences in the presence of multiple substitutions at a site. In analyzing polymorphism data, the finite sites model is most commonly needed when segregating sites are triallelic or quadrallelic (Figure 1.2A). In these cases, the number of segregating sites is no longer equal to the number of mutations in the history of a sample, as individual sites may have experienced two or more changes of state. Such changes can take the form of a new derived state, back-mutations to preexisting ancestral alleles, or recurrent mutations to a derived allele already present in the sample. Mutations in finite sites models can still be Poisson-distributed, but there must now be a chance for each new mutation to occur at sites that have already experienced mutations and to allelic states that already exist.

For both the infinite sites and finite sites models we have assumed perfect knowledge of the underlying DNA sequence. However, in the early days of molecular population genetics the DNA sequences themselves were not available. Allozyme technologies (e.g., Hubby and Lewontin 1966) only allowed one to distinguish sequences that differed at amino acid changes, causing proteins to run at different speeds through a gel. This meant that many amino acid mutations and all synonymous mutations (those that do not change the amino acid encoded by a codon) were undetectable. In order to model such a system—in which different alleles were identified as electrophoretically distinguishable variants—researchers used the *infinite alleles model* (Kimura and Crow 1964). This model posits that each new mutation at a locus creates a new allele that does not exist in the population, without necessarily having knowledge of the underlying DNA sequence and without dealing with the details of the sites that have mutated.

At this point we must clarify the relationship between the meaning of *alleles* in the infinite alleles model and its meaning in the infinite sites models (Figure 1.2A). As discussed above, we currently use *allele* to distinguish alternative states at a single nucleotide or codon position. The different alleles in the infinite alleles model, however, are equivalent to different haplotypes, where each haplotype may be distinguished by multiple nucleotide changes. The infinite alleles model does not consider the number of sites that differ between haplotypes/alleles and therefore does not take advantage of the full sequence information. In addition, although many historically important results have been derived for the infinite alleles model (e.g., Ewens 1972), they have generally ignored the effect of recombination in creating new haplotypes. For all of these reasons, the infinite alleles model is rarely used in modern molecular population genetics.

Related to the infinite alleles model is the *stepwise mutation model*, which was originally proposed in order to represent the effects of mutation moving between closely related alleles (Ohta and Kimura 1973). In the original model the alleles represented haplotypes, but the haplotypes were ordered such that mutations could take you between A_1 and A_2 or between A_2 and A_3, but not between A_1 and A_3 (**FIGURE 1.2B**). After its original usage in modeling allozyme loci, the stepwise mutation model has found much wider use as a model of microsatellites. Microsatellites—also called short tandem repeats (STRs) or variable number tandem repeats (VNTRs)—are made up of short repeat units (usually 1–6 bases in length) that can be repeated 1 to 50 times in a row (Goldstein and Schlötterer 1999; Ellegren 2004). There are often many alleles at polymorphic microsatellites, with each allele having a different number of repeat units. Mutations at microsatellites are thought to occur via polymerase slippage, resulting in more or fewer repeats. The stepwise mutation model allows allele sizes to change by adding or subtracting a single repeat unit, with symmetric probabilities of gain and loss and no relationship between allele size and mutation rate (Figure 1.2B). Real microsatellite data have revealed that there are multistep mutations that result in the gain or loss of more than one repeat at a time, and that the direction of mutation is often biased (Di Rienzo et al. 1994; Rubinsztein et al. 1995; Sun et al. 2012). For these reasons, alternative models that allow any allele to be reached from any other (also called infinite allele models) or that allow any particular distribution of step sizes and mutation rates, either symmetric or asymmetric, can be used (such models are known as *generalized stepwise models*; Kimmel and Chakraborty 1996).

Models of recombination

Modeling recombination is generally similar to modeling mutation. Recombination events are thought to occur in a Poisson-like manner along a chromosome, with the recombination rate given by the parameter c, defined as the probability of a crossing-over event occurring between two markers. In reality crossing over is more complicated than this, but we ignore most of these complications (e.g., cross-over interference) in population genetic models. We often measure the recombination rate per site per generation, so that the expected number of recombination events occurring in a single generation between two sites L nucleotides apart is cL. A prominent feature of crossing over in some organisms is the presence of so-called recombination hotspots (Jeffreys, Kauppi, and Neumann 2001; McVean et al. 2004). Recombination hotspots represent small regions where crossing-over events are clustered together, generally increasing the recombination rate many-fold above the background rate. Even organisms that appear to lack true hotspots (e.g., Singh, Aquadro, and Clark 2009; Comeron, Ratnappan, and Bailin 2012; Manzano-Winkler, McGaugh, and Noor 2013; Kaur and Rockman 2014) show a wide range of recombination rates along chromosomes. For recombination in general—and for scenarios including hotspots in particular—all of this means that events are often modeled in a "finite sites" manner, in which more than one recombination event can occur at the same position in the history of a sample.

One of the main differences between mutation and recombination is that recombination events do not change allele frequencies at single loci. By rearranging alleles along chromosomes, crossing over only changes the frequencies of DNA haplotypes (**FIGURE 1.3A**) while leaving the allele frequencies at each site unchanged. In addition, although crossing over is one outcome of recombination, another is *gene conversion*, the non-reciprocal exchange of genetic material (**FIGURE 1.3B**). In gene conversion (which does not have to involve a gene), one of the two allele copies in an individual acts as the donor, while the other acts as the acceptor. The donor's DNA "overwrites" the acceptor's, resulting in two identical

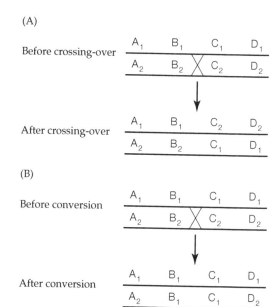

FIGURE 1.3 Different outcomes of recombination. (A) A diploid individual heterozygous at four loci, A, B, C, and D. A crossing-over event (indicated by the ×) occurs between the B and C loci, resulting in two recombinant haplotypes. (B) The same initial setting, with a conversion event affecting the B and C loci. Because the B_1 and C_1 alleles act as the donors, the result is an individual that is homozygous for the B_1 and C_1 alleles.

alleles. Sometimes this happens in a biased manner, such that one allele is more likely to be the donor (e.g., Galtier et al. 2001; Marais 2003). Whether biased or not, gene conversion can change allele frequencies. There can also be the non-reciprocal exchange of genetic material between duplicated loci within a genome, which is referred to as *non-allelic* or *interlocus* or *ectopic gene conversion* (Arnheim et al. 1980; Miyata et al. 1980; Scherer and Davis 1980). Here I consider only gene conversion between alleles; this is referred to as *allelic* or *intralocus gene conversion*.

The inclusion of gene conversion necessitates the introduction of two additional parameters to models of recombination. One is the rate of gene conversion per site per generation, g, and the other is the mean length of gene conversion tracts, q (Andolfatto and Nordborg 1998; Langley et al. 2000; Wiuf and Hein 2000). Because all recombination events result in a short gene conversion tract but not all events result in crossing over, the ratio of g/c (sometimes represented as its own parameter, f) is usually much larger than 1, ranging from about 4 in *D. melanogaster* (Comeron, Ratnappan, and Bailin 2012) to about 7 in humans (Frisse et al. 2001). Gene conversion tracts (i.e., the stretch of DNA converted) are on the order of 100–1,000 nucleotides (Chen, Cooper, et al. 2007), with a length distribution similar to a geometric distribution with parameter $1/q$. Gene conversion tracts are initiated with probability g at each site, and then the tract is modeled as extending to the right of the site of initiation (Wiuf and Hein 2000). Further details on both crossing over and gene conversion are considered in Chapter 4.

Models of natural selection

There are many aspects of natural selection that are relevant to molecular population genetics, not all of which can be discussed here. Perhaps the most useful introduction will simply be to define commonly used terms and to point out the ambiguities in the definitions as they are used in the field today.

With respect to its effects on organismal fitness, there are three consequences of a new mutation: it may be *advantageous*, *deleterious*, or *neutral* relative to the fitness of the ancestral allele. New mutations can of course have variable effects on fitness across genetic backgrounds and in heterozygous or homozygous form, but here we are interested in the average marginal effect of a mutation across backgrounds and population frequencies. If we define s as the selection coefficient of a new mutation—that is, the difference in relative fitness between individuals with one copy of the new allele and individuals homozygous for the ancestral allele—then for advantageous mutations, $s > 0$; for deleterious mutations, $s < 0$; and for neutral mutations, $s = 0$ because the two alleles have equal fitness. Mutations of large effect have larger $|s|$, either increasing or decreasing fitness, recognizing that in natural populations $s = 0.10$ (a 10% increase in fitness) is considered a very large effect. As we will see in Chapter 7, advantageous mutations with much smaller effects than this will quickly spread through a population.

As a way of summarizing the selective history of a gene or locus, we often use the phrases *negative* (or *purifying*) *selection*, *positive selection*, and *balancing selection*. These very useful, but often highly ambiguous, summaries of natural selection are not always aligned with the terms used to define the effects of single mutations. *Negative selection* refers to a history of a locus in which the vast majority of mutations that have arisen have been deleterious. A gene or noncoding region under negative selection is often therefore one that is conserved, with natural selection removing most mutations that change a functional DNA sequence. In some cases the primary sequence may not be conserved but the sequence can still experience purifying selection, as the length of a sequence or average biochemical property of a sequence is the selected trait (e.g., Podlaha and Zhang 2003; Lunter, Ponting, and Hein 2006). We also often say that such regions or nucleotide sites are *constrained* by selection to a limited region of sequence space. It should be clear from this definition that almost all protein-coding genes are under negative selection all the time, even if there is a large (or small) fraction of mutations that are neutral or advantageous. A region not under negative selection is not constrained and is under no selection. One ambiguity that arises when using the phrase negative selection is the meaning of a gene under "stronger" negative selection than another. This could mean that the average selection coefficients of alternative bases at constrained sites have become more negative; this is certainly what is implied by the phrase "weak selection," when s is close to 0. Alternatively, it could mean that more sites in a region (e.g., more amino acids within a protein) are under any constraint, and therefore that the average number of substitutions observed across the region is lower. Regardless, negative selection is a commonly used shorthand that has a clear meaning in most situations.

Positive selection refers to the history of a locus in which advantageous mutations have arisen and fixed or are in the process of fixing. Because advantageous mutations are much rarer than deleterious mutations, it is assumed that negative selection is acting on almost any conserved and functional region. By contrast, a locus with even a single detectable advantageous substitution is said to be under positive selection. Such substitutions may have happened in the distant past or may currently be passing through a population. Advantageous mutations can rapidly increase in frequency, and many patterns generated by this rapid rise are used to detect signatures of positive selection. As detecting positive selection has become a *raison d'être* of molecular population genetics, we will return to it many times throughout the book.

The final concept used to summarize the selective history of a locus is *balancing selection*. Multiple different selective models are subsumed under this term, but all have in common the maintenance of selected polymorphisms within a population or species. These models are in contrast to those with universally advantageous or deleterious mutations, as balanced polymorphisms consist of alleles whose relative fitness changes with time, space, or population frequency. As a result, balancing selection is often contrasted with the models of "directional" selection that include positive and negative selection. Commonly observed instances of balancing selection include *heterozygote advantage* (or *overdominant selection*), in which heterozygous genotypes have higher fitness than either homozygous genotype; *negative frequency-dependent selection*, in which rare alleles have higher fitness; and *spatially* or *temporally varying selection*, in which the fitness of alleles in a population is dependent on the environment or season, respectively, in which they are found. In many cases spatially varying selection is considered a form of local adaptation rather than balancing selection, although at the level of the whole species any such polymorphism is maintaining diversity and is therefore under balancing selection. For some forms of balancing selection, the balanced polymorphism may be biallelic or multiallelic. That is, there may be either two alternative alleles at a site that are balanced (e.g., Kreitman and Aguadé 1986) or multiple alleles—usually at multiple sites—that are balanced (e.g., Ségurel et al. 2012).

One last comment must be made on the language used to describe natural selection on quantitative traits, such as height, weight, and even some features of genomes (e.g., Kimura 1981). Quantitative traits are usually determined by the combined effect of alleles at many loci, and any one of these loci may be under negative, positive, or balancing selection. However, the phenotypes themselves are said to be under *stabilizing, directional*, or *disruptive selection*. Each of these processes is analogous to a form of selection on DNA, but the analogy only goes so far. For instance, stabilizing selection is not the same as negative selection—stabilizing selection acts to eliminate phenotypically extreme individuals, either by eliminating unconditionally deleterious alleles, maintaining balanced polymorphisms, or simply retaining the optimal number of alleles of near-equivalent fitness at many loci. Directional selection on phenotypes is most similar to positive selection on individual loci, although the strength of selection on any one allele that contributes to a quantitative trait can be very weak. Finally, disruptive (or *diversifying*) selection on phenotypes corresponds to selection against individuals with intermediate trait values but does not necessarily require balancing selection at any particular locus. Note, too, that the term *diversifying selection* is sometimes used to mean either multiallelic balancing selection at a locus (e.g., Foxe and Wright 2009) or rapid protein evolution across a phylogeny (e.g., Murrell et al. 2012). The take-home message should simply be that there are many forms of natural selection and many terms used to describe each of these, and that care must sometimes be taken in order to clearly communicate the model invoked in any particular case.

Models of migration

Models of migration in population genetics are largely just models of how populations—also called *demes* or *subpopulations*—are structured in an environment (more discussion of the definition of populations can be found in Chapter 5). These models require very little detail on how individuals actually move between populations and interbreed, and generally only the number and type of alleles being exchanged is of interest. However, there are several important details to the migration process that are common to all models of migration. These details are key in understanding how we infer patterns of gene flow across a species.

The migration rate, m, is defined as the fraction of all individuals (or chromosomes) in a population in the current generation that came from a different population in the previous generation. That is, the migration rate represents the proportion of individuals in a population that are migrants each genera-tion. These migrants are assumed to be a random sample of individuals from the source population, and it is assumed that neither the size of the source nor the receiving population changes as a result of migration. The first as-sumption means that we can easily calculate the expected change in allele frequency as a result of migration based on the allele frequencies in the source and recipient populations (and the migration rate). The second assumption also makes it easier to keep track of expected changes in allele frequency: be-cause any particular population can be both a source of migrants and a recipi-ent of migrants, we do not have to account for every individual's movements.

A large number of models of migration involve discretely organized popu-lations in which random mating occurs within each population and migrants may be exchanged between populations. These "island" models have inter-esting parallels with mutation models, the most obvious being that we can have *infinite island models* (**FIGURE 1.4A**) and *finite island models* (**FIGURE 1.4B**). The more assumption-laden of these is the infinite island model (generally

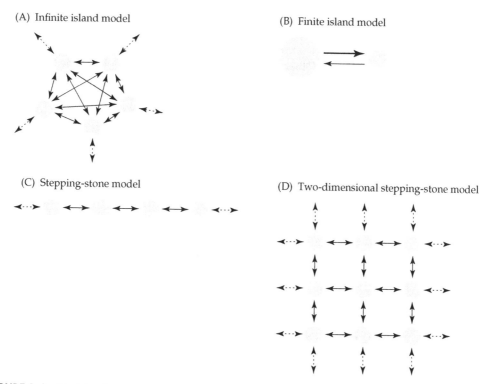

(A) Infinite island model

(B) Finite island model

(C) Stepping-stone model

(D) Two-dimensional stepping-stone model

FIGURE 1.4 Models of migration. (A) The infinite island model assumes an infinite number of popula-tions of equal size, all exchanging equal numbers of migrants with one another. Dashed lines represent migration with unseen populations. (B) The finite island model posits a limited number of populations, each with their own size and with a specific number of migrants entering and departing. The thickness of the arrow denotes differences in the migration rate. (C) The stepping-stone model assumes equally sized populations that can only exchange migrants with neighboring populations. (D) The two-dimensional stepping-stone model also only allows migration between neighboring populations.

ascribed to Wright 1931), which stipulates that there are an infinite number of populations of equal size, each of which exchanges migrants with every other population at the same rate, and assumes no selection and no mutation. These assumptions mean that this model has no geographic structure, as populations that are spatially proximate are no more likely to exchange migrants than those found at greater distances from one another. The infinite island model assumes that populations are at migration-drift equilibrium, so that there is also no trace of shared ancestry among populations that is not due to recent migration. For all of these reasons it has been called the "Fantasy Island" model (Whitlock and McCauley 1999), though many of its assumptions can be violated without large effects on the inferences regarding migration that are made (Neigel 2002). Nevertheless, more realistic migration models allow for more precise estimates of population parameters.

Finite island models include only a limited number of islands, each of which can have its own size and its own migration rate (Figure 1.4B). The simplest model involves two populations, of size N_1 and N_2, and two migration rates, m_1 and m_2. The migration rates specify the movement of alleles from population 2 into population 1 (m_1) and from population 1 into population 2 (m_2). It is common to see the assumption that migration rates are symmetric—that is, $m_1 = m_2$—so that only one parameter needs to be specified or estimated. Other versions of the two-island model assume that the two populations differ greatly in size, such that only migration from the larger population (the "continent") into the smaller population (the "island") has any effect on allele frequencies. This *continent-island model* is therefore simply a two-island model with one-way migration. For finite island models with more than two populations, we must specify the size of the ith population, N_i, and the migration rate between the ith and jth population, m_{ij}, for all i and j.

Continuing the parallel between migration and mutation models, just as the stepwise mutation model only allows changes to create neighboring alleles, the *stepping-stone model* of migration (Kimura 1953) only allows individuals to migrate between neighboring populations (or assumes that such migration is the most likely). Stepping-stone models are most commonly one-dimensional (**FIGURE 1.4C**) or two-dimensional (**FIGURE 1.4D**), but any number of dimensions can be accommodated (Kimura and Weiss 1964; Weiss and Kimura 1965). One issue with stepping-stone models of a finite number of populations is how the ends of the lattice are dealt with—that is, what to do when one arrives at the last population in a string of populations. For one-dimensional models, a simple fix is to arrange the populations in a circle so that all populations have two neighbors; for two-dimensional models, the equivalent shape is a torus (effectively, a donut). Interestingly, however, including true boundaries in the model can provide insight into the apportionment of diversity among an array of populations (e.g., Wilkins and Wakeley 2002).

Finally, an often more realistic model of migration is one that posits a continuously distributed population spread across a one- or two-dimensional surface. Such models are most often associated with Wright's concept of *isolation by distance* (IBD; Wright 1943), though the same pattern of IBD will be produced by stepping-stone models. In these models of continuously distributed populations (there is no common name for such models, though the term *continuum models* has been used; Felsenstein 1976), individuals proximal in space are more likely to be closely related because dispersal is limited, or at least much smaller than the scale of the habitat. Migration in the model is generally described by the variance in dispersal distance, σ^2, drawn from a normal distribution. These models have recently experienced a resurgence in popularity with the advent of "landscape genetics" (Manel et al. 2003) and will be discussed further in Chapter 9.

Models of mating

Much of population genetic theory is equally applicable to haploids and diploids, sexual and asexual organisms, and males and females. While this is not always true—and there are some very important exceptions—often the details make little difference in the predicted outcomes. One relevant area in which the details matter is when we must consider the frequency of diploid genotypes in sexual organisms. Because we often make inferences about evolutionary processes based on the observed frequencies of different genotypes (e.g., in identifying population structure; see Chapter 5), we must have a model to describe the expected genotype frequencies in the absence of these processes. The most widely used of such models is the *Hardy-Weinberg model* (Hardy 1908; Weinberg 1908).

In order to understand this model and its predictions, let us first consider what we are predicting. For a single locus with two alleles, A_1 and A_2, there are three possible diploid genotypes, A_1A_1, A_1A_2, and A_2A_2. The first and the third are homozygous genotypes, while the second is a heterozygote. We denote the frequencies of alleles A_1 and A_2 as p and q, respectively, and the frequencies of the three genotypes as p_{A1A1}, p_{A1A2}, and p_{A2A2}. The Hardy-Weinberg model gives us a way to calculate the expected genotype frequencies, assuming no selection, no mutation, no migration, no drift, and random mating between the sexes. Under these assumptions, the expected frequency of each genotype is:

$$E(p_{A1A1}) = p^2$$
$$E(p_{A1A2}) = 2pq \tag{1.5}$$
$$E(p_{A2A2}) = q^2$$

These expected genotype frequencies are equivalent to the expected outcome of choosing alleles from the population in proportion to their frequency in order to form the next generation, and are sometimes referred to as the *random union of gametes*. When a population matches the expected genotype frequencies, we say that it is at Hardy-Weinberg equilibrium (HWE).

There are a number of ways in which individuals can mate nonrandomly. Individuals may choose a mate with a similar phenotype, which we call *positive assortative mating*; if they mate with dissimilar individuals it is called *negative assortative* (or *disassortative*) *mating*. Because similar phenotypes are likely determined by similar alleles, positive assortative mating leads to more homozygous genotypes at the loci underlying the relevant traits, while the majority of the genome remains at HWE. An extreme form of assortative mating that affects all loci is *inbreeding*, in which relatives mate with each other more than is expected by chance. The inbreeding coefficient, F (Wright 1922), ranges from 0 to 1 and represents the fraction of loci in the genome that have alleles "identical by descent" as a result of matings between relatives. Inbreeding usually manifests as an excess of homozygous genotypes, such that more inbreeding (larger F) leads to more homozygosity in the population. Inbreeding can be due to multiple processes, including more mating within than between subpopulations (with an inbreeding coefficient denoted as F_{ST}) or more mating among relatives within the same subpopulation (denoted F_{IS}). Chapter 5 and Equation 5.3 deal with the excess of homozygotes that is due to F_{ST} in much more detail.

MODELS OF MOLECULAR EVOLUTION

The neutral theory of molecular evolution

There are relatively few models used as wide-ranging explanations for patterns of molecular variation, and only one body of work that has ever risen to the level of scientific theory in this field. That work—the neutral theory of molecular evolution—proposed independently by Motoo Kimura (1968) and Jack King and Thomas Jukes (1969), is one of the cornerstones of modern studies of molecular data. A wonderfully concise description of the neutral theory is given by Kimura himself (Kimura 1983, p. 306):

> The neutral theory claims that the great majority of evolutionary mutant substitutions are not caused by positive Darwinian selection but by random fixation of selectively neutral or nearly neutral mutants. The theory also asserts that much of the intraspecific genetic variability at the molecular level, such as is manifested in the form of protein polymorphism, is selectively neutral or nearly so, and maintained in the species by the balance between mutational input and random extinction or fixation of alleles.

The neutral theory is effectively making two claims about molecular evolution: (1) that differences between species at the molecular level are largely due to the substitution of one allele with another that is equivalent in fitness (i.e., neutral); and (2) that alternative alleles polymorphic within species are fitness neutral with respect to each other and have dynamics dominated by mutation-drift equilibrium. A consequence of these two claims is that polymorphism and divergence are simply two phases of the same process, with neutral alleles entering the population and eventually being lost or fixed by drift (Kimura and Ohta 1971). The neutral theory is therefore a framework for understanding both the causes of divergence between species and the processes that maintain variation within species.

There are a number of very valuable features of the neutral theory that have ensured its continuing use. One of the key insights of the neutral theory is its ability to explain differences in the rate of evolution among sites within genes, or among genes within a genome, by positing different levels of selective constraint. Although this point is often misunderstood (as we will see in the next section), the neutral theory does not deny a role for negative selection—in fact, it is variation in the strength of negative selection that is proposed to be the major determinant of the rate of evolution (as opposed to variation in the strength or frequency of positive selection). This insight can be summarized in the following equation (Kimura 1983, eq. 5.1):

$$\mu = \nu f_0 \tag{1.6}$$

Here the rate at which neutral mutations arise (μ) is determined both by the total mutation rate (ν) and by the fraction of all mutations that are neutral (f_0). Assuming a constant total mutation rate across a gene or genome, the neutral theory says that variation in the rate of evolutionary change is due to variation in the fraction of mutations that are neutral. This fraction, f_0, represents the amount of constraint, so that "the weaker the functional constraint, the larger the fraction of mutations that are selectively neutral and therefore, the higher the evolutionary rate" (Kimura 1983, p. 308). Consider levels of polymorphism and divergence at different positions within a codon. Fourfold degenerate sites evolve faster than twofold degenerate sites, which evolve faster than nondegenerate sites (e.g., Li 1997, fig. 7.2). The total (or "underlying") mutation rate will be very similar among sites within the same codon or within the same gene. What differs among sites is the amount of constraint (f_0), which in turn results in differences in the observed numbers of neutral polymorphisms or neutral substitutions.

It is important to understand what is meant by variation in the neutral mutation rate, μ. In the way Kimura intended it to be used, the neutral mutation rate can vary from site to site, codon to codon, among regions of the same protein, or among proteins. The neutral mutation rate varies because negative selection varies in strength across sites and genes. Nondegenerate sites within a codon (those at which mutations always result in nonsynonymous changes) may have a very different neutral mutation rate than fourfold degenerate sites (those at which mutations never result in a nonsynonymous change). Neutral rates for a single gene are themselves likely to be averages of different neutral rates among all the sites within a gene, as f_0 may, for instance, be lower for codons within functionally more important parts of a protein. Fourfold degenerate sites are generally assumed to have $f_0 = 1$—to have no constraint whatsoever—and are therefore thought to reflect the underlying rate of mutation (see Chapter 7 for more discussion of this assumption). But every gene and every site has its own neutral mutation rate—this is the rate at which neutral alleles arise, no matter how rarely. The distinction between total mutation rate and neutral mutation rate has important implications for many topics discussed throughout this book, and is a key concept underlying much of the way in which we understand molecular variation.

Another major contribution of the neutral theory has been its ability to help explain variation in levels of nucleotide polymorphism among species and populations. If most polymorphisms observed within species are neutral—not slightly deleterious polymorphisms kept at mutation-selection balance, not balanced polymorphisms maintained at intermediate frequency, and not advantageous mutations streaking through the population on their way to fixation—and there is no effect of selection on linked mutations, then the amount of variation expected within a population is determined by a balance between neutral mutation and genetic drift (**FIGURE 1.5**). Polymorphisms are added by (neutral) mutations and are removed by drift after being either lost completely or fixed. Assuming equivalent neutral mutation rates, larger populations experience less drift and therefore can maintain more polymorphism (Figure 1.5). The joint contributions of drift and mutation are one reason that we consider the product $N_e\mu$ to be the determinant of levels of polymorphism (though not the only one), and similarly why we focus on other compound parameters that include the contribution of drift (via N_e) in assessing levels of recombination ($N_e c$) and levels of migration ($N_e m$) within a population.

If the assumptions of the neutral theory are true and populations are at mutation-drift equilibrium, we have a large body of work describing the expected level and frequency of polymorphisms within species, much of it thanks to Kimura himself. The neutral theory provides a theoretical basis for understanding DNA variation with clear, testable hypotheses that enable the use of an array of statistical tools that can detect the action of natural selection. This corpus drives most of the expectations explained in this book, and ensures the continuing use of the neutral theory as a framework for studying molecular population genetics.

However, the neutral theory also has major difficulties in explaining much of the data being produced by the growing number of studies of molecular variation. To clearly grasp these problems, it is helpful to consider a weak version and a strong version of the neutral theory. The weak version states that most substitutions observed across the genome are in fact neutral and are fixed by random processes. The same is true for polymorphisms: across the genome most are neutral, representing the same alleles that will eventually be lost or fixed in a population. Because most mutations in a genome (at least in large,

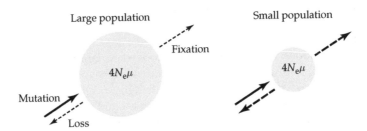

FIGURE 1.5 Equilibrium levels of variation. Equilibrium is reached when the input of variation into a population by mutation (solid arrow) is exactly equal to the combined output of variation through either loss or fixation of new alleles (dashed arrows). The population size (indicated by the size of the circle) determines the effects of drift, so that there is both more loss and more fixation in small populations (indicated by thicker arrows), such that the equilibrium level of variation is lower in small populations. Note that the rate of loss and the rate of fixation of alleles are not equal to each other, but the arrows have the same thickness for simplicity.

eukaryotic genomes) occur at noncoding, nonfunctioning sites, the weak version of the neutral theory continues to be accepted. The strong version says that most amino acid substitutions are also neutral, and that mutation-drift equilibrium is the major force determining levels and frequencies of neutral polymorphisms. Even if the vast majority of substitutions across a genome are neutral, what most researchers care about are substitutions that could have an effect on fitness—either those in coding regions or those in functional noncoding regions. And for these types of substitutions the neutral theory continues to be rejected (reviewed in Hahn 2008; Wright and Andolfatto 2008; Sella et al. 2009). Similarly, even if all observed polymorphisms were neutral—at coding, noncoding, and nonfunctional sites—observed patterns of polymorphism do not conform to the expectations of mutation-drift equilibrium, or even demographic perturbations of this equilibrium, likely because of selection on linked mutations. In summary, the weak version of the neutral theory is correct and invaluable; the strong version is almost certainly incorrect, but still sometimes useful. The tension between rejection of this theory and its remaining utility has still not been satisfactorily resolved (Kreitman 1996; Hahn 2008).

Misunderstanding and misuse of the neutral theory

As with any widely used scientific theory, the neutral theory has inevitably been misunderstood and, consequently, misused. Many of these problems recur even within molecular population genetics and lead to unnecessary confusion and miscommunication among authors. Most of the problems are caused because the neutral theory and the term *neutral* are both understood to mean "no selection," although Kimura stressed that "the theory does not assume that all the mutations at the time of their occurrence are selectively neutral" (Kimura 1983, p. 307).

In his formulation of the neutral theory Kimura was attempting to explain the DNA and protein sequence data accumulating in the early days of molecular biology, and the neutral theory is a statement about how he thought the world worked. He did not argue that the neutral theory should be a null model for evolution, though he certainly thought that it had fewer *ad hoc* assumptions than the models of pan-adaptationism proposed as alternatives. However, it is now common for *neutral model* to be used as a synonym for *null model*, with the implication that any model without positive or balancing selection is more parsimonious than one that includes them. It is of course useful to be able to specify a model with no positive or balancing selection, and commonly used phrases for such a model include *standard neutral model* or *neutral-equilibrium model*.

More confusing is the conflation of a generic "neutral model" with the idea of a model without any selection. Where does this sense of the term neutral come from? It likely started innocently enough with models of quantitative traits, for which researchers interested in the contributions of mutation and drift to the evolution of phenotypes coopted the term *selective neutrality* to mean a trait under no selection (Lande 1976). This understanding was even developed into a general model for such traits under the name *neutral theory of phenotypic evolution* (Lynch and Hill 1986). The moniker has been taken up many times since in many fields for models that do not include any sort of selection, including those found in studies of biodiversity (Hubbell 2001), gene expression (Khaitovich, Pääbo, and Weiss 2005), and cultural data (Lansing and Cox 2011).

A similar, and perhaps more continually vexing, conflation is between the neutrality of mutations and the neutrality of sites. To be clear, only mutations can be neutral: the term is explicitly a statement about the relative fitness of alternative alleles. There is no such thing as a locus that is neutral or "strictly neutral," but there can be loci under no selection. Unfortunately, a common shorthand in the field is to use *neutral locus* to refer to either a locus that produces only neutral mutations (i.e., is unconstrained), a locus with neutral polymorphisms that is linked to one with non-neutral polymorphisms, or a randomly chosen locus that is not known to be involved in a specific adaptive trait. The problem then arises not because this useful shorthand is deployed, but because many researchers do not understand that it *is* just shorthand for a more complex concept.

The misapplication of the term neutral to loci or positions in a sequence is more than simply an alternative usage of the word—it leads to a host of misunderstandings about both data and concepts in population genetics. Here are some of the misunderstandings that are commonly seen in the literature as a result of the application of the term neutral to both loci and mutations:

- It causes one to think that only unconstrained sites can have neutral mutations. Under the neutral theory all sites can have neutral mutations, and therefore all sites can have polymorphisms maintained by the balance of mutation and drift. The neutral mutation rate may be lower at sites that are constrained, but levels of diversity and rates of substitution can still be governed by population size and mutation rate. Understanding the expectations underlying, for instance, the McDonald-Kreitman test (Chapter 7) is dependent on understanding this distinction—that there are different neutral mutation rates at different types of sites. This seems to be an especially common conflation in comparative genomic studies, which aim to estimate the mutation rate in the absence of selection using analyses of substitution rates at unconstrained sites.

- It causes one to think that unconstrained sites present a pattern of nucleotide variation unaffected by the action of natural selection. This confusion is commonly seen in studies of demography and phylogeography, in which the use of markers thought to be under no direct selection (e.g., microsatellites, mitochondrial D-loop) leads researchers to believe they are studying patterns of variation that are not under the influence of selection. Instead, the fact that many unconstrained loci are linked to those loci under selection means that the level and frequency of variation is very much affected by selection—and can be "non-neutral" in the sense that they are not driven only by mutation-drift balance. It may be that unconstrained sequences offer the opportunity to examine the mutational process unimpeded by direct selection (e.g., Petrov, Lozovskaya, and Hartl 1996), but this is a very different use of the data.

- It causes one to think that genes or loci with constraint are always "non-neutral." We will discuss many tests of neutrality in this book and the many ways in which loci can be evolving neutrally (or not). But the main theme will be that we are almost never testing for the action of negative selection—again, this is assumed to act on functional sites. Instead, we will be asking whether there is evidence for positive selection, balancing selection, or the presence of segregating deleterious polymorphisms. A locus evolving neutrally is not one evolving without negative selection, but rather one for which there is no evidence for one of these alternative forms of selection.

- It causes one to think that loci or broad types of mutations that have segregating polymorphisms with $N_e s = 0$ are not under selection. This category of error is the converse of the ones described above: if we estimate selection on current polymorphisms (be they nonsynonymous amino acids, gene copy-number variants, or transposable elements) and find that they are all neutral, some researchers incorrectly infer that there is no selection on any mutations of these types. Obviously such a pattern would be perfectly consistent with the expectations of the neutral theory—which says that the polymorphisms we observe are neutral—and would in no way imply that all such mutations are neutral.

Alternative models of molecular evolution

If the neutral theory cannot explain current patterns of variation, what can? Though there is not a consensus answer to this question, we can at least begin to consider possible alternative models that are consistent with the data. An informative place to start may not be with theories that can explain the data, but rather with alternative models that are likely incorrect. By enumerating these possibilities, the hope is that we form a better idea of what a good theory will look like.

A set of related models that are almost certainly wrong are those in which the vast majority of polymorphisms and substitutions are not neutral (i.e., they are not even consistent with a weak version of the neutral theory). There are three obvious models of this type. A deleterious mutation model would propose that all variation is at least slightly deleterious and is maintained at mutation-selection balance. Polymorphism would be dominated by low-frequency variants, and all substitutions between species would be made up of the few deleterious variants that are able to fix. An advantageous mutation model would posit a constant influx of advantageous mutations. In this model all polymorphisms (or at least those at possibly functional sites in the genome) would represent adaptive alleles on their way to fixation, likely at high frequencies in the population. All substitutions between species in this model correspond to the fixation of advantageous alleles. A balanced mutation model would propose the maintenance of balanced polymorphisms throughout the genome. Polymorphism would comprise these intermediate-frequency variants, with substitutions made up of the rare balanced allele that drifts to fixation. There are many reasons to doubt each of these models, based on both empirical and theoretical grounds.

The two models that are largely consistent with current data both include a large role for the effect of selection on linked neutral variation (see Chapter 8 for more discussion). While neither has reached the critical mass of theoretical results and empirical support that generally accompany a "Theory," both are well-developed models that provide detailed predictions about molecular variation. The *background selection model* (Charlesworth, Morgan, and Charlesworth 1993; Charlesworth, Charlesworth, and Morgan 1995) proposes that the constant influx of deleterious mutations that must be removed from a population reduces neutral polymorphism at linked sites. This model is not the same as the deleterious mutation model described above, as background selection does not propose that the polymorphisms we observe are themselves non-neutral. Though the background selection model does not always explain patterns of divergence very well (Stephan 2010; but see McVicker et al. 2009), the ubiquity of deleterious mutation means that it almost certainly influences levels of variation across all genomes (Charlesworth 2012). The *hitchhiking model* (Maynard Smith and Haigh 1974; Kaplan, Hudson, and Langley 1989) proposes that the rapid fixation of advantageous alleles reduces polymorphism at linked sites. Again, this model is quite different from the advantageous mutation model, as most polymorphisms are assumed to be neutral; it is the effect of rapidly fixing substitutions on levels of linked neutral polymorphism that is the major feature of the model. In addition, the hitchhiking model is consistent with the high rates of adaptive substitution observed in many species (Hahn 2008; Sella et al. 2009). It is almost certainly the case that both background selection and hitchhiking are acting in all species—what remains to be determined is the relative importance of each process across regions of the genome and across species.

Finally, it is also worth briefly mentioning the *nearly neutral theory of molecular evolution* (Ohta 1972a, 1972b), as it is often considered to be an alternative to the neutral theory. The nearly neutral theory posits a large number of slightly deleterious (and/or slightly advantageous) mutations, and was initially proposed to explain unexpected constancy in the observed rate of amino acid substitutions (Ohta 1992). However, rates of amino acid substitution are now known to vary widely across species, and the nearly neutral theory does not explain observed variation in levels of diversity across the genome. While there undoubtedly are many slightly deleterious mutations produced each generation, and—as is predicted under the nearly neutral theory—patterns of protein evolution are weakly correlated with population size (Akashi, Osada, and Ohta 2012), the theory remains explanatory for only a small slice of data.

EXPERIMENTAL DESIGN

2

In order to make inferences about the evolutionary process, molecular population genetic approaches require DNA sequences from a random sample of chromosomes. Although this seems like a relatively straightforward task to accomplish, deciding how many samples to collect, how to ensure that they are "random," and exactly how to obtain the sequence data are all important issues that deserve attention. These uncertainties are especially relevant now that large datasets can be collected relatively easily and cheaply using next-generation sequencing (NGS) technologies. Because the optimal experimental design largely depends on the questions being asked, I discuss a number of different approaches below.

POPULATION SAMPLING

Datasets in molecular population genetic studies generally use between 10 and 50 individual chromosomes (i.e., 5 to 25 diploid individuals). This is because many inferences can be made about population parameters from even very small samples. For instance, if we are not overly concerned with allele frequencies but simply need a count of segregating sites, then two chromosomes may be sufficient as long as there is enough variation between them. Sequencing more and more individuals always results in diminishing returns to the number of segregating sites discovered, and for many purposes sequencing a longer region rather than more individuals may be optimal (Pluzhnikov and Donnelly 1996). If we are applying methods that are exquisitely tuned to allele frequencies, then sample sizes larger than 50 may be required. However, most of our tests and statistics are intended for use with small samples, and there is often little increase in statistical power with sample sizes larger than 50 (Simonsen, Churchill, and Aquadro 1995). In fact, interpreting sequence data when the sample size n is close to the population size N can be problematic (e.g., Wakeley and Takahashi 2003; see Chapter 6). To some extent the "optimal" experimental design for molecular population genetics will always be dominated by the costs and constraints of the latest sequencing technologies, rather than the availability of samples. Because of these anticipated constraints it will be more helpful not to give prescribed numbers of individuals to use, but instead to provide general guidelines as to how studies can be judiciously conducted. When discussing particular methods used to infer evolutionary histories throughout this book I will be sure to note when sample sizes outside the range of 10 to 50 chromosomes can or must be used.

The samples collected must be a set of unrelated individuals, or at least chosen without respect to relatedness. Of course all individuals in a population are related to some degree, but avoiding the selection of siblings as well as parents and their offspring is especially preferred. If there is known to be some microstructuring of a habitat—such as relatives living in the same nest or seeds falling not far from the parental tree—then it is recommended that samples be chosen from sites spaced at least as far as the structuring occurs. Many sequencing datasets in human population genetics are now collected as "trios," comprising a mother, father, and offspring. If data from trios are to be used for molecular population genetics, then only the mother and father sequences should be used since they usually represent unrelated individuals.

Most, but not all, inference in molecular population genetics requires a set of sequences sampled from a single population. We define a *population* as a group of randomly mating individuals. This definition is clearly not very helpful in dealing with asexual or primarily selfing species; we will see that making inferences about the evolutionary history of such species can often be difficult exactly because it is hard to define populations. But the theoretical basis for most population genetic methods is derived assuming a single panmictic population. As a result, these methods can be extremely sensitive to the sampling procedure. Sampling 1 individual from each of 10 subpopulations rather than 10 individuals from just 1 subpopulation can provide a very different set of results. Researchers may have no idea about possible population structure within their study species, and we will discuss how to test for the presence of multiple subpopulations in Chapter 5. Obviously, many questions revolve around differentiation between subpopulations, in which case multiple subpopulations will purposely be sampled. In these scenarios the general guidelines on sample sizes given above apply to each subpopulation. It may also be the case that chromosomes are sampled from different subpopulations in order to provide the maximal amount of variation—for instance, to obtain molecular markers for crossing studies. There are a number of applications that can use these types of datasets, and I will point out each of them as they are discussed.

In addition to sampling sequences from within a species of interest, it is common to sample one or a few homologous sequences from a closely related species. Such *outgroup sequences* can provide estimates of divergence between species and can be used to "polarize" segregating alleles into ancestral and derived states. Again, different population genetic methods require slightly different numbers of outgroup sequences, or outgroups at different degrees of divergence, and so are discussed as each method is introduced.

DNA SEQUENCING

One of the most important requirements of the DNA sequences used in molecular population genetic studies is that they be obtained by direct sequencing, with full ascertainment of all variation present in the sample. In other words, variation in the sample of interest should not be confined only to polymorphisms that were previously known, but should include all mutations. There are a number of important exceptions to this rule (see the final section of this chapter), but for the vast majority of cases we still require full sequence data from each sampled chromosome. This ensures that our sample corresponds to the assumptions underlying our models.

Decisions about which locus to sequence, how many loci to sequence, or whether to simply sequence a whole genome will depend on the specific aims of each experiment. Assuming a set cost for sequencing a single nucleotide, for some purposes it is better to sequence more individuals rather than more loci, and vice versa (Pluzhnikov and Donnelly 1996). Some statistical methods require multiple loci (e.g., Hudson, Kreitman, and Aguadé 1987), and therefore application of these methods constrains experimental design. Different technologies also differ in their ease of application to one or multiple loci: for instance, next-generation sequencing technologies are wonderful for whole-genome studies but less useful for targeted studies of a limited number of loci. Additionally, next-generation technologies have small but nonnegligible costs related to the preparation of each individual library to be sequenced, and therefore a more complex calculation must be made about the exact sequencing approach taken (i.e., there is not a fixed cost per nucleotide). In the next two sections I discuss many of the technical aspects of sequencing DNA for a single locus using traditional Sanger sequencing. Extensions to multiple loci using this technology follow directly from these single-locus approaches. Although these methods have largely been surpassed by the use of other technologies, in order to understand the older literature it is important to understand what has been done previously and why. Following this review, I attempt to distill the most important considerations common to all next-generation sequencing approaches and the implications of these ideas for studies of population genetics.

Sanger sequencing

The most common, direct method for obtaining sequence data uses polymerase chain reaction (PCR; Saiki et al. 1985) to amplify the locus of interest, followed by chain-termination DNA sequencing (Sanger, Nickleu, and Coulson 1977). This combined approach is usually simply referred to as *Sanger sequencing*, but within this relatively straightforward approach are a number of variations that will determine the nature and quality of sequence data. Consider the gene shown in **FIGURE 2.1A**. There are multiple exons and introns, and the entire coding region covers more than 20 kilobases (kb). While some long-amplification (LA-) PCR methods may be able to amplify the entire region in one reaction, it is more likely that multiple primer pairs will have to be designed to amplify multiple amplicons (**FIGURE 2.1B**). If the latter is the case, then researchers must decide whether they need sequence data from both coding and noncoding regions or whether they are only interested in coding regions. One important thing to remember is that variation in the primer sequences themselves will not be detected, except indirectly through amplification failure. This means that placing primers inside exons

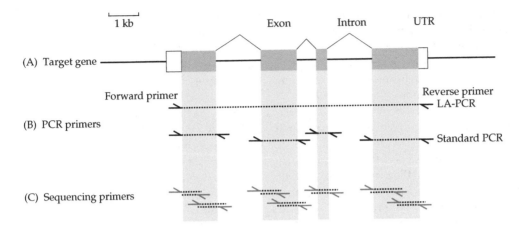

FIGURE 2.1 DNA sequencing. (A) Target gene to be sequenced. Coding exons (colored boxes), untranslated regions (UTRs; open boxes), and introns are shown. (B) Primer pairs for PCR, either long-amplification PCR (LA-PCR) or standard PCR. (C) Sequencing primers, using existing PCR primers when possible.

(when the aim is to sequence only coding sequences) will reduce the number of sites considered. (Needless to say, primer sequences should not be included in resulting alignments.) Of course if a genome sequence is not available then primer design may have to rely on relatively more conserved coding regions in order to work at all, and so there will be no option to place primers in flanking intronic or noncoding regions.

Sanger sequencing methods provide only about 700 bases of high-quality DNA sequence per read. Because the current standard is that DNA should be sequenced on both strands (i.e., that the forward and reverse primers should be used to generate overlapping sequences), any amplicon longer than this will require multiple internal sequencing primers (**FIGURE 2.1C**). There do not appear to be any inherent advantages to using either fewer, longer amplicons that use different internal sequencing primers or more, shorter amplicons that do not require internal primers. So other than the fact that all PCR primers can be reused for sequencing if each amplicon is smaller than 700 bp, the optimal choice will depend on the length of the target sequence and the cost of PCR reagents.

Another important choice to be made is whether PCR products will be cloned into DNA vectors before sequencing. If sequencing is done directly from amplified PCR products, then primers will actually be sequencing off of thousands of individual DNA strands. For non-inbred diploid individuals this means that both chromosomes will be sequenced for any locus, and therefore that some sites will be heterozygous (a point that will be elaborated below). In contrast, if PCR products are first cloned into a DNA vector, then sequencing proceeds from clones of only a single DNA template, and no heterozygotes are possible. Although a large number of companies produce easy-to-use kits for cloning, it is a relatively time-consuming process. One reason it is so time-consuming follows from the fact that the error rates of DNA polymerases used for PCR are relatively high: approximately 10^{-4} for *Taq* and 10^{-6} for *Pfu* (Hengen 1995). This implies that there can be errors introduced during PCR, and that any single DNA clone may therefore represent an incorrect sequence. For these reasons, *singleton* mutations—those found only on a single chromosome—are routinely resequenced from multiple clones, increasing the amount of work that must be done to collect even a single sequence. Because data obtained directly from PCR products gives the average DNA sequence from among the millions present in the reaction, it is unlikely that errors of this sort are introduced using this approach.

The results of chain-termination sequencing reactions are obtained as fluorescently labeled nucleotides passed across a laser, resulting in peaks of intensity corresponding in color to one of the four dyes attached to the four bases. The depiction of these raw data of peak heights and identities is called either a *chromatogram*, an *electropherogram*, or a *trace*. High-quality DNA sequences will have peaks of approximately equal intensity for each base, with equal area under each peak and regular spacing of peaks. Automated algorithms that call bases from raw chromatograms are the key to collecting high-confidence DNA sequences. There are a number of methods for doing this that come with the most commonly used proprietary software packages (e.g., Sequencher, CodonCode, Vector NTI), but the most popular software is the free *phred* package (Ewing et al. 1998; Ewing and Green 1998). (*Phred* stands for *Phil's read editor*—"Phil" being Phil Green, one of the original developers of the program.) One of the most useful aspects of phred is that it attaches a quality score to each base, colloquially called a *phred score*, based on the shape of peaks in the chromatogram. The phred score, Q, equals $-10*\log_{10}(p)$, where p is the error probability for each base. This means that a base with a score of 30 should have a probability of being incorrect of less than 1 in 1,000. In practice, phred scores greater than 20 are generally accepted (0.01 error probability); most proprietary software packages will represent bases with phred scores lower than this (or below some equivalent proprietary algorithm's score) with an "N" in the final sequence. When using phred users can choose to accept basecalls with whatever minimum quality score they wish, replacing low-quality bases with Ns.

Although PCR error may not present a large problem when clones are not used, distinguishing homozygotes from heterozygotes when direct sequencing from an outbred diploid individual can be challenging. This is because both bases will be represented at a single position in the chromatogram, resulting in overlapping peaks. Fortunately, a number of methods have been created specifically for the purpose of identifying heterozygous positions in samples of sequenced individuals (e.g., *PolyPhred* [Nickerson, Tobe, and Taylor 1997], *novoSNP* [Weckx et al. 2005], *PolyScan* [Chen, McLellan, et al. 2007]). The newest version of *PolyPhred* (Stephens et al. 2006) provides a measure of statistical confidence for each potentially heterozygous position and is reported to have 99.9% genotyping accuracy. All of these methods appear to miss a relatively higher number of singleton polymorphisms because the minor allele is not observed in other samples, though this is still a small number overall. One feature that *PolyPhred* does not yet have but that can be found in *novoSNP*, *PolyScan*, and some proprietary packages is the ability to account for heterozygous indels. The effect of heterozygous indels on chromatograms can be severe: in this case sequences from the two strands may overlap one another such that even high-quality sequences can appear to be unreadable. Algorithms that disambiguate the two strands have relatively good success when indels are short and there are not multiple indels in a single read (Weckx et al. 2005).

Haplotypes

If multiple heterozygous sites are present in a DNA sequence, then there is another major difference between direct sequencing and sequencing from cloned PCR products: direct sequencing does not give us the *gametic phase* of the heterozygous alleles. As a simple example, imagine that a single individual has two heterozygous sites 100 bases apart from each other, one an A/G polymorphism and one a C/T polymorphism. In this case there are two possible configurations of alleles on the two homologous chromosomes carried by the individual, either AC/GT (with the A and C alleles inherited from one parent and the G and T from the other) or AT/GC (**FIGURE 2.2**). The two configurations of alleles on chromosomes are also called the individual's *haplotypes*.

Sequencing from cloned PCR products gives us the phase of the alleles directly, as each sequence is read from only one of the two chromosomes at a time (ignoring the possibility of PCR-mediated recombination). Of course, if the heterozygous sites lie far apart and cannot be captured in a single amplicon, then there is also no information about the phase of the alleles. One reason to sequence from inbred individuals is that there are no heterozygotes, and therefore the phase is known regardless of the sequencing method used. This is also the reason that so many early studies in molecular population genetics focused on genes found on the X chromosome—sequencing hemizygous males avoided the problems associated with obtaining the phase.

For many applications we do not need the gametic phase of alleles, and in these cases it is perfectly appropriate to randomly assign alleles to haplotypes if necessary (e.g., many software packages expect sequences to be represented as distinct haplotypes even if the haplotypic information is never used). But what if haplotypes are needed? There are some experimental techniques that allow us to phase genotypes—such as cloning—but these are generally limited to small genomic regions that can be inserted into the cloning vector. There are also some technologies that allow for the isolation of individual chromosomes (e.g., hybrid cell lines), but these are often extremely time-consuming and not at all amenable to applications involving a population of individuals. Other promising experimental technologies are being developed (reviewed in Snyder et al. 2015) but so far have not been widely applied.

Because of experimental limitations and cost, the most common methods used to phase alleles (i.e., to turn genotypes into haplotypes) have been computational (Browning and Browning 2011). Given only genotypic information, there are a huge number of possible haplotypic configurations: 2^S for S biallelic segregating sites. Luckily there

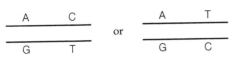

FIGURE 2.2 Haplotype configurations. Given two heterozygous sites, A/G and C/T, the two possible haplotypes are shown.

is not free recombination between all polymorphisms, and consequently there are always many fewer than 2^S combinations found in natural populations. This means that it is possible to infer haplotypes. Many computational methods have been developed for inferring the phase of alleles, including basic parsimony methods (Clark 1990), more advanced maximum likelihood methods using expectation-maximization (Excoffier and Slatkin 1995), hidden Markov model (HMM)–based methods (Browning and Browning 2007), and Bayesian methods (Stephens, Smith, and Donnelly 2001; Stephens and Donnelly 2003). The Bayesian methods implemented in the software package *PHASE* (Stephens, Smith, and Donnelly 2001; Stephens and Donnelly 2003) have been found to be most accurate for smaller samples (Marchini et al. 2006), while *BEAGLE* (Browning and Browning 2007) and *fastPHASE* (Scheet and Stephens 2006) are much faster and possibly more accurate for very large samples. These methods are all applied to samples of unrelated individuals from a population, although alternative methods can be used with groups of related individuals (e.g., Kong et al. 2008). While experimental phasing is the generally preferred method, because of recombination during PCR it is sometimes the case that computational methods can be more accurate than experimental methods (Harrigan, Mazza, and Sorenson 2008).

All computational methods are based on a similar idea—the haplotypes for some pairs of sites in some individuals from a population are known because they either are doubly homozygous or have only a single heterozygous site. Following from the example above, if we observe an individual whose genotype is A/A at the first site and T/T at the second site, we know that the two haplotypes must be AT/AT. All of the computational methods take advantage of individuals with known phase in order to infer the phase of doubly heterozygous individuals, essentially assuming that there are few recombination events that will break up allelic combinations. Once all genotypes in a population have been phased, researchers will often treat these haplotypes as if they were experimentally obtained. However, some programs will also report confidence scores for the placement of individual alleles (e.g., *PHASE*), explicitly incorporating uncertainty into the estimated haplotypes.

Next-generation sequencing

To a larger and larger extent, studies relying solely on Sanger sequencing are being overtaken by whole-genome (or at least genome-wide) datasets that use next-generation sequencing technologies. Even though PCR followed by Sanger sequencing is still considered the gold standard, the high throughput and relatively low cost per nucleotide of next-generation approaches allows large amounts of population genetic data to be collected quickly and cheaply. For many whole-genome studies this means that only the most interesting loci will be further Sanger sequenced in a targeted manner, possibly among a larger number of individuals (and even this approach is becoming less common all the time). There are still some species for which a reference genome sequence is not available; for these species many NGS applications to genomic DNA will not be appropriate, and Sanger sequencing will have to be used. However, alternative approaches using RNA sequencing or reduced representation sequencing may be a way to collect large amounts of data on sequence variation even from non-model organisms (see below).

Along with the enormous datasets produced by NGS technologies come a number of important issues, including both technology-specific biases and general data-handling problems. Discussing the technical details of any particular technology in a book is a losing proposition: the ever-evolving nature of the field means that the specific approaches taken now will become obsolete in a short period of time. In a little more than a decade, high-throughput alternatives to Sanger sequencing have included Affymetrix sequencing microarrays (Chee et al. 1996; Hacia et al. 1996), 454 pyrosequencing (Margulies et al. 2005; Wheeler et al. 2008), Solexa/Illumina sequencing-by-synthesis (Bentley et al. 2008), ABI SOLiD sequencing-by-ligation (McKernan et al. 2009), and Complete Genomics nanoball sequencing (Drmanac et al. 2010). Not included in this list is the growing number of even newer technologies, most of which have not yet been applied to the sequencing of population samples (see Chapter 10). Some of the technologies that have made major contributions to studies of variation—most notably array sequencing (e.g., Cargill et al. 1999; Cutler et al. 2001; Hinds et al. 2005; Kim et al. 2007)—are already obsolete. (This latter approach should not be confused with arrays used for genotyping, which are discussed later in the chapter.)

Despite the important technical differences among commonly used next-generation sequencing platforms, we can identify two key features shared by all of them that are likely to be challenges for most analyses. These two features are (1) short sequence reads, and (2) high error rates.

For most NGS technologies the reads are getting both longer and more accurate, so to some extent both of these problems will be mitigated soon. But it appears that at least some concerns will remain for the foreseeable future, and therefore it is best to understand how these features affect downstream analyses. To understand the effects of these problems, it is first important to understand a few more general characteristics of next-generation sequencing approaches.

NGS technologies typically produce millions of individual sequence reads from a single run. The reads are derived from the source DNA, which has been fragmented into many small pieces in order to serve as an efficient template for the sequencing reaction. The random sequencing of the input DNA ("shotgun sequencing") results in a wide distribution of read depths for each position in the genome (**FIGURE 2.3**). The read depth at each position—in other words, the number of sequence reads originating from a specific portion of the genome—is expected to be Poisson-distributed under simplifying assumptions (Lander and Waterman 1988). In practice there are a number of different features that bias read depth, including GC-content of the template sequence and the exact DNA preparation protocol used. Most importantly for studies of variation, the read depth will vary from site to site and will not necessarily be consistent between samples. In other words, some positions in the genome will be sequenced many times over, and some will be sequenced very rarely, but the coverage of any particular position can differ between one individual and the next. All of this means that we will have variable sequence quality along chromosomes, and that we cannot expect that every sequenced sample will have a genotype call at every position. We must therefore use appropriate methods that are able to deal with large amounts of missing data (e.g., Ferretti, Raineri, and Ramos-Onsins 2012). In what follows I generally use the terminology associated with Illumina sequencing, but most of the issues described arise in conjunction with all NGS technologies.

The effect of short sequence reads

Depending on the technology and the precise settings used within a technology, NGS platforms used in population studies generate reads between 35 and 600 nucleotides long. Reads can often be produced as either single-end or paired-end. *Paired-end* means that two reads of equal length come from the same larger piece of DNA, the length of which is approximately known. These sequences can be quite useful in effectively increasing the number of sequenced bases from any particular genomic location because the paired reads must have originated from the same template, though some DNA in the middle may not be sequenced. Paired-end sequences are also crucial to identifying copy-number variants (CNVs), or indeed any form of structural variation in a sampled genome (for a review see Schrider and Hahn 2010; Alkan, Coe, and Eichler 2011).

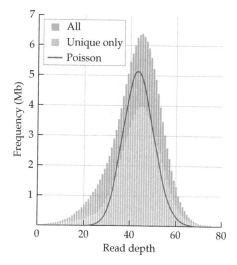

FIGURE 2.3   Read depth for the human X chromosome using Illumina sequencing technology. The read depth sampled at every 50th position along the chromosome is plotted for uniquely mapped reads and all mapped reads. Average read depth across the chromosome was 40.6X, meaning that most 50-bp windows were covered by 40.6 sequence reads. The Poisson expectation for a distribution with the same mean as the uniquely mapping reads is also plotted. (After Bentley et al. 2008.)

The most important consequence of collecting relatively short sequence reads is that we cannot make *de novo* genome assemblies for each of the samples sequenced. Instead, with this method we "map" individual reads against an existing genome assembly, one that has been constructed to be used as a reference sequence for the whole species. The need to map millions of reads to a genome—that is, to find the genomic location corresponding to the source of each read—means that we must use approximate, fast methods that do not rely on performing millions of individual alignments. There are many software packages that can carry out this mapping step, using a number of different computational shortcuts. The most popular programs include *BWA* (Li and Durbin 2010), *Bowtie* (Langmead et al. 2009), *SOAP* (Li, Kristiansen, and Wang 2008), and *Stampy* (Lunter and Goodson 2011). There are a large number of new mapping programs being introduced all the time, each with their own advantages and disadvantages, and each often best for a specific NGS technology.

The general approach taken to discover polymorphisms using NGS reads is to find high-quality differences between sequenced samples and the reference genome (**FIGURE 2.4**). At any genomic position the sequenced samples can be either homozygous for the base that matches the reference, homozygous for an alternative base, or heterozygous. (I frame the discussion in terms of SNPs, but similar logic applies to any kind of variation.) This approach is often called *resequencing* because it requires a preexisting genome from an individual of the same species. There are some cases in which multiple *de novo* assemblies have been made from a single eukaryotic species (e.g., Gan et al. 2011; Li et al. 2011), but this is still rare (see Chapter 10).

The use of a reference genome means that we are highly dependent on this sequence, and that to some extent this sequence constrains the variation we see. Reference genomes are produced in a variety of ways, from a variety of material. Reference genomes can be made from single inbred lines (e.g., Adams et al. 2000), single outbred individuals (e.g., Mikkelsen et al. 2005), a group of partially inbred individuals (e.g., Holt et al. 2002), or a mixture of a limited number of outbred individuals (Lander et al. 2001). The individuals to be sequenced are often chosen because of the availability of high-quality DNA, not because they signify any particular genotype or the consensus genotype. Regardless of who is chosen as the reference line or individual, we represent genome sequences as haploid in online databases; this means that any variation present within the individual or pool must be collapsed down into a haploid state, whether or not that particular haplotype exists in nature. All of this implies that the genome sequence we use to map NGS reads against represents only a single sequence taken from nature and cannot possibly include all of the variation present in a species or population. As a result, we will be unable to map sequence reads that are very different from the reference, or that include sequences not present in the reference (either because of an insertion in our sample or because of a

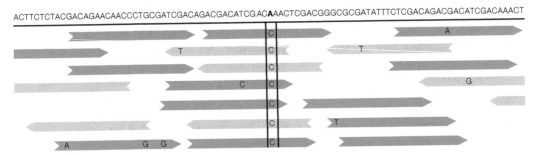

FIGURE 2.4 True polymorphisms are supported by multiple sequence reads. The reference genome is represented at the top, with reads from both strands of the sequenced sample presented below (indicated by arrows facing opposite directions and different shading). Only bases that differ from the reference are shown. The site at which a homozygous polymorphism exists is in bold and highlighted by vertical bars; all other varying sites are intended to represent sequence errors. (After Thorvaldsdóttir, Robinson, and Mesirov 2012.)

deletion in the individual used to make the reference). This is also an important reason why using a reference genome from even a very closely related species can result in biased read coverage: regions with high divergence will have fewer reads mapping to them. In general, resequencing approaches limit the type and level of variation that we detect.

A further consequence of short sequence reads is that they may map equally well to multiple places in a genome, especially after possible sequencing errors are taken into account. Longer reads (or paired-end reads) mean more confidence in the placement of individual sequences, but some repetitive elements are quite long and can be several thousand nucleotides in length. Because of this problem, we often assign a *mapping quality score* to individual reads (Li, Ruan, and Durbin 2008). Uniquely mapping reads are those in which we place the most confidence, and the mapping quality score of all reads is inversely related to the number of places in the genome a read can map. Often nucleotide variants are inferred based solely on uniquely mapping reads, of which there are many fewer in any sequencing run (Figure 2.3). The amount of nucleotide variation within repeated sequences in reference genomes is especially hard to assess, as reads with allelic variants are difficult to assign to a particular repeat element. This means that it is very difficult to identify SNPs within repeats.

The effect of high error rates

Recall that for Sanger sequencing we had more confidence in the sequence derived from PCR products because of the high error rates associated with any particular cloned fragment. Similarly, for next-generation sequencing we must be wary of the sequence associated with any single read, instead gaining confidence only when many independent reads support a particular genotype (Figure 2.4). The problem of error is more severe for NGS technologies compared with Sanger sequencing: even though error rates are dropping quickly, per base error rates are still about 0.3% in a single read using the Illumina sequencing technology (Schirmer et al. 2016). This means that there will be a high false-positive rate for nucleotide mismatches if only single reads are used. Error rates often vary in a predictable manner along a read—for instance, getting higher toward the end of reads—and there may be systematic errors that result in repeated mismatches at particular positions or associated with particular sequence motifs (Meacham et al. 2011). Sequence errors resulting in indels can occur at different rates than nucleotide mismatches, but depending on the specific NGS technology used, they may be lower (Illumina) or higher (454) in comparison (Loman et al. 2012).

As with Sanger approaches, NGS technologies aim to assign a phred-like *base quality score* to each nucleotide in a sequence read (these scores are "phred-like" because the numerical value given to each base is intended to reveal the same associated error rate as a phred score does). This base quality score represents the probability that the called base is in error and is generally fairly accurate (e.g., Hu et al. 2012). The *consensus quality score* of a position represents the final error probability associated with the specific nucleotide (or nucleotides, if heterozygous) at a site and is again a phred-like score. There are multiple ways in which consensus quality scores are determined by different software packages or in different studies, but they all generally use a combination of base quality score, mapping quality score, and the base quality scores at nearby nucleotides. These values—and several others, including those indicating the presence of neighboring indels—are then leveraged so that more reads provide increasing consensus quality scores for a single genotype. Popular programs used to identify SNPs in NGS data are *SAMtools* (Li et al. 2009) and *GATK* (McKenna et al. 2010). One problem with combining scores from different reads is that these scores do not always provide independent data on nucleotide sequences: systematic errors resulting in the same nucleotide mismatch can occur on multiple reads. These systematic errors are consistently associated with reads derived from one template strand (1000 Genomes Project Consortium 2010; Meacham et al. 2011), so the consensus quality score can be further improved by including reads from both strands supporting non-reference bases (Figure 2.4). In general, and regardless of the specific technology used, increased coverage results in increased confidence in the inferred sequence.

Diverse experimental designs for next-generation genome sequencing

The above discussion begs the question: How much coverage is enough? The answer to this question for any particular sample will be determined by factors including the heterozygosity of the individual being sequenced, the genetic distance between the individual and the reference, and the repeat content of the genome. The Lander-Waterman model can be used to calculate a rough estimate of the fraction of the genome that will have sufficient coverage, though actual coverage may differ from these expectations (e.g., Figure 2.3). As a general guideline, for completely inbred lines of species with relatively low numbers of repeats, 10X coverage of Illumina single-end sequences results in a highly accurate genome sequence (e.g., Langley et al. 2012). In contrast, for a single human individual, we likely need at least 30X paired-end coverage to achieve low genotyping error rates, largely because many heterozygous sites may be missed with lower coverage (e.g., 1000 Genomes Project Consortium 2010). Because we require multiple reads from both strands to support any allele that differs from the reference genome, the coverage requirements for heterozygous individuals is disproportionately high relative to what is needed for homozygotes. Also note that the coverage values used above refer only to uniquely mapping reads: for genomes with very high repeat content (or high amounts of contamination in the sample preparation) the total amount of sequence data that needs to be collected may be significantly larger.

The need for high read depth to ensure accurate genotypes means that whole-genome sequencing for large numbers of samples can be prohibitively expensive. Even with the huge amount of data produced by single runs of most NGS platforms, some genomes are so big that current costs do not allow a large number of individuals to be sequenced deeply (as in **FIGURE 2.5A**). One solution to this problem is to sequence multiple individuals to low depth (**FIGURE 2.5B**), leveraging the fact that true polymorphisms will appear

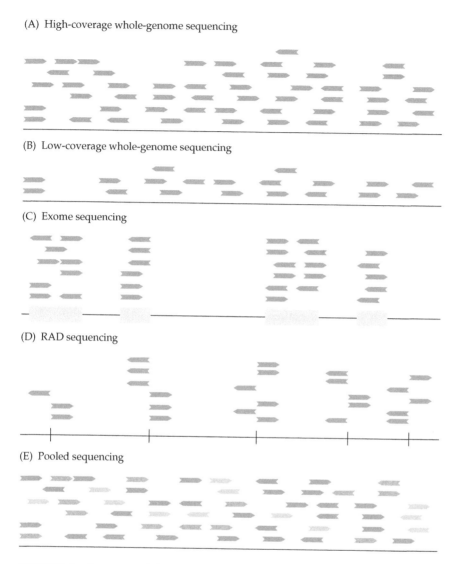

(A) High-coverage whole-genome sequencing

(B) Low-coverage whole-genome sequencing

(C) Exome sequencing

(D) RAD sequencing

(E) Pooled sequencing

FIGURE 2.5 Experimental design for next-generation sequencing. (A) For high-quality genotype calls in a single sampled individual, high read coverage is needed (each read is represented by a blue arrow, with the reference genome below). (B) Low coverage for each individual can be used when multiple individuals from a population are sampled. (C) Exome sequencing is a reduced representation approach that provides high read depth at targeted regions using sequence capture (exons are indicated by yellow boxes). Transcriptome sequencing targets the same regions but does so by sequencing mRNAs. (D) RAD sequencing is a reduced representation approach that provides high read depth at restriction enzyme cut sites. At each cut site (indicated by vertical lines) reads generated from alternative template strands extend in opposite directions. (E) Pooled sequencing combines multiple unlabeled sample individuals in a single sequencing lane so that the source of each read cannot be determined. For demonstration purposes, reads from three hypothetical individuals pooled together are represented.

many times in the total sample to find high-quality SNPs (e.g., Liti et al. 2009; Sackton et al. 2009; 1000 Genomes Project Consortium 2010). This approach does not necessarily ensure accurate genotypes for each individual—and will miss many low-frequency alleles—but it does have the advantage of quickly identifying a large number of variants. With very low coverage many sites will not even be sequenced in every sample, leading to different sample sizes at each base in the genome (e.g., Begun et al. 2007). Programs made specifically to estimate genotypes and allele frequencies from this type of error-prone data should be used in these cases (e.g., *ANGSD*; Korneliussen, Albrechtsen, and Nielsen 2014).

In order to obtain accurate calls for each sampled individual, several *reduced representation* approaches have been used to sequence a smaller portion of each genome at higher coverage. The simplest reduced representation method is PCR: by amplifying only those regions of interest, one can get high coverage of a targeted set of loci. But PCR can produce only relatively short amplicons—even when long-amplification PCR is used—and therefore a very large number of reactions must be performed in order to justify the use of next-generation methods to obtain sequence reads.

Instead of PCR, three alternative reduced representation technologies are now in wide usage. The first uses *sequence capture* to target specific loci for sequencing. The capture is often accomplished by hybridization to a microarray that has been constructed with thousands of oligonucleotides complementary to the targeted loci (Albert et al. 2007; Hodges et al. 2007; Okou et al. 2007). After the sample has been hybridized to the array and the non-complementary DNA has been washed away, the remaining DNA can be eluted and used as a template for next-generation sequencing. A similar approach using oligonucleotides suspended in a solution can also be used (Gnirke et al. 2009), even with non-model organisms (Jones and Good 2016). The most common application of sequence capture technologies is so-called *exome* sequencing (**FIGURE 2.5C**), in which every exon in an annotated genome is represented by capture probes, allowing deep coverage of coding regions within single samples to be achieved. This approach has now been used to find rare variants in samples of thousands of individuals (e.g., Tennessen et al. 2012).

Similar to exome sequencing, *transcriptome sequencing* (or *RNA-seq*) targets coding regions of the genome by sequencing mRNAs (Figure 2.5C). This reduced portion of the genome can be queried in multiple individuals cheaply, and using the same tissues in each individual ensures that similar genes are recovered. Transcriptome sequencing does not require a reference genome and therefore can be used in many non-model organisms (e.g., Vera et al. 2008; Parchman et al. 2010; Gayral et al. 2013; De Wit, Pespeni, and Palumbi 2015). The extraction of RNA and the necessary reverse transcription steps make transcriptome sequencing slightly more expensive per library than DNA sequencing, but it has become a powerful approach for population genomics (e.g., Romiguier et al. 2014; Pease et al. 2016).

A third reduced representation approach ensures that a relatively limited number of positions in the genome will be sequenced many times by using restriction enzymes. This approach, called *restriction-associated DNA (RAD) sequencing* (Baird et al. 2008), uses oligonucleotide adaptors to capture and sequence only those template fragments that have been cut by a specific restriction enzyme (**FIGURE 2.5D**). Although multiple different variants on this approach are in use (e.g., Andolfatto et al. 2011; Elshire et al. 2011; Luca et al. 2011; Peterson et al. 2012), they all rely on the fact that there are usually only hundreds to thousands of sites that can be cut by any particular restriction enzyme in a given genome (collectively they are sometimes called *genotyping-by-sequencing [GBS] methods*). While the cut-sites obtained via these methods are not necessarily specific targets of sequencing, each site can be reliably resequenced in each sample. By sequencing only the cut-sites, RAD-based approaches provide deep coverage across many loci for single individuals. The resulting sequence data can provide an accurate view of genome-wide patterns of diversity within and between populations (e.g., Hohenlohe et al. 2010).

At the other end of the spectrum of genome sizes, for many species current NGS platforms produce unnecessarily high read depth. Individual samples can be labeled with barcodes so that they can be combined in a single sequencing lane (i.e., in a single run of a particular platform), but each labeling reaction remains a moderate expense. Therefore, as an alternative method to both discover high-quality SNPs and to estimate the sample allele frequencies of these SNPs, DNA from multiple individuals can be pooled together in a single lane without barcodes (**FIGURE 2.5E**). Pooled sequencing cannot provide individual genotypes, instead leveraging deep read depth to estimate accurate allele frequencies genome-wide (reviewed in Schlötterer et al. 2014). Pooled approaches were first applied to individuals hybridized to microarrays (Turner, Hahn, and Nuzhdin 2005) or to the DNA sequencing of reduced representation samples (e.g., Altshuler et al. 2000; Van Tassell et al. 2008; Druley et al. 2009). They have now been used to sequence the whole genomes of multiple organisms using NGS technologies (e.g., Burke et al. 2010; Turner et al. 2010; Kolaczkowski et al. 2011; Cheng et al. 2012).

Pooled sequencing estimates allele frequencies by counting reads supporting each alternative allele. The aim is to have each read come from a different sampled chromosome from among the input DNAs; for this reason, it is important to have a larger number of chromosomes than the average number of reads at a locus (Futschik and Schlötterer 2010). If there are too few chromosomes used in the pool some will be sequenced multiple times, leading to bias in allele frequency estimates (Futschik and Schlötterer 2010; Kolaczkowski et al. 2011). In any case, the minimum variance achievable in estimates of the allele frequency occurs when the number of reads equals the number of chromosomes in the pool (Futschik and Schlötterer 2010; Kolaczkowski et al. 2011; Lynch et al. 2014). Although there are some valid concerns about the accuracy of pooled sequencing—especially regarding the false-positive rate for low-frequency alleles (Cutler and Jensen 2010)—empirical validation has provided remarkable support for the approach overall (Zhu et al. 2012). In addition to SNPs, pooled sequencing has been used to detect both transposable elements (Kofler, Betancourt, and Schlötterer 2012) and copy-number variants (Schrider, Begun, and Hahn 2013).

While the optimal sampling design for NGS projects likely does not exist, the main goal must be to collect enough reads to ensure an accurate set of polymorphisms. Depending on the size and repeat content of the sampled genome, the experimental protocols described here attempt to find a balance between inadequate coverage per sample and too much coverage per sample. Further consideration must be contingent on the need for accurate allele frequency estimates versus accurate genotypes of particular strains. And if individuals or strains can be sampled again in the future to improve coverage—or can be phenotyped for association studies—this knowledge can change the decisions one makes (e.g., on pooled sequencing versus low-coverage whole-genome sequencing). Of course I have assumed that an investigator has only finite sequencing capabilities and that library costs are nontrivial. All of these considerations may change with the *next* next-generation of sequencing technologies.

GENOTYPING

A major issue in dealing with the high error rates associated with NGS platforms is an unnecessarily high false-negative rate. In the above discussion we considered the read depth needed to avoid false-positive calls that are due to error, but just as important may be the number of true SNPs that are missed because of conservative criteria for calling non-reference bases. For many applications the missing genotype data and skewed allele frequencies that result from false negatives will cause larger problems than any false positives. Especially when the reference base happens to represent the ancestral state (which it often does; see Chapter 6), false negatives can cause a rather extreme downward shift in the estimated frequency of derived alleles. This is why it is preferable to treat unreliable base calls as missing data rather than defaulting to the reference allele.

As alluded to in the previous section, we can often take advantage of having a large sample of low-coverage genomes to make more accurate genotype calls by separating the variant *discovery* process from the variant *genotyping* process. The distinction between these two steps can be made clearer by considering the sequence evidence we would be willing to accept in order to call an individual heterozygous for a site at which we are sure there is a SNP in the population (say, an A/G polymorphism). Many people would readily accept even one high-quality A read and one high-quality G read as evidence of heterozygosity if this was already a known variant. Without independent knowledge of the A/G polymorphism, however, it is unlikely that one read supporting each base (or, more importantly, one read supporting the non-reference base) would tip the scales in favor of genotyping the individual as a heterozygote.

In the above hypothetical scenario we have relaxed our stringency in the variant genotyping process because the variant discovery process has identified a variant site. We have effectively started the genotyping process with a prior probability of there being a non-reference allele present and have changed our posterior probability of the genotype accordingly; this Bayesian approach is in fact how several programs achieve high genotyping accuracy (e.g., McKenna et al. 2010; DePristo et al. 2011; Le and Durbin 2011; Li 2011; Nielsen et al. 2012). In practice we may go through several iterations of the variant discovery and genotyping processes in order to more accurately identify high-quality polymorphisms, updating our prior probability at each iteration. This *allele-aware* approach to genotyping (which is also referred to as *genotyping-by-sequencing*) is especially effective in genetic crosses both because linkage disequilibrium in the recombinant population is high, and therefore missing genotypes can be imputed easily, and because there are no low-frequency alleles. Extremely low read depth can be used in recombinant populations designed to be homozygous (e.g., recombinant inbred lines), and accurate genotyping can be accomplished with even 0.1X coverage per individual (Huang et al. 2009; Xie et al. 2010).

Putting aside NGS technologies, there are a number of fast, cheap, and accurate approaches to genotyping natural populations (for a review, see Gibson and Muse 2009, pp. 178–185). For most of these methods we must know not only the position of the SNP ahead of time, but also the identity of the two alternate alleles. Given this information we can then apply both low- and high-throughput methods that enable us to genotype individuals more cheaply than sequencing. The simplest genotyping method couples PCR with restriction digestion to genotype polymorphisms found in restriction enzyme cut-sites. This method—referred to as *Snip-SNP, PCR-RFLP*, or *cleavable amplified polymorphic sequences (CAPS)*—is dependent on the polymorphism being located within a nondegenerate position in a cut-site, with the two alternate alleles producing bands that run at different speeds when visualized on an agarose gel. A fluorescently labeled primer that enables automated scoring of individual genotypes can also be used. High-throughput methods for genotyping can use array-based technologies to query hybridization intensity to alternative alleles; these are often referred to as *SNP-chips*. The first such arrays contained 500 SNPs (Wang et al. 1998), while current Affymetrix arrays (v. 6.0) contain more than 900,000 probes for known SNPs and more than 900,000 probes for known CNVs in humans (McCarroll et al. 2008). Medium-throughput methods can query many fewer SNPs but can genotype multiple individuals at a time (e.g., Sequenom iPlex assays). Such technologies can be useful for conducting follow-up experiments to whole-genome studies for a smaller number of interesting polymorphisms in a much larger sample (e.g., Jones et al. 2012).

One final caveat must be mentioned in regard to genotyped samples: we should remember that all the SNPs being used were previously ascertained. The original sample used to discover the SNPs was likely smaller than the one being genotyped—it may even have consisted of a single outbred individual sequenced to construct a reference genome (so $n = 2$). As a result, we have introduced an *ascertainment bias* to our study. We are biased because we are disproportionately sampling intermediate-frequency polymorphisms in the small sample, and this bias can have important downstream effects on inferences from population genetic data if it is not taken into account (e.g., Kuhner et al. 2000; Wakeley et al. 2001; Nielsen, Hubisz, and Clark 2004). We will consider the effects of ascertainment bias on various inferences in molecular population genetics throughout the book, with Figure 6.6 providing an illustrative explanation of the general problem.

DESCRIBING VARIATION

3

Once we have obtained a sample of sequences from a population of interest, we must describe the variation we observe. There are many ways to summarize molecular variation for individual nucleotides, microsatellites, and full sequences, and therefore many different *statistics* that describe the data. A statistic is simply a summary of the sample of observations (the sequences). In molecular population genetics we tend to focus on those statistics that are *estimators* of the theoretical parameter $\theta\,(\equiv 4N_e\mu)$, which represents the amount of variation found at autosomal loci in hypothetical populations in which all variation is neutral and at mutation-drift equilibrium. This focus is motivated by a desire to connect our empirical observations to theoretical predictions of the neutral model about both the amount of diversity found in nature and the expected dynamics of selected mutations in populations with varying values of θ. It is important to note, however, that the statistics discussed below are only estimators of θ under relatively restrictive assumptions—namely, idealized populations without selection and at demographic equilibrium, with additional restrictions specific to each mutational model. There are alternative ways to understand these statistics, and later in the chapter I will discuss several different interpretations of summaries of variation.

There are also a growing number of maximum likelihood (ML) and Bayesian methods that can be used to estimate the parameter θ from data (e.g., Kuhner, Yamato, and Felsenstein 1995; Nielsen 2000; Beerli 2006). Some likelihood-based methods can even estimate θ from panels of pre-ascertained SNPs rather than full sequence data, as long as the ascertainment scheme is known exactly (e.g., Kuhner et al. 2000). While these methods are growing in popularity—and have some advantages over moment-based estimators—they also have some important limitations. Most of these limitations are simply computational: the size of the datasets that can be analyzed, in terms of both the number of samples and the length of sequences, is limited. These methods do not scale up to whole-genome (or even whole-chromosome) datasets. In addition, many likelihood-based methods (both ML and Bayesian) do not do particularly well with missing data, a limitation especially relevant to whole-genome resequencing projects that may have variable coverage across nucleotides. Many of these problems will be overcome as computational tools and resources improve, but here I will only discuss some of the most commonly used methods for summarizing sequence diversity from simple summaries of the data.

MEASURES OF SEQUENCE DIVERSITY

Common estimators of θ for SNPs under the infinite sites model

For individual nucleotide polymorphisms, the ith allele at a site has sample frequency p_i, such that the sum of all allele frequencies is equal to 1. If we consider only biallelic sites, then obviously $p_1 + p_2 = 1$. A useful way to summarize the variation at a single polymorphic site is by calculating the sample *heterozygosity*, which is given by:

$$h = \frac{n}{n-1}\left(1 - \sum p_i^2\right) \tag{3.1}$$

where n is the number of sequences in the sample. Although this value can be calculated from data obtained from outbred individuals, inbred lines, or haploid chromosomes, it is referred to as *heterozygosity* (or, more properly, *expected heterozygosity*) because it is an estimate of the proportion of all individuals that would be heterozygotes if there was random union of gametes—that is, if chromosomes were randomly paired to create diploid individuals. In fact, it is common to refer to many measures of nucleotide diversity as measures of heterozygosity, regardless of the origin of the data.

Given this measure of diversity at a single site, we now consider measures of diversity across an entire sequence. There are two methods that are most commonly used to summarize nucleotide diversity, as they do not require assignment of derived and ancestral alleles (additional measures that use this information are described below). Following our above definition of heterozygosity at a single site, we can define the sum of site heterozygosities as:

$$\pi = \sum_{j=1}^{S} h_j \tag{3.2}$$

where S is the number of segregating sites and h_j is the heterozygosity as defined in Equation 3.1 for the jth segregating site (Nei and Li 1979; Nei and Tajima 1981a; Tajima 1983). Under the infinite sites model for a diploid Wright-Fisher population at equilibrium, $E(\pi) = \theta$, which is why this statistic is sometimes called θ_π. Statistics that are estimators of parameters are often denoted with a caret above them (e.g., $\hat{\theta}_\pi$), but I will not use this notation here. Because heterozygosity at a monomorphic site is 0, we could also sum across all sites in an alignment with no change in result. Calculating π for the alignment shown in Figure 1.1, we find that the site heterozygosities are 0.5, 0.667, 0.5, 0.667, 0.5, and 0.5 for the six polymorphisms, resulting in $\pi = 3.33$. We often give this value per site, in which case $\pi = 3.33/15 = 0.222$.

This measure of diversity gives us the average number of pairwise nucleotide differences between any two sequences, and can therefore also be calculated as:

$$\pi = \frac{\sum_{i<j} k_{ij}}{n(n-1)/2} \tag{3.3}$$

where k_{ij} is the number of nucleotide differences between the ith and jth sequences in the sample and the denominator represents the number of unique comparisons being made between n sequences. If we were to calculate the number of differences in each of the six possible pairwise comparisons of the four sequences in Figure 1.1, we would again find that the average number of differences is 3.33. If the number of sequences is large, calculating π using Equation 3.2 is much more efficient than calculating it using Equation 3.3.

As discussed earlier in the book, we often ignore indel variation when calculating differences between sequences. When applying Equation 3.2, individual sequences with indels are treated as missing data, and therefore the value of n will differ across positions in calculating site heterozygosities. This will also be the case in resequencing datasets with varying coverage—and consequently a substantial amount of missing data—across sites. When applying Equation 3.3, it is common to simply ignore positions in the alignment with a gap character in any of the n sequences. This approach will make a difference in the calculations of π produced by Equations 3.2 and 3.3, as we can demonstrate using the modified alignment shown in **FIGURE 3.1**.

In this case $\pi = 3.49$ using Equation 3.2 ($h = 0.66$ instead of 0.5 at the first segregating site), but $\pi = 2.83$ using Equation 3.3 because we have completely excluded positions 1, 2, 9, and 10 from all of the pairwise comparisons. Even if there were no segregating sites within positions with gaps (as there is in the second site in Figure 3.1), the different methods would also give different results when calculated per base pair. This is because all 15 sites are included in the first calculation, but only 11 are included in the second. Especially for genome-scale datasets, it is much more useful to apply methods (such as those based on Equation 3.2) that can easily deal with large amounts of missing data (e.g., Begun et al. 2007; Ferretti, Raineri, and Ramos-Onsins 2012).

We can also express the expected amount of variance in π as a function of the parameter θ. We express the variance in terms of our theoretical models because we know there is a large historical component to the structure of variation in our sample—that is, the structure of the genealogy that relates the sampled chromosomes. Although the theoretical variance of π under the neutral-equilibrium model will not always be ideal to use—for example, in cases where populations do not meet our assumptions—the alternative would be to use highly inaccurate expectations that do not take the hierarchical structure of the underlying genealogy into account and are much lower than the

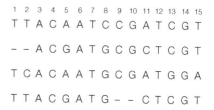

FIGURE 3.1 Alignment of four sequences with gaps.

values observed in most datasets. For the case of no recombination, the expected variance in π under this idealized model is:

$$Var(\pi) = \frac{n+1}{3(n-1)}\theta + \frac{2\left(n^2+n+3\right)}{9n(n-1)}\theta^2 \tag{3.4}$$

As Equation 3.4 shows, there is a large amount of variance associated with π, and even with a large number of samples the variance does not approach zero (Tajima 1983). The expected variance under increasing levels of intralocus recombination is given in Pluzhnikov and Donnelly (1996).

An alternative method to π for summarizing nucleotide variation uses the total number of segregating sites in the sample, S. However, because larger sample sizes will result in larger values of S, we must adjust the statistic to be (Ewens 1974; Watterson 1975):

$$\theta_W = \frac{S}{a} \tag{3.5}$$

where a is equal to:

$$a = \sum_{i=1}^{n-1}\frac{1}{i} \tag{3.6}$$

The θ_W statistic is often referred to as *Watterson's theta*, as it is also an estimator of the parameter θ assuming a Wright-Fisher population at equilibrium and the infinite sites mutational model. We can also rewrite the above equations to express the expected number of segregating sites as a function of θ: $E(S) = \theta a$. Given the alignment of sequences in Figure 1.1, $\theta_W = 6/(1/1 + 1/2 + 1/3) = 3.28$. We can see that this value is very similar to the one given by π. Once again, differences in the exact value of θ_W can arise in alignments with gaps depending on whether we include or exclude positions with gaps. If we include positions with gaps, we need to calculate separate values of θ_W for regions with different numbers of sequences, so that $\theta_W = 5/(1/1 + 1/2 + 1/3) = 2.73$ for positions with four sequences and $\theta_W = 1/(1/1 + 1/2) = 0.667$ for positions with three sequences in Figure 3.1. The total value of the statistic is then the sum of these two, $\theta_W = 3.40$, and per site measures would have to weight the number of positions with different numbers of samples accordingly.

We can again give the expected variance of S in terms of θ, assuming a Wright-Fisher population at equilibrium and the infinite sites mutational model with no recombination:

$$Var(s) = \sum_{i=1}^{n-1}\frac{1}{i}\theta + \sum_{i=1}^{n-1}\frac{1}{i^2}\theta^2 \tag{3.7}$$

When there is free recombination, the variance reduces to just the first term in Equation 3.7, as there is no longer any evolutionary variance (see Chapter 6). Pluzhnikov and Donnelly (1996) give the variance in S for intermediate levels of recombination. Equation 3.7 can also be used to find the expected variance in θ_W, which is given by $Var(\theta_W) = Var(S)/a^2$, with a as defined in Equation 3.6.

Comparing the expected variances of our two estimators of θ, we see that the variance in θ_W is lower than the variance in π and approaches zero with larger sample sizes (albeit very slowly). This would seem to suggest that θ_W is the "better" of the two estimators. However, it is also much more sensitive to both the presence of slightly deleterious alleles and sequencing error, even in equilibrium populations (see below). It is therefore common to see both statistics used side by side, as well as a comparison of the two (i.e., Tajima's D statistic; see Chapter 8).

The frequency spectrum of alleles

So far we have considered descriptions of variation using only a single summary statistic like π or θ_W. But there are also useful graphical ways to describe variation, and these in turn will lead us to more measures of nucleotide diversity. Consider the alignment shown in **FIGURE 3.2A**. Among these 10 chromosomes are 9 segregating sites, each of which can have a frequency between $1/n$ and $(n-1)/n$ (remember that if an allele is present on all n chromosomes it is not polymorphic). If we do not know which allele is ancestral and which is derived, a simple way to describe the variation at each site would be the *minor allele frequency* (*MAF*), or the frequency of the less common allele. The minor allele frequencies in a sample therefore range from $1/n$ to 0.5. We can visually summarize the MAF of all the segregating sites in our sample by graphing the *allele frequency spectrum*, in which each bin in the histogram represents the number of sites (or proportion of sites) in the alignment with a given minor allele frequency (**FIGURE 3.2B**). If we define S_i^* to be the number of segregating sites with alleles found on i chromosomes, with the * specifying that this count is without regard to ancestral/derived relationships, and allow the value of i to range from 1 to $n/2$ (rounding down for odd-numbered n), then the allele frequency spectrum is simply a plot of the various values of S_i^*. The

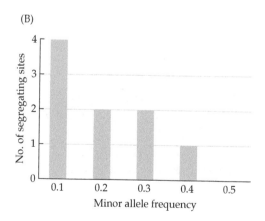

(A)

```
  1 2 3 4 5 6  7  8 9 1011121314 15 16 17 18 1920212223 2425 2627 2829
GCTCACCGGAATTATCCGATATGCTAGTA
GCTTACCGGAATTATGCGATATGCTTGTA
GCTCACCGGAATTATGCGATATGGTAGAA
GCTCACCGGAATTATGCGATATGGTAGAA
GCTCACCGGGATGATGCGATATGCTAGTA
GCTCACCGGAATTATGCGATATGCTAGAA
GCTTACCGGAATTATCCGATATGCTAGTA
GCTCACAGGGATTATGCGCTATGCTAGTA
GCTCACCGGAATTATGCGATATGGTAGAA
GCTCACCGGAATTATCCGATATGCTAGTA
```

FIGURE 3.2 (A) For each segregating site the minor allele is blue and the major allele is black. For this sample, $n = 10$, $L = 29$, and $S = 9$. (B) The "folded" allele frequency spectrum for the alignment shown in (A).

allele frequency spectrum is also referred to as the *site frequency spectrum* (*SFS*), and when it shows only minor allele frequencies it is called the *folded spectrum*.

In order to infer the ancestral and derived states of each allele it is common to use the sequences of one or more closely related species. It is generally assumed that the allele matching these species represents the ancestral state, although parallel mutations in each lineage can lead to incorrect inferences; this is why using two outgroup species is often preferred—the chance of parallel mutations in all three lineages should be vanishingly small. If the outgroup species sequences do not match either allele, or if they are contradictory, then no assignment of states as ancestral and derived can be made. **FIGURE 3.3A** shows the same alignment as in Figure 3.2A, but one allele has been assigned as the derived state. We can now describe variation at each site using the *derived allele frequency* (*DAF*), which can fall between $1/n$ and $(n - 1)/n$. Again, we summarize the DAF of all segregating sites by the allele frequency spectrum, this time in the *unfolded* state (**FIGURE 3.3B**). If we define S_i to be the number of segregating sites with *derived* alleles found on i chromosomes—with i now ranging from 1 to $n - 1$—then the allele frequency spectrum is simply a plot of the various values of S_i. For clarity I have plotted just S_i, but it is also common to see the allele frequency spectrum plotted as S_i/S, or the proportion of all segregating sites at each frequency (see Chapter 6 for further discussion of this point). It should also be clear that some alleles with MAF < 0.05 may actually have DAF > 0.95, since no distinction can be made without assigning derived and ancestral states. This implies that disregarding sites with MAF < 0.05 (as might be done if these are considered to be either sequencing errors or deleterious mutations) may lead to the loss of evolutionarily important variation. In addition to being a nice way to summarize variation, in later chapters we shall see that theoretical

(A)

```
        1  2 3 4  5  6 7  8  9 10 11121314 15 16 17 181920212223 24 25 262728 29
GCTCACCGGAATTATCCGATATGCTAGTA
GCTTACCGGAATTATGCGATATGCTTGTA
GCTCACCGGAATTATGCGATATGGTAGAA
GCTCACCGGAATTATGCGATATGGTAGAA
GCTCACCGGGATGATGCGATATGCTAGTA
GCTCACCGGAATTATGCGATATGCTAGAA
GCTTACCGGAATTATCCGATATGCTAGTA
GCTCACAGGGATTATGCGCTATGCTAGTA
GCTCACCGGAATTATGCGATATGGTAGAA
GCTCACCGGAATTATCCGATATGCTAGTA
```

(B)

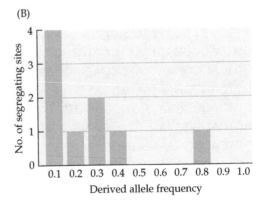

FIGURE 3.3 Same alignment as in Figure 3.2, but with alleles now polarized as ancestral and derived. (A) For each segregating site the derived allele is blue. (B) The "unfolded" allele frequency spectrum for the alignment shown in (A).

predictions about the shape of the allele frequency spectrum can be used to infer the history of a set of sequences.

We now consider several statistics summarizing sequence diversity that use information about the frequency of derived alleles, as these capture more information about our data. Fu and Li (1993a) defined a statistic, ξ_1, based on the number of derived singletons in a sample, which can be written in the terms used here as:

$$\xi_1 = S_1 \tag{3.8}$$

where S_1 is the number of segregating sites with derived alleles found on only one chromosome. This statistic is sometimes referred to as θ_e (with e referring to external branches of genealogies). From the alignment in Figure 3.3A, we can calculate the value of $\xi_1 = 4$. We cannot calculate ξ_1 from Figure 3.1 or 3.2A, as we did not identify the derived alleles. However, we can also define a statistic, η_1, based on all singletons in a sample—whether or not they are ancestral or derived—which in the terms used here is written as:

$$\eta_1 = S_1^* \frac{n-1}{n} \tag{3.9}$$

with S_1^* as defined above (Fu and Li 1993a). From the alignment in Figure 3.2A, the value of $\eta_1 = 3.6$. The $(n-1)/n$ term corrects for the fact that counting all singletons in a sample regardless of ancestral/derived relationship will tend to slightly overcount the expected number of derived singletons in a neutral-equilibrium population.

A second summary statistic of diversity that uses ancestral state information is θ_H:

$$\theta_H = \frac{\sum_{i=1}^{n-1} i^2 S_i}{n(n-1)/2} \tag{3.10}$$

where S_i is again the number of segregating sites where i chromosomes carry the derived allele (Fay and Wu 2000). (According to Justin Fay, the H in θ_H refers to the weight this measure gives either to high-frequency derived alleles or to the homozygosity of the derived variant.) From the alignment in Figure 3.3A, the value of θ_H can be calculated as $([1^2 * 4] + [2^2 * 1] + [3^2 * 2] + [4^2 * 1] + [8^2 * 1])/45 = 2.36$. We can see that this is slightly lower than the value of ξ_1 because of the relatively large number of derived singletons in the sample.

All of these statistics—ξ_1, η_1, and θ_H—are estimators of θ in a Wright-Fisher population at equilibrium under an infinite sites mutational model; specifically, $E(\xi_1) = E(\eta_1) = E(\theta_H) = \theta$. These relationships arise because we know the expected shape of the allele frequency distribution under our standard neutral assumptions (see Equation 6.8 and Figure 6.5). Knowledge of the expected spectrum and summary statistics of this spectrum will be especially useful in detecting natural selection (Chapter 8) and nonequilibrium demographic histories (Chapter 9).

Incidentally, we can also define θ_W and π in terms of derived allele frequencies as:

$$\theta_W = \frac{\sum_{i=1}^{n-1} S_i}{a} \tag{3.11}$$

with a as defined in Equation 3.6, and:

$$\pi = \frac{\sum_{i=1}^{n-1} i(n-i)S_i}{n(n-1)/2} \tag{3.12}$$

The values of θ_W and π from the alignment in Figure 3.3A are 3.18 and 2.98, respectively. See Chapter 8 for further discussion of the differences between estimators of θ.

Estimators of θ for SNPs under finite sites models and incorporating sequencing error

The statistics described above are estimators of the population mutation parameter θ only under several restrictive assumptions. One of these assumptions is that mutations at a given site have only occurred once—that is, the infinite sites mutational model (Chapter 1). It is easy to see why these statistics are biased downward if there are multiple mutations at a site. For example, the θ_W statistic is an underestimate of diversity when polymorphic sites are not biallelic. In this case the total number of segregating sites is smaller than the number of separate mutations, and therefore θ_W underestimates the heterozygosity of the sample. In populations with low diversity this will not be much of a problem: in human samples there are relatively few triallelic sites and almost no quadrallelic sites (Hodgkinson and Eyre-Walker 2010). However, in species with high diversity this can be a real problem (e.g., Wang et al. 2010).

Conversely, in the presence of sequencing errors these estimators will be biased upward. This is because errors will be mistaken for true segregating sites, and consequently many of these sites will be included in calculations of diversity statistics. Sequencing error is bound to be a large problem in genome-scale datasets, simply because most next-generation sequencing technologies are more error-prone and many sites will have low sequencing coverage within a single sample, resulting in incorrect base calls.

In both cases—multiple true mutations at a single position or high sequencing error—the statistic θ_W will be more affected than π (Tajima 1996; Achaz 2008; Johnson and Slatkin 2008). There is less of an effect on π in the case of finite sites because the third (or fourth) allele at a site will contribute less to pairwise differences as a result of its lower frequency. Likewise, we expect sequencing errors to be present on only one or a few sampled chromosomes, and these will therefore contribute little to the average number of differences between chromosomes. Nonetheless, all of the statistics described above will be affected by these complications to some degree (Clark and

Whittam 1992; Tajima 1996; Achaz 2008; Johnson and Slatkin 2008; Knudsen and Miyamoto 2009).

Tajima (1996) provided multiple ways to adjust the calculations of θ_W and π to account for a finite sites mutation model by using the Jukes-Cantor correction (Chapter 7) and assuming that there is no sequencing error. To adjust θ_W we first need to count the minimum number of mutations in the sample, S', rather than the number of segregating sites, S. If there are j different alleles at a site in our sample, then there have been at least $j-1$ mutations at this site; summing the values of $j-1$ across all positions in an alignment gives S'. (Note that if all sites are biallelic, then $S' = S$.) Using the number of mutations per site (i.e., dividing by the length of the sequence, L), Tajima showed that a new version of the statistic θ_W that takes into account the possibility of multiple mutations at a site is:

$$\theta_{W(finite)} = \frac{S'/L}{a - b(S'/L)} \tag{3.13}$$

with a as defined in Equation 3.6 and b defined as:

$$b = \frac{4a}{3} - \frac{3.5\left(a^2 - \sum_{i=1}^{n-1}\frac{1}{i^2}\right)}{3a} \tag{3.14}$$

For the alignment shown in Figures 3.2A or 3.3A, $S' = S = 9$ and $\theta_{W(finite)} = 0.125$, which across the whole locus equals $0.125 * 29 = 3.62$ (compared to $\theta_W = 3.18$).

Likewise, a new version of π that takes into account the possibility of multiple mutations at a site using the Jukes-Cantor correction is given by:

$$\pi_{finite} = \frac{\pi/L}{1 - 4(\pi/L)/3} \tag{3.15}$$

with π calculated according to Equations 3.2 or 3.3 (Tajima 1996). For the alignment shown in Figure 3.2A or 3.3A, $\pi/L = 2.98/29 = 0.103$. Following Equation 3.15, $\pi_{finite} = 0.119$, which over the whole locus would equal 3.45. We can see that the values of both statistics corrected for the possibility of multiple mutations at a site are slightly larger than the uncorrected versions, which is consistent with our intuition that the effect of ignoring finite sites is to underestimate nucleotide diversity.

Sequencing errors will occur using any sequencing technology, from traditional Sanger-based methods to modern next-generation methods. As discussed in Chapter 2, there are steps researchers can take to minimize errors, including sequencing multiple clones when doing Sanger sequencing and using strict filters—including minimum read depth and minimum quality scores—when using next-generation technologies. However, increasing the minimum acceptable quality score or read depth can also lead to bias: as more and more true polymorphisms are eliminated, the estimated levels of heterozygosity will inevitably go down (Lynch 2008). Instead, most corrections for errors in estimators of θ take advantage of the fact that errors will likely only be found on a single chromosome and will therefore appear as derived singletons in a sample (Achaz 2008; Johnson and Slatkin 2008; Knudsen and

Miyamoto 2009). Other methods are based on deep sequencing of only a single outbred individual, and errors are recognized at sites with more than two alleles (Lynch 2008). Because we do not know which singletons are errors in larger samples, we must ignore all singletons, adjusting the statistics that use the expected numbers of segregating sites (θ_W) and pairwise differences (π) after the expected number of singletons have been removed (statistics based on only the number of singletons in a sample obviously cannot be corrected in this way). Multiple methods have been proposed for carrying out this, or similar, corrections (e.g., Achaz 2008; Johnson and Slatkin 2008; Knudsen and Miyamoto 2009), but here I only present the error-corrected estimators given in Achaz (2008).

Let us define S_{multi} as the number of nonsingleton (*multiton*) segregating sites, that is, the number of segregating sites not including derived singletons: $S_{multi} = S - S_1$. We can now introduce a new statistic based on S_{multi} that adequately corrects for sequencing errors:

$$\theta_{W(multi)} = \frac{S_{multi}}{a - 1} \qquad (3.16)$$

with a as defined in Equation 3.6. Using the alignment shown in Figure 3.3A, $\theta_{W(multi)} = 2.73$ (compared to $\theta_W = 3.18$ for the same dataset).

Similarly, if we define π_{multi} as the sum of site heterozygosities (Equation 3.2) for only those polymorphic sites that do not have derived singletons, then we can give another new statistic that takes errors into account:

$$\theta_{\pi(multi)} = \pi_{multi} \left(\frac{n}{n - 2} \right) \qquad (3.17)$$

Using the alignment shown in Figure 3.3A, $\theta_{\pi(multi)} = 2.72$ (compared to $\pi = 2.98$ for the same dataset). Achaz (2008) also shows how to correct θ_W and π for errors if the derived and ancestral states are not known.

Both $\theta_{W(multi)}$ and $\theta_{\pi(multi)}$ will accurately estimate the parameter θ in a Wright-Fisher population at equilibrium under an infinite sites mutation model, assuming that all errors are present in the sample as singletons. Errors present on more than one chromosome, possibly because of a consistent bias in sequencing technologies, will not be accounted for by such approaches. Regardless of whether the assumptions of the standard neutral model are true, both $\theta_{W(multi)}$ and $\theta_{\pi(multi)}$ represent statistics that summarize nucleotide diversity without counting derived singletons and may therefore be preferred when error rates are high. Note that while our corrections for finite sites and sequencing error are both special cases of the standard θ_W and π, they assume very different things about the process of mutation. Corrections for finite sites explicitly allow three alleles at a position but no error, while the framework for correcting sequencing errors assumes there are infinite sites and therefore no positions with multiple hits that are not errors. Finally, although we saw earlier that the theoretical variance for θ_W is lower than that for π, this expectation assumes no sequencing error. In the presence of sequencing error, π becomes a "better" measure of diversity as it is less affected by these errors (Achaz 2008; Johnson and Slatkin 2008). Of course both of these statistics are more robust than estimators that use only singletons (ξ_1 and η_1), as these are expected to be the most affected by errors in sequencing.

Interpreting measures of nucleotide variation

Under a neutral model—in which all alleles at segregating sites are selectively equivalent and there is no effect of selection on sites linked to the neutral variants—the level of variation at a locus is governed solely by a balance between the input of mutation and the loss of alleles by genetic drift. If these assumptions hold and populations are at equilibrium (and even often when they are not), the expected average number of nucleotide differences between sequences per site, π, is (Li 1977; Tajima 1983):

$$E(\pi) = \theta = 4N_e\mu \tag{3.18}$$

In this case, levels of polymorphism at a locus should be linearly proportional to both the effective population size (which determines the effect of genetic drift) and the neutral mutation rate.

However, diversity is only weakly correlated with apparent population size. This early observation was called the *paradox of variation* (Lewontin 1974) and still remains even after many more datasets have been collected (Leffler et al. 2012; Corbett-Detig, Hartl, and Sackton 2015). Measurements of nucleotide variation from hundreds of species across the tree of life continue to show that even though population sizes vary across many orders of magnitude (from ubiquitous bacteria to exceedingly rare vertebrates), the mean difference in nucleotide diversity between bacteria and vertebrates spans less than two orders of magnitude (**FIGURE 3.4**). Among eukaryotes, levels of variation in mitochondrial DNA show no correlation with population size (Bazin, Glemin, and Galtier 2006), and there is only a weak relationship between nuclear genes and what we assume is the census population size of different organisms (Figure 3.4).

In addition to the lack of a linear relationship between the assumed population size and levels of variation, there is an unexpected positive correlation

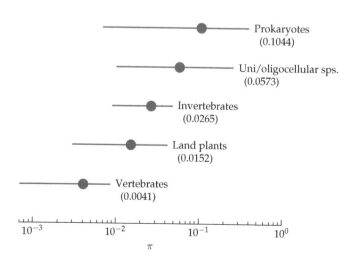

Prokaryotes
(0.1044)

Uni/oligocellular sps.
(0.0573)

Invertebrates
(0.0265)

Land plants
(0.0152)

Vertebrates
(0.0041)

10^{-3} 10^{-2} 10^{-1} 10^{0}

π

FIGURE 3.4 The range of π values per site for synonymous variants from a range of prokaryotic and eukaryotic organisms. The mean values are reported in parentheses. (From Lynch 2006.)

between recombination rates and levels of variation (reviewed in Hahn 2008; Cutter and Payseur 2013). This correlation is unexpected under a neutral model but is expected under models that include the effects of linked selection (see Chapters 1 and 8). The two most complete models of linked selection predict either a weakly positive relationship (background selection; Charlesworth, Morgan, and Charlesworth 1993) or no relationship between diversity and population size (hitchhiking; Maynard Smith and Haigh 1974), and both predict a positive correlation between recombination rates and levels of variation. The background selection model proposes that the constant influx of deleterious mutations that must be removed from a population reduces polymorphism at linked sites. The hitchhiking model proposes that the rapid fixation of advantageous alleles reduces polymorphism at linked sites. Although distinguishing between these two models is not straightforward (Stephan 2010; Coop and Ralph 2012), we can write down the expected average number of nucleotide differences between sequences per site under both models.

Under the background selection model, the expectation for π is:

$$E(\pi) = 4N_e\mu f \tag{3.19}$$

where f is the fraction of homologous chromosomes in the population that carry no deleterious mutations (Charlesworth, Morgan, and Charlesworth 1993). The fraction of mutation-free chromosomes is dependent both on the effective population size and on the recombination rate, in a relatively complex way depending on assumptions about dominance, recombination, and mutation (Charlesworth, Morgan, and Charlesworth 1993). But it is easy to see that with decreasing rates of recombination, sites with neutral polymorphisms will become increasingly linked to neighboring deleterious alleles, leading to a decrease in f and a concomitant decrease in π.

Under the hitchhiking model with recombination, one expression for the expected average number of nucleotide differences between sequences per site is:

$$E(\pi) = 4N_e\mu \frac{\rho}{\rho + \alpha} \tag{3.20}$$

where $\rho = 4N_e c$ (the population recombination parameter; see Chapter 4) and α is a complex parameter that depends on the rate at which advantageous alleles arise, the population-scaled strength of selection ($N_e s$), and the distance of selected alleles from the neutral polymorphisms being assayed (Wiehe and Stephan 1993). As can be seen, increasing levels of recombination result in levels of diversity that are quite close to the expectations with no selection. Increasing the effect of hitchhiking (i.e., increasing α) lowers levels of diversity. The details of different models of hitchhiking give slightly different expectations for π, but there is almost always a positive relationship between levels of recombination and diversity.

These considerations imply that while by definition $\theta = 4N_e\mu$, it is not necessarily the case that the statistics π, θ_W, θ_H, η_1, or ξ_1 are equal to $4N_e\mu$, even though they are unbiased estimators of θ in a Wright-Fisher population at equilibrium under the infinite sites model. In the presence of linked selection—which can have a substantial influence even in regions of normal recombination (e.g., Loewe and Charlesworth 2007)—the statistics used to measure levels of polymorphism will tell us at least as much about the stochastic effects of natural selection on linked variation (*genetic draft*; Gillespie 2000) as they will about drift and mutation. This also implies that the value of the effective population size (N_e) that is calculated by dividing π by 4μ may not indicate an interpretable biological value (and certainly not a number of "individuals"; Chapter 1), as it only represents the effective population size under a neutral model. The coalescent effective population size definition of this quantity may still represent the timescale of coalescence, but no rescaling of the value of N_e will capture all aspects of the linked selection model (Neher 2013). Instead, the observation that π is correlated with apparent census population size—however weakly (Figure 3.4)—suggests that using this or other summaries of nucleotide variation directly may be just as fruitful as using inappropriate estimates of N_e. Moreover, it is likely that the census population size is more important in determining the rate of adaptation than any effective population size, as the census size plays a large role in determining the rate of input of adaptive mutations (e.g., Karasov, Messer, and Petrov 2010). The rate at which the selective environment changes is also likely to play a larger role in adaptive evolution than any conception of population size (Gillespie 2001; Lourenço, Glémin, and Galtier 2013).

ADDITIONAL MEASURES OF DIVERSITY

Microsatellite variation

Variation at a single microsatellite locus can be measured in much the same way as variation at a single nucleotide site, though there are usually many more than two alleles (often >10 alleles). These alleles will also differ quite a lot in terms of repeat lengths, with individual alleles having anywhere from 1 to >30 repeat units depending on the structure of the microsatellite. Variation does not have to be normally distributed about a mean repeat length and in fact is often bi- or even trimodal (see **FIGURE 3.5** for examples).

There are multiple methods for summarizing diversity at a microsatellite locus, including methods based on simple summary statistics (see below) and those based on maximum likelihood (e.g., Nielsen 1997; Wilson and Balding 1998; RoyChoudhury and Stephens 2007). As with summarizing nucleotide variation, the most commonly used statistics are also estimators of the population mutation parameter θ in a Wright-Fisher population at equilibrium. In this context the mutation rate, μ, measures the rate of mutation to new neutral alleles at the microsatellite locus, and we must assume a mutation model appropriate to microsatellites. The most commonly used model is the stepwise mutation model (SMM), which allows only single repeat unit changes for each mutation (Chapter 1).

A straightforward way to summarize variation at a microsatellite locus is to report the number of different alleles in the sample, K (sometimes denoted j, n_e, or n_a). While this statistic can be used as an estimator of θ (e.g., Haasl and

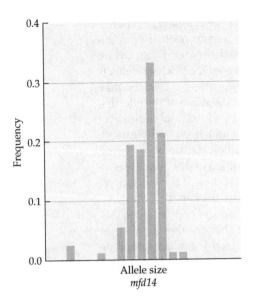
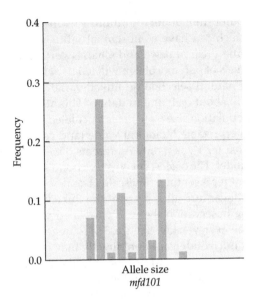

FIGURE 3.5 Examples of allelic variation at microsatellite loci in humans. Both loci have only dinucleotide repeats, with adjacent bars indicating the frequencies of alleles that differ by one repeat unit. (From Valdes, Slatkin, and Freimer 1993.)

Payseur 2010), in practice it is usually reported alone, without even a correction for the number of chromosomes sampled.

There are two common statistical summaries of microsatellite variation that do provide an estimation of θ under a stepwise mutation model and in Wright-Fisher populations at equilibrium. The first uses the expected homozygosity of the sample, F, which can be calculated as:

$$F = \sum_{i=1}^{K} p_i^2 \tag{3.21}$$

where p_i is the frequency of the ith allele at a locus (out of K total alleles). Confusingly, unlike the calculation for nucleotide heterozygosity shown in Equation 3.1, commonly used calculations for homozygosity do not use Bessel's correction for bias (the $n/[n-1]$ term). However, because the proportions of heterozygotes and homozygotes must sum to 1, we also must define uncorrected heterozygosity, H, as $1 - F$. Given this method for calculating the sample homozygosity, we can define an estimator of the parameter θ for microsatellite loci as (Ohta and Kimura 1973):

$$\theta_F = 0.5\left(\frac{1}{F^2} - 1\right) \tag{3.22}$$

Despite the fact that θ_F is a slightly biased estimator of θ, this bias can be reduced by using improvements based on simulation (Xu and Fu 2004).

The second estimator of θ for microsatellite loci under a stepwise mutation model uses the variance in repeat number among alleles observed at a single locus, V. Because mutations in microsatellites generally produce new alleles one or a few repeat units away from the original state, the variance in repeat number can be a good indicator of the amount of variation maintained at a locus. If we assume that only single-step mutations can occur, with an equal probability of mutations adding or subtracting repeat units, then $E(V) = \theta$ (Moran 1975; Goldstein et al. 1995; Slatkin 1995). A straightforward estimator of θ is therefore given by:

$$\theta_V = V \tag{3.23}$$

While this estimator is unbiased, it has a large variance (Zhivotovsky and Feldman 1995).

All of the summary statistics presented here are only estimators of the parameter θ under the stepwise mutation model and all of the other assumptions of the standard neutral model. However, it is possible to derive estimators of θ under a generalized stepwise model that has arbitrary properties with respect to the number of average repeat units added or subtracted and any asymmetry in gain or loss of repeats (Kimmel and Chakraborty 1996). Although these estimators do require specification of the specific mutation model to be used—and many of the parameters in such a model will be unknown for any particular locus—it appears that even some of the estimators that assume the SMM may be relatively robust to model mis-specification (Xu and Fu 2004; Haasl and Payseur 2010).

Haplotype variation

If nucleotide sequence data are phased—either experimentally or computationally—then the 4 sequences in Figure 3.1 represent 4 haplotypes and the 10 sequences in Figure 3.2A represent 10 haplotypes. In Figure 3.1 there are also four unique haplotypes, with all sequences representing a different combination of nucleotides (these are also sometimes referred to as *alleles*; see Chapter 1). Figure 3.2A has eight unique haplotypes. With no recombination, new haplotypes can only be created by mutation, and the number of unique haplotypes, K (also sometimes denoted n_e or n_a), is at most $S + 1$. With high recombination, new haplotypes can be created by both recombination and mutation, and there can therefore be up to 2^S different haplotypes. Assuming a Wright-Fisher population at equilibrium and the infinite alleles mutation model with no recombination (Chapter 1), there is a relationship between the expected number of unique haplotypes at a locus, K, and the population mutation parameter, θ:

$$E(K) = \frac{\theta}{\theta} + \frac{\theta}{\theta + 1} + \frac{\theta}{\theta + 2} + \cdots + \frac{\theta}{\theta + n - 1} \tag{3.24}$$

where n is again the number of sampled chromosomes (Ewens 1972). It is generally necessary to use numerical optimization in order to solve for θ here, and I will refer to this estimator of θ as θ_K. Remember that in this context the mutation rate, μ, measures the rate of mutation to new neutral alleles (i.e., unique haplotypes) at the locus.

Other methods for summarizing haplotype variation use an analog of Equation 3.21 to calculate the haplotype homozygosity: the only change is that here p_i represents the frequency of the ith unique haplotype out of K unique haplotypes. Once again, this value is often referred to as F, with $H = 1 - F$ used to calculate haplotype heterozygosity (also called *haplotype diversity*; Depaulis and Veuille 1998). Haplotype heterozygosity and the number of unique haplotypes are obviously extremely dependent on the inclusion of singleton nucleotides, as each singleton will create a new haplotype. It is therefore common to see both of these statistics calculated without including these mutations.

While it is useful to consider haplotypes as the unit of variation when working with (largely) non-recombining sequences such as mitochondrial DNA or the Y chromosome, many studies of recombining nuclear loci focus on nucleotide site variation. It is important to note that the statistics for summarizing nucleotide diversity described above do not depend on accurate phasing of haplotypes: the values of π, θ_W, and all the other nucleotide-based statistics will not change even if haplotypes are constructed at random from the same set of segregating sites. This being said, there are a number of uses specifically for haplotypes in detecting natural selection that will be discussed later (see Chapter 8).

RECOMBINATION

4

In the previous chapter we considered multiple ways of describing the amount of diversity in a sample of DNA sequences, from summaries of the number of segregating sites, to the number of alleles at a microsatellite locus, to the number of unique haplotypes. These statistics—including those based on phased haplotypes—were all intended to summarize the effect of mutation in generating diversity. In this chapter I will introduce multiple ways of summarizing the effect of recombination in generating molecular diversity, both between pairs of sites and in a recombining region. While our ability to detect recombination requires the presence of nucleotide variation, the effects of recombination in reassorting these variants among haplotypes are quite distinct from the effects of mutation. I start by describing methods for examining the effect of recombination on two-locus genotypes before moving on to regional and genome-wide summaries of recombination.

MEASURING LINKAGE DISEQUILIBRIUM

Linkage and linkage disequilibrium

Collecting two haplotypes from a single individual implies that we are observing the two parental gametes that combined to create the individual. That is, we have the maternally and paternally derived chromosomes that came together in the initial zygote. Of course these sequences are not themselves necessarily exactly the haplotypes present in either the maternal or paternal genomes, but instead may represent a new recombinant haplotype that is a combination of the two haplotypes present in the parental genomes and is passed on as a gamete. *Linkage* is the co-inheritance of sites together in gametes because they are physically linked on chromosomes.

FIGURE 4.1 gives an example of an individual who is heterozygous at two loci (A_1/A_2 and B_1/B_2) and who has two haplotypes, A_1B_1 and A_2B_2. The four possible gametes produced by such an individual are A_1B_1, A_2B_2, A_1B_2, and A_2B_1. If the two loci are closely linked, then almost all gametes will be either A_1B_1 or A_2B_2. If the two loci are completely unlinked, then all four gametes are expected to be produced during meiosis with equal frequency. We define the frequency of recombination between two loci, c, as the fraction of A_1B_2 and A_2B_1 gametes produced

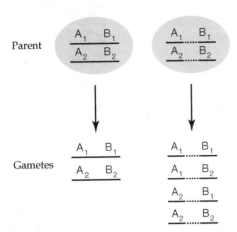

Parent

Gametes

FIGURE 4.1 Linkage. The example on the left shows two closely linked loci such that only the two parental gametes are produced ($c = 0$), while that on the right shows two unlinked loci with all four gametes produced at equal frequency ($c = 0.5$).

by the individual such that $c = 0.5$ implies that the two loci are unlinked and $c < 0.5$ implies that they are linked.

Just as linkage is defined as the association of alleles through meiosis, *linkage disequilibrium* (LD) is defined as the nonrandom association of alleles in a population. The term is a bit misleading, as linked loci are not necessarily in linkage disequilibrium, and even unlinked loci can be in linkage disequilibrium. But physical linkage is the major cause of LD—and many closely linked sites are in fact in LD—leading many people to conflate the two terms (*association disequilibrium* might have been a better choice for this phenomenon).

If two loci are randomly associated with one another in a population, then having an A_1 allele at the first locus tells us nothing about whether the second locus will be B_1 or B_2. But if they are nonrandomly associated (i.e., in linkage disequilibrium), then individuals with an A_1 might also be more likely to have a B_1. If we define p_i to be the frequency of allele A_i at the first locus and q_j to be the frequency of allele B_j at the second locus, and if there is random association between alleles, we can simply multiply individual allele frequencies at the two loci to get the expected haplotype frequencies, g_{ij}. These loci are said to be in *linkage equilibrium*. However, if there is an association between alleles, then these expectations will be off by some amount, D, which we define as the *linkage disequilibrium coefficient* (Robbins 1918; Lewontin and Kojima 1960). We can see this more clearly if we consider the four haplotypic combinations that can be generated by two loci that each have two alleles. The haplotypes occur with frequencies:

$$g_{11} = p_1 q_1 + D$$
$$g_{12} = p_1 q_2 - D$$
$$g_{21} = p_2 q_1 - D$$
$$g_{22} = p_2 q_2 + D$$

(4.1)

When there is no linkage disequilibrium ($D = 0$), then the haplotypic frequencies are exactly equal to the product of the constituent allele frequencies. But in the presence of LD, all haplotype frequencies are off by amount D.

Note that our choice of adding D to the $A_1 B_1$ and $A_2 B_2$ haplotypes and subtracting it from $A_1 B_2$ and $A_2 B_1$ was arbitrary: D itself can be either positive or negative and must be the same for each haplotype in this scenario, so that our inference of the magnitude of linkage disequilibrium will not be affected by this choice. However, it is standard to add D to the haplotypes in "coupling" phase (i.e., $A_1 B_1$ and $A_2 B_2$) and to subtract D from the haplotypes in "repulsion" phase (i.e., $A_1 B_2$ and $A_2 B_1$). Although the assignment of the coupling and repulsion types is also arbitrary in a population sample, Langley, Tobari, and Kojima (1974) suggested that the haplotypes containing the two more common alleles and the two less common alleles be considered coupling, and this nomenclature has become relatively standard. Alternative naming systems, such as the assignment of ancestral-ancestral and derived-derived pairs of mutations to be in coupling phase, can also be useful (e.g., Langley and Crow 1974; Machado et al. 2002; Takahasi and Innan 2008).

Measures of pairwise linkage disequilibrium

Given the above definitions, it is easy to see that the coefficient of disequilibrium, D, can be calculated as:

$$D = g_{ij} - p_i q_j \tag{4.2}$$

for any combination of alleles. Because the deviation from the observed haplotype frequency can only be as large as the product of the marginal allele frequencies, the range and value of D are dependent on p_i and q_j. For instance, when $p_1 = 0.5$ and $q_1 = 0.5$, D can range between -0.25 and $+0.25$, and when $p_1 = 0.1$ and $q_1 = 0.7$, D is between -0.07 and $+0.03$. This means that differences in the magnitude of D among pairs of sites can reflect both differences in the strength of the association among alleles and differences in allele frequencies. In order to be able to compare levels of linkage disequilibrium across pairs of sites, Lewontin (1964) suggested the following normalized measure:

$$
\begin{aligned}
D' &= \frac{D}{\max(-p_1 q_1, -p_2 q_2)}, \quad \text{for } D < 0 \\
D' &= \frac{D}{\min(p_1 q_2, p_2 q_1)}, \quad \text{for } D > 0
\end{aligned}
\tag{4.3}
$$

where $\max(x,y)$ denotes the larger of x and y and $\min(x,y)$ the smaller. The value of D' is now constrained between -1 and $+1$, though usually only the absolute value is reported. Note that while the range of D' is no longer dependent on allele frequencies, the value of this statistic—and in fact all measures of linkage disequilibrium—still is (Lewontin 1988).

A second commonly used measure of linkage disequilibrium is based on the squared correlation of alleles at two loci and was first introduced by Hill and Robertson (1968):

$$r^2 = \frac{D^2}{p_1 p_2 q_1 q_2} \tag{4.4}$$

Here the statistic r^2 is used instead of just the correlation of allelic states, r $(=D/\sqrt{[p_1 p_2 q_1 q_2]})$, in order to remove the sign associated with calculating D. The r^2 statistic is therefore bounded by 0 and 1. In addition to the value of r^2 being slightly less dependent on marginal allele frequencies than is D', it is a useful statistic because it is related to the population recombination parameter, ρ (see below).

The two measures of linkage disequilibrium, D' and r^2, also provide very different kinds of information about the association between alleles. Put simply, while D' tells us about recombination between two loci, r^2 tells us about our

A_1 B_1	A_2 B_1
A_1 B_1	A_2 B_1
A_1 B_1	A_2 B_1
A_1 B_1	A_2 B_1
A_1 B_1	A_2 B_2

FIGURE 4.2 Sample of 10 chromosomes from a population. There are three unique haplotypes observed.

	B_1	B_2
A_1	5	0
A_2	4	1

FIGURE 4.3 Counts of haplotypes in a sample, based on the chromosomes shown in Figure 4.2.

power to predict the identity of alleles at one locus given allelic states at another locus. To see this, consider the example given in **FIGURE 4.2**. There are only three unique haplotypes observed in this sample (A_1B_1, A_2B_1, and A_2B_2), and D' will therefore be equal to 1 (this is sometimes called *complete LD*). In fact, D' will always be 1 unless all four haplotypic combinations are observed, which is only possible if recombination has occurred between the two loci (except in the case of recurrent mutation, when all four haplotypes can be present without recombination). These constraints are the basis for the *four-gamete test* for recombination (Hudson and Kaplan 1985; also see below). In contrast, the value of r^2 calculated from the sample in Figure 4.2 is 0.111. The value is so low because there is very little power to predict the allele at the B locus with knowledge of the allele at the A locus, even though no recombination has occurred. Unless there are only two haplotypes in a population and they each contain alternative alleles at the two loci (often called *perfect LD*), r^2 will always be less than 1. This aspect of r^2 makes it very useful in association (linkage disequilibrium) mapping of phenotypic traits, in which the questions revolve less around the presence of recombination than around the prediction of the genotypes that underlie different phenotypic states.

Calculating the significance of any observed linkage disequilibrium is straightforward, regardless of the statistic used. One method is to arrange the counts of the four possible haplotypes in a 2 × 2 contingency table, as shown in **FIGURE 4.3**. Any test for independence (e.g., Fisher's exact test) can then be applied to these counts to ask whether D is significantly different than 0 ($P = 1.0$ in the example shown). A second method results from the fact that nr^2 is χ^2-distributed with 1 degree of freedom (where n represents the number of phased haplotypes sampled). In this case the χ^2 statistic is $10 * 0.111 = 1.11$, which is also not significant. Because it is difficult to have significant values of r^2 when minor allele frequencies are low (without very large sample sizes), it is common to ignore calculations involving polymorphisms with minor allele frequencies of <0.05.

There are several nuances in calculating linkage disequilibrium that must be considered when dealing with genome-scale datasets (see Chapter 2 for a discussion of the issues surrounding the experimental and computational phasing of haplotypes). One common problem will be missing data: there may often be variation among sites in the number of chromosomes with accurate genotype information. Because no haplotype can be defined in an individual for a pair of sites where the state of one of the sites is not known, the sample size for each pair of sites may differ. Correct calculation of either D' or r^2 in the presence of missing data requires that any individual with at least one undefined site be removed from both the count of haplotypes and the counts of allele frequencies at both sites, with the sample sizes and allele frequencies adjusted accordingly. This requirement obviously only applies to pairwise comparisons involving the site with missing data; the same sampled chromosome can be used in other comparisons for which genotypes at both sites are defined. Missing data will occur whether obtained via sequencing or a genotyping platform, and linkage disequilibrium can be calculated from either data source. However, the ascertainment bias inherent to genotyping methods (such as SNP-chips) means that there can also be a bias in overall levels of linkage disequilibrium (Akey et al. 2003). To be clear: individual measures of linkage disequilibrium will not be biased for any pair of sites, but the average level of disequilibrium among many pairs will be biased relative to the expectation for idealized populations (see below for these expectations). Perhaps unsurprisingly, the effects of ascertainment bias are also quite different for D' and r^2 because of the different aspects of the data they are using (Nielsen and Signorovitch 2003). The fact that ascertainment bias results in an overrepresentation of intermediate-frequency polymorphisms means that D' will tend to be downwardly biased and r^2 will tend to be upwardly biased, although both of these biases can be corrected if the original ascertainment scheme is known (Nielsen and Signorovitch 2003).

Summarizing pairwise linkage disequilibrium

Calculating coefficients of linkage disequilibrium for many pairs of segregating sites—whether within a region or from across the genome—provides a huge amount of data. The question then becomes: How do we make sense of all these data? There are a number of approaches to summarizing pairwise LD, differing largely in whether the aim is to provide a local or global view of linkage disequilibrium. Local views attempt to summarize patterns of LD within a single region, while global views may summarize LD across many regions, or even across the genome.

There are two straightforward ways to summarize pairwise linkage disequilibrium within a region, one graphical and one simply using the average of all the pairwise measures. Kelly (1997) defined the average r^2 value among S segregating sites in a sample of n sequences as:

$$Z_{nS} = \frac{2}{S(S-1)} \sum_{i=1}^{S-1} \sum_{j=i+1}^{S} r_{ij}^2 \tag{4.5}$$

(The Z was chosen because it is John Kelly's favorite letter.) Like r^2, the value of Z_{nS} varies between 0 and 1 and represents the average pairwise LD at a locus. Without recombination, the mean value of Z_{nS} at equilibrium is expected to be about 0.15 (Kelly 1997). Intra-locus recombination causes the value of Z_{nS} to be lower, while certain forms of selection can increase the value of Z_{nS}. Rozas et al. (2001) suggested a modification of Z_{nS} that could be used to detect whether recombination has occurred by measuring the decline in LD among pairs of sites located farther and farther apart. However, Z_{nS} is most often used in the context of detecting natural selection (see Chapter 8).

A graphical summary of pairwise linkage disequilibrium in a region is shown in **FIGURE 4.4** for the $su(w^a)$ locus in *D. melanogaster* (Langley et al. 2000). Significant pairwise comparisons are denoted by colored boxes, with

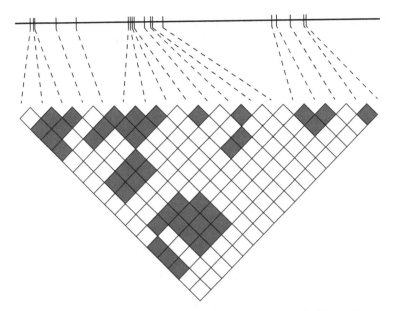

FIGURE 4.4 Pairwise linkage disequilibrium at the *suppressor of white apricot* (*su* [*w*ᵃ]) locus in *D. melanogaster*. Pairwise comparisons with *D* that are significantly different from 0 are indicated by colored boxes ($P < 0.01$; Fisher's exact test). Only those polymorphisms that occurred on at least 2 of the 51 sampled chromosomes are compared. The dashed lines connect the position of the segregating sites to their corresponding columns in the matrix. (From Langley et al. 2000.)

low *P*-values indicating evidence for linkage disequilibrium (equivalent to the test shown in Figure 4.3). We can see in this example that many (but not all) neighboring polymorphisms are significantly associated with one another, and that there are also some significant associations between polymorphisms almost 2,000 bases apart. Otherwise, there is not a lot of structure to the patterns of linkage disequilibrium in this region, and LD does not extend very far, as is typical for *D. melanogaster*. By contrast, **FIGURE 4.5** provides a graphical summary of pairwise LD in a 500-kb region in humans (International HapMap Consortium 2005). In this representation only pairs of polymorphisms with $D' = 1$ are indicated by colored boxes; many more pairs have $D' > 0$. A striking feature of regional LD in this figure is due to the punctate nature of recombination in humans: because recombination largely occurs at a limited number of locations along the chromosome (Jeffreys, Kauppi, and Neumann 2001; McVean et al. 2004), there are large blocks of very high LD that can be almost independent from neighboring blocks. These so-called *haplotype blocks* can include hundreds of polymorphisms in complete linkage disequilibrium with one another, with recombination "hotspots" separating individual blocks. Although the criteria used to define a haplotype block can differ among studies, such blocks are a common feature of mammalian populations studied thus far (e.g., Guryev et al. 2006; Villa-Angulo et al. 2009). Simple representations of regional LD such as those shown in Figures 4.4 and

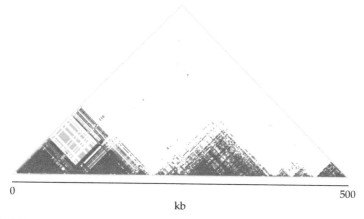

0

500

kb

FIGURE 4.5 Pairwise linkage disequilibrium in the 2q37.1 region in *H. sapiens* from East Asia. Pairwise comparisons with $D' = 1$ are indicated by colored boxes. (From International HapMap Consortium 2005.)

4.5 can be a relatively straightforward way to visualize differences in patterns of LD and recombination among loci and species.

In order to gain a more global view of linkage disequilibrium across a genome, it is clear that we cannot use the type of graphs shown in Figures 4.4 and 4.5—we would need a separate graph for each chromosome, and each triangular matrix would have to be huge to encompass a whole chromosome. Of course we could give a single statistic that averages LD across all pairs of sites, but this sort of summary does not provide a lot of information about the way in which LD is structured across a genome. One common summary of LD across a genome (or across a smaller region) that is intended to reveal the scale of LD is the distance over which LD drops below some threshold. Often this threshold is an arbitrary value of r^2 (say, $r^2 < 0.2$) or the distance at which the average r^2 falls below 50% of the initial value (i.e., the value between polymorphisms at adjacent bases).

FIGURE 4.6 shows just this sort of display of the decay in linkage disequilibrium. Pairwise values of r^2 are plotted for about 350,000 comparisons between polymorphisms less than 20 kb apart in the genome of the model legume *Medicago trunculata* (Branca et al. 2011). The whole genomes of 26 *M. trunculata* inbred lines ($n = 26$) have been sequenced, and the distribution of r^2 values for all comparisons between SNPs separated by less than 20 kb—across all eight assembled chromosomes—are shown. (LD can be calculated between indels, or indeed between any variants, but was not done so in this case.) We can see that there is a huge spread in r^2 values, with many pairs having $r^2 = 0$ at very short distances and $r^2 = 1$ at distances up to 7 kb. The figure reveals that the scale of LD in *M. trunculata* is approximately 5 kb; that is, mean LD falls below $r^2 = 0.2$ for SNPs separated by about 5 kb, and declines to less than half of its initial value by 2 to 3 kb. This figure is of course averaging over many different regions, each of which possibly has its own recombination rate. The decay in observed LD may therefore be different within each region, and the summary

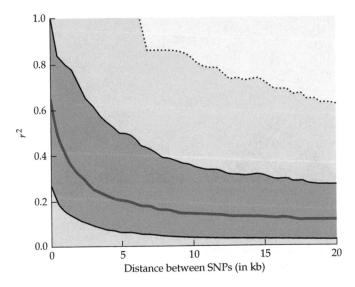

FIGURE 4.6 The decay of linkage disequilibrium across the *M. truncatula* genome. The mean value of r^2 between all pairs of polymorphisms up to 20 kb apart is shown by the thick red line (individual points are not shown). The 50% range of values is shown in darker blue, and the 90% range is in lighter blue. (From Branca et al. 2011.)

presented in Figure 4.6 represents the decline in pairwise LD for the genome-wide average recombination rate; Equations 4.11 and 4.12 below provide the theoretical expectation for this decay given an average recombination rate and population size among loci. Although this figure does not show r^2 values between polymorphisms separated by more than 20 kb, we expect the average r^2 to eventually decline to $1/n$ for unlinked loci.

Extensions to multiple alleles, multiple loci, and genotypes

The measures of linkage disequilibrium discussed so far have considered only comparisons between two loci, each with two alleles, or the summary of such pairwise measurements. We have also assumed that we know the phase of alleles on chromosomes and therefore have also considered only disequilibrium between loci that are on the same chromosome. Here I briefly discuss measures of LD that relax each of these constraints in turn; more details can be found in Weir (1996).

When there are multiple alleles at one or both loci, separate disequilibrium coefficients must be calculated for every pair of alleles:

$$D_{ij} = g_{ij} - p_i q_j \tag{4.6}$$

Unlike the two-allele case, the coefficients calculated for more than two alleles are not equivalent between all pairs. However, as in the two-allele case, the sum of all D_{ij} values must be 0, and the sum of all values of D_{ij} for any particular i or j must also equal 0 (e.g., if there are three alleles at the second locus, then $D_{11} + D_{12} + D_{13} = 0$).

In addition to pairwise linkage disequilibrium, we can also consider disequilibria between more than two loci. All higher-order measures of LD are easily derived from the three-locus case, so I only discuss three-locus disequilibrium here. Given three loci A, B, and C with allele frequencies p_i, q_j, and r_k, we want to know whether there is a nonrandom association among sets of alleles at all three loci that is not explained by nonrandom associations among pairs of alleles. To answer this question we include the pairwise disequilibrium coefficients in our calculation:

$$D_{ijk} = g_{ijk} - p_i D_{jk} - q_j D_{ik} - r_k D_{ij} - p_i q_j r_k \tag{4.7}$$

Similar calculations are done for disequilibrium among more than three loci, always taking into account all lower-order disequilibria. It is important to realize that these measures are not summarizing regional LD, but rather are testing for higher-order interactions among more than two loci.

All of the methods for calculating linkage disequilibrium discussed thus far have assumed that all individuals have known phase: that the arrangement of alleles on chromosomes are known. For this reason, it is common to refer to the previous statistics as measures of *gametic linkage disequilibrium*. But if phase is not known—either because the two sites are far apart on a single chromosome or because they are located on different chromosomes—we can still ask whether there are nonrandom associations between genotypes at two loci (e.g., **FIGURE 4.7**), a state that is referred to as *genotypic linkage disequilibrium*. Given two loci with two alleles each, there are three different possible genotypes at each locus ($A_1/A_1, A_1/A_2, A_2/A_2$, and $B_1/B_1, B_1/B_2, B_2/B_2$) and nine different combinations of genotypes (**FIGURE 4.8**). If we denote the frequency of $A_1 A_1 B_1 B_1$ individuals as g_{1111}, the frequency of $A_1 A_1 B_1 B_2$ individuals as g_{1112}, the frequency of $A_1 A_2 B_1 B_1$ individuals as g_{1211}, and the frequency of $A_1 A_2 B_1 B_2$ individuals as g_{1212}, then the genotypic disequilibrium coefficient can be calculated as (Weir 1979):

$$\Delta = 2g_{1111} + g_{1112} + g_{1211} + 0.5g_{1212} - 2p_1 q_1 \tag{4.8}$$

For the sample of nine diploid genotypes shown in Figure 4.7, $\Delta = 0.136$. Assuming that the population is in Hardy-Weinberg equilibrium,

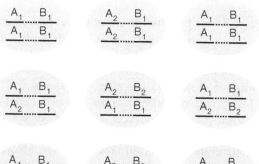

FIGURE 4.7 Sample of nine diploid genotypes from a population. Linkage relationships (and therefore gametic phase) between the A and B locus are unknown.

	B_1/B_1	B_1/B_2	B_2/B_2
A_1/A_1	2	0	0
A_1/A_2	2	2	0
A_2/A_2	1	1	1

FIGURE 4.8 Counts of two-locus genotypes in a sample, based on the individuals shown in Figure 4.7.

significance can again be assessed using a test of independence on the 3 × 3 table of genotypes (Figure 4.8). This estimate of genotypic disequilibrium is also equivalent to the gametic disequilibrium coefficient, D, if there is random mating among known haplotypes. Assuming each of the 18 haploid genotypes (e.g., $A_1 \ldots B_1$) shown in Figure 4.7 represents a phased haplotype, $D = 0.123$ for this sample, similar to our genotypic disequilibrium coefficient.

ESTIMATING RECOMBINATION

Finding the minimum number of recombination events in a sample

The statistics discussed thus far have largely been limited to pairwise comparisons among segregating sites and therefore can only tell you about the effects of recombination between a specific pair of sites. However, we would like to be able to make more general inferences about the effects of recombination in individual regions and across the genome. In this and the following section I describe multiple approaches to summarizing the effects of recombination. Here we start with methods for estimating the minimum number of recombination events that must have occurred in a sample of sequences from a single region.

As alluded to earlier in explaining the differences between D' and r^2, definitive evidence for recombination in population genetic datasets can be inferred using the four-gamete test (Hudson and Kaplan 1985). The essence of this test is that with two loci that each have two alleles, and under an infinite sites assumption, the only way to generate all four possible gametes is to have a recombination event somewhere between the two loci. The infinite sites assumption is important because recurrent mutation of an allele can create all four gametes without recombination. For instance, Figure 4.2 shows a sample of sequences in which only three haplotypes are present: A_1B_1, A_2B_1, and A_2B_2. However, if there is a recurrence of the $B_1 \rightarrow B_2$ mutation, *and* it occurs on an A_1B_1 background, then all four gametes will be present in the population, and possibly in a sample of chromosomes.

Of course simply observing all four gametes does not tell us how many recombination events have occurred in the history of a sample, only that a minimum of one event must have occurred (*modulo* recurrent mutation). Estimating the number of recombination events in a sample, R, is in fact quite difficult: unlike the number of segregating sites in a sample, S, R is not directly observable. Many recombination events can occur in the history of a set of sequences and not be observed, either because they did not occur between two sites that were both in a heterozygous state in the ancestor or because they generated a recombinant haplotype already present in the sample. We can, however, estimate the *minimum* number of recombination events that must have occurred in a sample, a number that is typically much smaller than R because most recombination events will not be detected (Hudson and Kaplan 1985; Stephens 1986; Posada and Crandall 2001).

There are two common statistics used to put a bound on the minimum number of recombination events in a sample: R_m (Hudson and Kaplan 1985) and R_h (Myers and Griffiths 2003). R_m determines the maximum number of non-overlapping intervals in a region in which recombination must have occurred based on the four-gamete test. R_h is based on the observation that in any region with K unique haplotypes and S polymorphisms, the number of recombination events must satisfy $R \geq K - S - 1$. Therefore, the minimum number of recombination events can be estimated by $K - S - 1$. This relationship breaks down for longer sequences, particularly when the number of segregating sites becomes greater than the number of distinct haplotypes (which is of course limited by the number of chromosomes sampled, n). But by combining together minimum recombination estimates from many subsets of sites each with $K \geq S + 1$, R_h can provide a bound on the global minimum number of recombination events across a region. The algorithms used to find R_m and R_h also provide the locations of the inferred recombination events.

In general, R_h will be a more accurate estimate than R_m, such that $R \geq R_h \geq R_m$ (Myers and Griffiths 2003). To see why this is the case, consider the example given in **FIGURE 4.9**. In this example we have four sites $(1 - 4)$ and eight sampled chromosomes $(a - h)$, each of which has a unique haplotype $(K = 8)$. The algorithm to calculate R_m proceeds as follows (with explicit results for the example in Figure 4.9):

1. For all pairs of sites i and j, determine if all four gametes are present.

   - Figure 4.9: (1,2), (1,3), (1,4), (2,4), and (3,4)

2. Remove intervals that completely contain other intervals, including intervals with the same start or end points.

   - Figure 4.9: remove (1,3), (1,4), and (2,4)

3. If any remaining intervals are overlapping, go to step 4; if not, go to step 5.

   - Figure 4.9: go to step 5

4. For the first pair of sites—listed according to increasing starting points—(i_1,j_1), remove any interval (m,n) with $i_1 < m < j_1$. For the next non-overlapping pair (i_2,j_2), remove any interval (m,n) with $i_2 < m < j_2$, repeating this procedure until no overlapping intervals remain.

   - Figure 4.9: no such intervals

5. The number of remaining intervals indicates the minimum number of recombination events, and one event is assigned to have occurred in each interval.

   - Figure 4.9: recombination events in the intervals (1,2) and (3,4)

	Site			
	1	2	3	4
a	0	0	0	0
b	0	1	0	1
c	1	1	0	0
d	0	1	1	0
e	1	1	1	1
f	1	1	0	1
g	1	1	1	0
h	1	0	0	1

FIGURE 4.9
Example of haplotypes from four segregating sites in eight sampled chromosomes $(n = 8)$. Each polymorphism has two alleles, arbitrarily denoted 0 or 1. Each four-locus haplotype is unique, so $K = 8$. For this example, $R_m = 2$ and $R_h = 3$ (see text for details). (After Myers and Griffiths 2003.)

In this example, $R_m = 2$ and the Hudson-Kaplan algorithm says that at least one recombination event must have occurred between sites 1 and 2, and at least one between sites 3 and 4. In contrast, $R_h = 3$ ($= 8 - 4 - 1$), and the Myers-Griffiths algorithm (see Myers and Griffiths 2003) says that at least one recombination event must have occurred between all pairs of sites. We can get an intuitive feeling for why the Hudson-Kaplan algorithm has missed a recombination event (between sites 2 and 3) by noting that this *pair* of sites only has three haplotypes. However, the recombination event that did occur between these sites generated a new four-locus haplotype that the Myers-Griffiths algorithm takes into account, and so the inferred minimum number of recombination events is larger. Many other algorithms exist for finding more accurate bounds on the minimum number of recombination events (e.g., Song and Hein 2005), although they become computationally intractable for large numbers of samples and polymorphisms.

The population recombination parameter

Many evolutionary processes affect estimated levels of linkage disequilibrium in a particular region: recombination, natural selection, drift, demographic history, and even mutation (the last because without sufficient polymorphism many recombination events will go undetected). However, as we saw in Chapter 1, under the standard neutral model the only factors that contribute to the amount of recombination in a sample of sequences—whether detectable or not—are the per site recombination rate, c, and the effective population size, N_e, which is proportional to the magnitude of genetic drift. We define the population recombination parameter as $\rho \equiv 4N_e c$ for autosomal loci. (ρ is sometimes referred to as C or R, and c is often called r. For consistency in naming parameters with Greek letters, and to avoid confusion between r and r^2, we will use the symbols given here.) Under the assumptions of the standard neutral model, ρ is positively correlated with the number of recombination events in a sample and negatively correlated with the amount of linkage disequilibrium. In particular, the expected number of recombination events in a sample of size n is given by (Hudson and Kaplan 1985):

$$E(R) = \rho a \tag{4.9}$$

where

$$a = \sum_{i=1}^{n-1} \frac{1}{i} \tag{4.10}$$

This expectation for R is of exactly the same form as the expectation for the number of segregating sites, S, given in Chapter 3, with θ replaced by ρ. Just as θ is not equivalent to the mutation rate, ρ is not equivalent to the recombination rate but is expected to be highly correlated with it across a chromosome. Indeed, estimates of c from pedigree studies in humans are highly correlated with estimates of ρ (e.g., McVean et al. 2004), though localized departures between these measures may indicate the action of natural selection (O'Reilly, Birney, and Balding 2008).

Multiple relationships between the expected value of r^2 and ρ have been derived under slightly different assumptions about whether ρ is large or small, the size of the sample, and whether or not low-frequency alleles are removed before calculating r^2. The classic relationship given between r^2 and ρ, assuming infinite sample size and small values of ρ, is (Sved 1971):

$$E(r^2) \approx \frac{1}{1 + \rho} \tag{4.11}$$

This formula does not correct for small sample sizes and incorrectly predicts a mean value of $r^2 = 1$ when $\rho = 0$, that is, between completely linked sites (recall that r^2 can only equal 1 when the marginal allele frequencies are exactly the same at a pair of sites). An alternative relationship (Weir and Hill 1986; Hill and Weir 1988) corrects for finite sample sizes and provides the correct expectation for small ρ when low-frequency alleles ($<10\%$) are excluded:

$$E(r^2) \approx \left[\frac{10 + \rho}{(2 + \rho)(11 + \rho)} \right]\left[1 + \frac{(3 + \rho)(12 + 12\rho + \rho^2)}{n(2 + \rho)(11 + \rho)} \right] \tag{4.12}$$

This formula is not correct for large values of ρ—in particular, it does not predict a value of $1/n$ for r^2 when ρ is large enough that sites are effectively unlinked. For $\rho = 0$ and very large sample sizes it predicts a value of $r^2 = 5/11$, similar to other results for infinite sample sizes (Ohta and Kimura 1971; McVean 2002).

The relationships given in Equations 4.11 and 4.12 give us a way to tie together levels of recombination and linkage disequilibrium. In terms of comparisons among populations of different size, these results predict lower levels of LD in populations with larger sizes (assuming equivalent per site recombination rates). Comparisons between African and non-African human populations bear out this prediction, with much less LD in Africans than non-Africans as a result of the bottleneck that reduced effective population sizes as humans migrated out of Africa (International HapMap Consortium 2005). In terms of a pair of segregating sites, the appropriate value of c is a multiple of the number of nucleotides separating the sites, such that $\rho = 4N_ec$ for polymorphisms at neighboring nucleotides and $\rho = 4N_ec(L-1)$ for polymorphisms L nucleotides apart. This means that ρ increases for sites farther and farther apart, and, plugging in larger and larger values of ρ into Equation 4.12, that r^2 is expected to decay approximately exponentially along a chromosome. Figure 4.6 shows just this sort of decreasing relationship between r^2 and the distance between polymorphisms. In fact, we can use Equation 4.12 and data plotted as in Figure 4.6 to find the best-fit value of ρ for any particular dataset (e.g., Brown et al. 2004). While this estimate of ρ represents an average value across all the regions plotted, there are more accurate methods for estimating ρ. Next I describe some of these methods.

Estimating the population recombination parameter

There are two important differences between the common summary statistics used to describe nucleotide variation—all of which are estimators of the parameter θ under the appropriate assumptions—and the statistics used to estimate the recombination parameter, ρ. For one thing, the statistics used to describe nucleotide variation (e.g., Equations 3.2, 3.5, 3.8, 3.9, and 3.10) are all straightforward descriptions of sequence diversity and are easy to interpret even when the assumptions of the equilibrium Wright-Fisher model do not hold. While summaries of the effects of recombination such as K, Z_{nS}, R_m, and R_h are more or less easy to interpret, none of these are estimators of ρ, though they can be used indirectly to estimate this parameter (see below). Instead, the most accurate, most commonly used estimators employ approximate likelihood methods that assume the standard neutral model to calculate the likelihoods, and it is not always obvious to researchers what these methods are doing. While offering intuition into their function is not a requirement of accurate methods, statistics should be interpretable in cases in which the assumptions of the standard neutral model are violated. A second, more important, difference is that estimators of ρ are much less accurate than estimators of θ. That is, even when we are dealing with a population that matches the assumptions of the standard neutral model, these estimators will do a much poorer job at correctly estimating the population recombination parameter than will the equivalent estimators of θ, especially in cases in which ρ is low and there are few segregating sites (Wall 2000a; Hudson 2001). The dependence on levels of polymorphism is especially troubling, as it means that there will be biased estimates of ρ in regions with low mutation rates, or in regions that have reduced levels of diversity as a result of selection or recent changes in population size. In spite of all these caveats, estimating these statistics has proven to be useful in many applications.

As with estimators of θ, there are both moment-based and likelihood-based estimators of ρ. Full likelihood methods (e.g., Griffiths and Marjoram 1996; Kuhner, Yamato, and Felsenstein 2000; Nielsen 2000; Fearnhead and Donnelly 2001) are highly computationally intensive, and the increased accuracy—if any—that they offer over approximate likelihood methods is likely to be small given the limits on the size of the datasets that can be analyzed. I therefore focus on describing moment-based estimators and multiple forms of approximate likelihood estimators.

The moment-based estimators of ρ are based on the insight that the variance in pairwise differences among sampled chromosomes (i.e., the variance in π) goes down with increasing recombination (Hudson 1987; Wakeley 1997). To see why this is the case, imagine a sample with maximal LD between all pairs of segregating sites: such a sample would contain only two haplotypes, regardless of how many polymorphisms there were. As a result, there would be maximal variance—all comparisons between the two different haplotypes would entail differences at every polymorphic nucleotide, while all comparisons among samples of the same haplotype would have no nucleotide differences. The effect of recombination would be to break up the sample into many more unique haplotypes, thereby reducing the variance in pairwise differences. This description also highlights another difference between estimating ρ and θ: unlike the moment-based estimators of θ, which all use the first moment of the distribution of the number of segregating sites (i.e., the mean), the estimators of ρ use the second moment (i.e., the variance).

Wakeley (1997) slightly improved the estimator first proposed in Hudson (1987); this improved estimator, denoted ρ_π, is now the most widely used moment-based estimator. We saw earlier (Equation 3.3) that by using the number of nucleotide differences (k_{ij}) between all pairs of sequences i and j, we can calculate π as:

$$\pi = \frac{\sum_{i<j} k_{ij}}{n(n-1)/2} \tag{4.13}$$

Based on this we can calculate the variance in π as:

$$S^2 = \frac{\sum_{i<j} \left(k_{ij} - \pi\right)^2}{n(n-1)/2} \tag{4.14}$$

Wakeley (1997) showed that the expectation of S^2 in a sample with recombination depends on both the parameters ρ and θ, in a complex way, but can be used to estimate ρ_π.

Unfortunately, ρ_π is a more biased estimator than those based on various likelihood methods; the earlier moment-based method of Hudson (1987) is even worse (Wall 2000a; Hudson 2001). Fortunately, there are four different approaches to finding likelihood-based estimators of ρ. None of these would be recognized as a full-likelihood method, as they all use approximations of one kind or another, but in general they offer improvements in accuracy over ρ_π and improvements in computational efficiency and speed over full-likelihood methods. Two of these methods can be summarized briefly: Hey and Wakeley (1997) showed that a likelihood estimate of ρ can be calculated exactly for samples of $n = 4$ and $S = 3$. For samples larger than this, their estimator, ρ_γ (what they referred to as γ), takes the average estimate for every set of three adjacent polymorphisms across all possible subsets of four sequences. In general, ρ_γ seems to give a slightly downwardly biased estimate of ρ (Hey and Wakeley 1997; Hudson 2001). Another method, proposed by Li and Stephens (2003), takes advantage of the fact that the probability of seeing any particular haplotype is conditional on all other haplotypes in a sample as well as the value of both ρ and θ. This conditional probability is different for each haplotype in the sample, but one can find the value of ρ that maximizes the likelihood (what I will refer to as ρ_{LS}) by averaging over all haplotypes in the entire sample using Markov chain Monte Carlo (MCMC) methods. In the program *PHASE* (Li and Stephens 2003), ρ_{LS} can be estimated at the same time that haplotypes are reconstructed from un-phased data.

A third method (Wall 2000a) uses what is sometimes called an *approximate likelihood approach* (see Chapter 9 for more details). This method proceeds by first calculating several summary statistics for a sample, such as K and R_m, and then simulating equivalent datasets in terms of number of samples and number of segregating sites for many values of ρ. The probability of the observed summary statistic given any particular value of ρ is the proportion of trials that have the same value of the statistic. The maximum likelihood estimator is defined as the value of ρ that maximizes this proportion. The estimators ρ_K and ρ_{KRM}—in effect the ones that find the value of ρ that maximizes the observed number of haplotypes (K) or both the number of haplotypes and the minimum number of recombination events (K and R_m)—were found to work best (Wall 2000a; note that this paper refers to the number of haplotypes as H).

As an example of an application of Wall's method, **FIGURE 4.10** shows the outcome of one such simulation run, for a single value of ρ. If in a sample we observe $K = 10$, then the results in this figure say that the probability that $K = 10$ given $\rho = 0.03$ (per site) is approximately 0.079. That is, out of 1,000 random simulations with $\rho = 0.03$ and the same sample size and number of segregating sites, only 79 of them will have produced $K = 10$. This is clearly not the most likely outcome when $\rho = 0.03$, so it is likely that a smaller parameter value would maximize the probability of seeing $K = 10$. Carrying out similar simulations for a range of values for ρ will allow us to find the maximum likelihood estimate, ρ_K. When finding the maximum likelihood (ML) estimator for ρ_{KRM}, we need to find the value of ρ that maximizes the product of the probabilities for the observed values of both K and R_m; we therefore also have to record the probability of R_m given specific values of ρ.

The approximate likelihood method is straightforward to implement and can be used to find maximum likelihood estimators for many different types of parameters without the need for a likelihood function (e.g., Andolfatto 2007). In fact, it is quite easy to see how, given an observed number of segregating sites, one could also simulate many different values of θ to find the estimate of this parameter that maximizes the likelihood of the observed S. However, simple moment-based methods provide equally accurate (and much faster) estimates of θ, and likelihood methods are therefore generally unnecessary in this case.

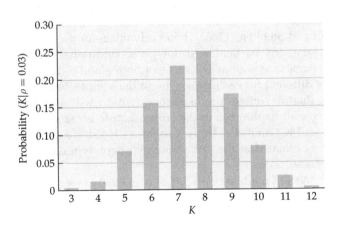

FIGURE 4.10 Finding the probability of observing K haplotypes given $\rho = 0.03$. One thousand coalescent simulations were run with $n = 20$, $S = 10$, and $\rho = 0.03$, and the number of unique haplotypes, K, was recorded for each simulation run.

The final estimator of ρ that I consider here is also now one of the most widely used, partly because of its accuracy and partly because of the ease of use of software packages that implement it. The original idea for this estimator is due to Hudson (2001), who showed that two-locus sampling distributions can rapidly be calculated given different values of ρ, and that consequently a "composite" likelihood estimate, ρ_{CL}, can be made (Chapter 9 also discusses composite likelihood methods in more detail). A two-locus sampling distribution is in some ways analogous to the allele frequency spectrum described in the previous chapter. The expected allele frequency spectrum represents the probability density function for a single polymorphism—that is, the probability that a single variant will be found at any particular frequency in a sample. The two-locus sampling distribution represents the probability density function for a pair of sites, specifically the probability that all four possible two-locus haplotypes— A_1B_1, A_2B_2, A_1B_2, and A_2B_1—will be found in a specific combination of frequencies (such that $g_{11} + g_{22} + g_{12} + g_{21} = 1$). The probability of a particular two-locus sample configuration is dependent on ρ, and we can therefore find the maximum likelihood estimate of ρ by simulating many two-locus samples with different values of ρ (assuming small θ), similar to the approach of Wall (2000a) described above. The key insight of Hudson (2001) was to show that the individual ML estimates for many pairs of linked sites—each of which could be different distances apart and therefore could have different recombination rates between them—can be combined to generate a single composite likelihood estimate for the entire dataset, ρ_{CL}. Because of the fact that this is a composite likelihood method, the mean of this estimate will reflect the true mean, but the confidence intervals will be narrower than expected because of the non-independence among the individual ML estimates. Bootstrapping of the data has been used to generate accurate bounds on the estimated value of ρ_{CL}, but this can be computationally costly. A newer method for accurately estimating confidence intervals in composite likelihood calculations uses the Godambe information matrix (Coffman et al. 2016), which allows for rapid and efficient calculation of uncertainty.

Improvements to the composite likelihood estimator have been made, including allowing the possibility of multiple mutations at the same site (i.e., the finite sites model; McVean, Awadalla, and Fearnhead 2002) and allowing the value of ρ_{CL} to vary along a chromosome (McVean et al. 2004). Missing data are also easily dealt with in the composite likelihood framework. The ability to allow varying values of ρ_{CL} has been particularly powerful, especially when working with species that have widely disparate recombination rates within a region (e.g., **FIGURE 4.11**). Although the programs *LDhat* (McVean, Awadalla, and Fearnhead 2002) and *LDhelmet* (Chan, Jenkins, and Song 2012) use MCMC methods to estimate variable recombination patterns along a chromosome, and such methods are generally more computationally intensive, the information gained can often outweigh the slightly longer run times required.

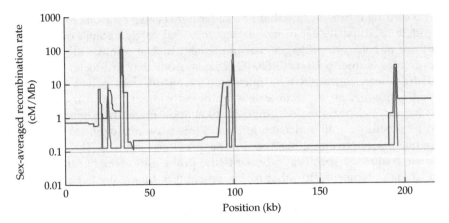

FIGURE 4.11 A comparison of estimates of recombination rates from population genetic data (red) and direct gamete genotyping (blue) in the HLA region in humans. The estimates of ρ_{CL} (the red line) have been converted to recombination rates by assuming a single value of the human effective population size and that $\rho_{CL} = 4N_ec$. (After McVean et al. 2004; based on Jeffreys, Kauppi, and Neumann 2001.)

One last comment should be made about all of the estimators of ρ discussed here. As with the estimators of θ discussed in the previous chapter, all of these statistics only estimate the quantity $4N_ec$ under the assumptions of a Wright-Fisher population at equilibrium and the appropriate mutational model. If these assumptions do not hold, then our statistics will certainly be related to the amount of recombination in a sample, but we will not know the exact relationship. This problem is especially acute when comparing estimators of θ and ρ. Under the assumptions of the standard neutral model, $\theta/\rho = 4N_e\mu/4N_ec = \mu/c$, so the ratio of these parameters tells us about the relative importance of mutation and recombination at a locus. However, unless all the assumptions hold, then statistics such as θ_W and ρ_{CL} will not be accurate estimators of θ and ρ, and therefore θ_W/ρ_{CL} will not equal μ/c. Even when violations of our model seem to offer a straightforward way to estimate the equivalent equilibrium values—as with a population bottleneck—the idiosyncratic behaviors of individual estimators make it hard to ensure that we are estimating μ/c. For example, the estimators θ_W and π recover to their equilibrium values at different rates after a population bottleneck (see Chapter 9), so the choice of statistics in nonequilibrium populations will strongly affect the result. This means that the rate at which various estimators of ρ recover to their expected value can also differ widely. In sum, while a comparison of estimators of the population recombination and population mutation parameter offers us some insight into the relative magnitudes of recombination and mutation, this ratio will generally not equal μ/c.

GENE CONVERSION

In our examination of the effects of recombination thus far we have only considered the role of crossing over in breaking up associations between sites. As we have seen earlier, however, gene conversion is an alternative outcome of recombination (Chapter 1).

The main difference between crossing over and gene conversion in terms of patterns of linkage disequilibrium is the scale of each one's effect. The effects of crossing over increase over increased physical distances, with LD decaying according to the relationships given in Equations 4.11 and 4.12. However, because gene conversion tracts (i.e., the stretches of DNA converted) are relatively short—on the order of 100 to 1,000 nucleotides (Chen, Cooper, et al. 2007)—this process will only have effects on closely linked sites and should be independent of distance for markers that are sufficiently far apart (Andolfatto and Nordborg 1998; Wiehe et al. 2000). To see why this is the case, consider the example given in **FIGURE 4.12**, in which gene conversion is limited to only a relatively small stretch of sequence, affecting only one of the three polymorphisms shown. If we imagine a population where solely $A_1B_1C_1$ and $A_2B_2C_2$ haplotypes exist prior to the conversion event shown, the creation of the $A_1B_2C_1$ haplotype clearly lessens the association between alleles at the A and B loci and between alleles at the B and C loci. However, there is no effect on linkage disequilibrium between the A and C loci, as the conversion event does not change the relationship between loci on either side of the conversion tract. In this way the outcome of conversion for loci flanking the tract is similar to that of a double crossover, differing only in the length of the affected sequence.

Gene conversion is therefore expected to have larger effects on closely linked sites relative to more distant sites, with crossing over dominating at large distances. Patterns of linkage disequilibrium in natural populations are consistent with an important role for conversion, with lower LD between tightly linked variants than would be expected if crossing over were the only mechanism of recombination (Langley et al. 2000; Ardlie et al. 2001; Frisse et al. 2001). In fact, many methods for estimating the effects of gene conversion rely on the contrast between small-scale and large-scale LD (e.g., Padhukasahasram et al. 2006). Hudson's composite likelihood estimator ρ_{CL} and similar methods (Wall 2004) can easily be altered to additionally estimate both the ratio of gene conversion to crossing over (f) and the average tract length (q), which is assumed to be geometrically distributed (Hilliker et al. 1994). The main problem for all methods used to detect gene conversion is that without a large amount of data many such events will be missed (Wiuf and Hein 2000); even with

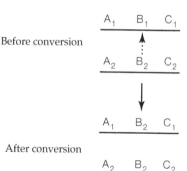

FIGURE 4.12 The effect of allelic gene conversion. The upper panel shows a diploid individual heterozygous at three loci, A, B, and C. A gene conversion event occurs during meiosis (indicated by the dashed arrow in the upper panel), with allele B_2 as the donor and B_1 as the acceptor. The result is an extra B_2-carrying gamete.

large amounts of data, if the average tract length is much smaller than the average inter-polymorphism distance, then it will be almost impossible to accurately estimate the tract length distribution (but see Miller et al. 2012 for a possible way to do this). Alternative methods that are based on the distribution of polymorphisms along a sequence—and do not attempt to estimate ρ—provide an explicit test for gene conversion but will also miss shorter tracts and will not detect any events if conversion is so rampant that the entire sequence is homogenized (Sawyer 1989).

POPULATION STRUCTURE

POPULATION DIFFERENTIATION

Populations, subpopulations, and demes

As mentioned in Chapter 2, inferences from molecular population genetic datasets can be very sensitive to the precise individuals sampled, especially to whether individuals are all taken from the same population. By *population* I mean a group of freely interbreeding diploid individuals; the term's definition for selfing or asexual lineages is much more contentious. Although they can sometimes be used to indicate slightly different things, the terms *subpopulation* or *deme* often mean the same thing as *population*, and I will use them interchangeably. The more loaded terms *races* and *subspecies* are generally avoided, except in specific instances.

Species can be subdivided into any number of populations, whether they are wholly separated from one another (i.e., they exchange no migrants) or only partially separated (i.e., they exchange some migrants). Just as there is not a particular threshold of genetic distance for separating different species, there is also not one particular threshold for calling two groups of individuals separate populations or not. As will be discussed below, the problem of actually identifying populations from molecular data can be reduced to the problem of identifying freely interbreeding groups of individuals—but the theoretical ideal of complete panmixia is rarely, if ever, met in real organisms. In addition, populations may have been separated for varying amounts of time, or there may be some phylogenetic structure to how they are related (e.g., populations A and B are more closely related to each other than either is to C). If subpopulations have not yet reached migration-drift equilibrium, many of the assumptions of models used for inference may be violated.

If subpopulations are not completely interbreeding, any of the evolutionary forces that change allele frequencies within a population (mutation, selection, drift, migration) can lead to differences in these frequencies among them (**FIGURE 5.1**). These populations are then said to be *differentiated* (I will generally reserve the term *diverged* for differences between species). Differentiation leads to *population structure*

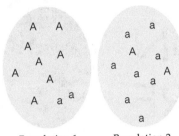

FIGURE 5.1 Example of two populations, each polymorphic for alleles
A and a. The frequency of the A
allele in population 1 is $p_1 = 0.8$
and in population 2 is $p_2 = 0.2$
(note that p_1 and p_2 do not have to
sum to 1).

Population 1   Population 2

among groups, which simply means that individuals
within populations tend to be more closely related than
individuals between populations. Obviously, populations that completely ceased to exchange migrants only
yesterday are not freely interbreeding, but neither will
they show any structure.

This simple discussion of populations begs two very
important questions: (1) How do we measure differences
among populations? (2) How do we define populations
in practice? Although I will discuss many approaches
to inferring the demographic history of a population in
Chapter 9, the importance of the above questions to the
analysis of population genetic data requires that these
issues be addressed here. In addition, the study of population structure preceded the availability of molecular
data by many decades and has developed somewhat independently of other methods for inferring population
history. It therefore merits a full discussion of its own, especially given the
many methods that are currently in use for measuring population differences
and defining populations. We tackle each of these topics in turn; however, the
answers to both of our questions can be better understood by first understanding the *Wahlund effect*.

The Wahlund effect

The simple expectations for genotypic frequencies under Hardy-Weinberg
equilibrium (Chapter 1) provide a strong null hypothesis against which we
can test for departures from the assumption of random mating. There are
many reasons why a locus or population may be out of Hardy-Weinberg equilibrium, not all of which involve nonrandom mating or even natural selection (Waples 2015). However, population structure can cause departures from
HWE because it generates strong nonrandom mating among populations.
Departures from HWE occur when multiple subpopulations are sampled
without knowledge of the underlying structure: even if all of the individual
subpopulations are themselves in HWE, the aggregate "population" can be
far from the HWE expectations. The form of this deviation from expectations is always the same—there are fewer heterozygous individuals observed
than are expected under HWE. This deficit of heterozygotes is known as the
Wahlund effect after the Swedish geneticist Sten Gösta William Wahlund, who
first described the consequences of this type of sampling (Wahlund 1928).

To see why there is a deficit of heterozygotes, consider the two populations
in Figure 5.1. We will denote the frequency of the A allele in population 1 as
p_1 and the frequency of the a allele in population 1 as q_1 (we use this notation here so as not to conflate alleles A_1 and A_2 with populations 1 and 2). In
population 1, $p_1 = 0.8$ and $q_1 = 1 - p_1 = 0.2$, while in population 2, $p_2 = 0.2$
and $q_2 = 1 - p_2 = 0.8$ (note that $p_1 + p_2$ does not have to equal 1, as these are
the frequencies of the same allele from two different populations). If we have
sampled an equal number of individuals from each population, the average

allele frequency across both populations, $\bar{p}$, is 0.5, and therefore 50% ($=2\bar{p}\bar{q}$) of all individuals are expected to be heterozygous under HWE. However, the observed proportion of heterozygotes in the sample will be approximately 32% ($=2p_1q_1$ or $2p_2q_2$), which is much lower than the expected value.

What you will notice by doing a few more such calculations is that the magnitude of the deviation from the Hardy-Weinberg expectations is proportional to the difference in allele frequencies: populations with similar allele frequencies will show very small deviations from HWE when grouped together relative to populations with very different allele frequencies. Because we will often be dealing with more than two subpopulations, a useful measure of allele frequency differences among populations will be the variance in frequencies, σ^2. In general, then, the magnitude of the Wahlund effect on the expected frequency of each genotype can be denoted as:

$$E(p_{AA}) = \bar{p}^2 + \sigma^2$$
$$E(p_{Aa}) = 2\bar{p}\bar{q} - 2\sigma^2 \qquad (5.1)$$
$$E(p_{aa}) = \bar{q}^2 + \sigma^2$$

where again $\bar{p}$ and $\bar{q}$ represent the average allele frequencies across populations, and p_{AA}, p_{Aa}, and p_{aa} represent the frequencies of the three different diploid genotypes at a locus. A larger variance in allele frequencies leads to more homozygotes and fewer heterozygotes, to the limit at which the maximum variance ($=0.25$), obtained when populations are fixed for alternative alleles, results in no observed heterozygotes. These relationships suggest that the variance in allele frequencies among populations might be a good way to quantify deviations from HWE, and therefore a good way to measure the amount of differentiation among populations. We will see that a slight modification of the variance is an ideal way to measure population differentiation, one that connects theoretical predictions to data.

MEASURING POPULATION DIFFERENTIATION

Measuring population differentiation using F_{ST}

Given samples of multiple individuals from multiple populations within a species, we would like to be able to say how different the populations are. Putting aside for a moment the exact way in which we calculate these differences, there are actually a number of alternative ways to summarize differentiation. We can ask about the average amount of subdivision among multiple populations, or about pairwise relationships between all populations. We may also have very different kinds of data, from allele frequencies at single polymorphic sites (or multiple unlinked sites) to full haplotypic sequences in coding or noncoding regions. Methods for dealing with two or more than two subpopulations are largely the same, but the many different ways of measuring differentiation at single sites versus whole sequences are often distinct; we therefore first discuss differentiation at single polymorphic sites.

In the above discussion of the Wahlund effect we saw that the variance in allele frequencies among populations was a natural way to measure differentiation, and that it connected nicely with theoretical expectations of genotype frequencies. However, the variance in allele frequencies is not itself a very useful statistic owing to the fact that its value is highly dependent on the average allele frequency. Given binomial sampling of alleles every generation, a locus with an average allele frequency of $p = 0.5$ will show a much higher variance among populations than a locus sampled from exactly the same individuals but with $p = 0.01$. This means that it is hard to compare variances between different loci or between different sets of populations. In order to standardize the variance by the maximal variance possible, we can define a new statistic of differentiation as:

$$F_{ST} = \frac{\sigma^2}{\overline{p}\,\overline{q}} \tag{5.2}$$

where σ^2 is the observed sample variance in the frequency of allele A among populations, $\overline{p}$ is the average frequency of allele A among populations, and $\overline{q}$ is the average frequency of allele a among populations ($= 1 - \overline{p}$). The value of F_{ST} from the populations shown in Figure 5.1 is $0.09/0.25 = 0.36$. This measure was first introduced by Wright (1931, 1943, 1951), who defined it as the amount of differentiation among subpopulations and showed that it had the same expected value for neutral alleles at any frequency. For our purposes the subscript "ST" is not really necessary, but it differentiates this measure from similar measures of deviation from HWE not covered here (specifically, F_{IS} and F_{IT}, where the "I" stands for "individual," the "S" for "subpopulation," and the "T" for "total population" in Wright's original notation).

F_{ST} has some very nice properties as a measure of population differentiation. It ranges between 0 and 1, with 0 indicating no differentiation and 1 indicating complete fixation of alternative alleles in different subpopulations. There are many different ways to understand what F_{ST} represents exactly, but we can relate it back to the Wahlund effect via simple algebra:

$$
\begin{aligned}
E(p_{AA}) &= \overline{p}^2 + \overline{p}\,\overline{q}\,F_{ST} \\
E(p_{Aa}) &= 2\overline{p}\,\overline{q} - 2\overline{p}\,\overline{q}\,F_{ST} \\
E(p_{aa}) &= \overline{q}^2 + \overline{p}\,\overline{q}\,F_{ST}
\end{aligned}
\tag{5.3}
$$

This equation says that when there is no differentiation ($F_{ST} = 0$) there is no deficit of heterozygotes, but when there is complete differentiation ($F_{ST} = 1$) there is a complete lack of heterozygotes. So F_{ST} can be thought of as a measure of the magnitude of the Wahlund effect. Alternatively, one can consider F_{ST} to be a measure of the amount of total variation in a set of populations explained by among-population variance as opposed to within-population variance, or as the correlations between alleles within subpopulations relative to the total population (Holsinger and Weir 2009).

F_{ST} has become a very popular way to quantify population differentiation using molecular data. However, many different ways of calculating "F_{ST}" are used. As with the many different statistics that estimate the nucleotide diversity parameter θ (Chapter 3), there are also many different ways of calculating F_{ST}-like statistics. Another similarity with measures of nucleotide diversity is that F_{ST} is commonly, and confusingly, used to denote both a parameter and a statistic—that is, both a theoretical construct related to models of populations and a precise way of estimating this parameter from data. Even more confusing is the fact that a common F_{ST}-like statistic is denoted θ (Cockerham 1969; Weir and Cockerham 1984)! This conflation of parameters and statistics dates back to Wright's original definition of F_{ST}, and I will not attempt to sort out the notations or to introduce new notations of my own to distinguish among the competing usages (I have also had to use the term in both ways in this chapter). However, one should be aware that many different statistics and parameters for quantifying the relative amount of variation within and among populations are all called F_{ST}.

If the number of sampled chromosomes is the same across subpopulations, Equation 5.2 is an unbiased way to estimate F_{ST} under a "fixed"-effect model (explained further below). If sample sizes are unequal, then a better method to calculate this statistic is:

$$F_{ST} \;=\; \frac{\sum_i \dfrac{(p_i - \bar{p})^2}{K - 1}\,\dfrac{n_i}{\bar{n}}}{\bar{p}\bar{q}} \tag{5.4}$$

where for all K subpopulations, each subpopulation i has a sample size n_i and a sample allele frequency p_i, and there is an average sample size $\bar{n}$ and average sample allele frequency $\bar{p}$ for the entire dataset (Weir 1996). Note that if all subpopulations have the same sample size (i.e., all n_is are equal), this equation reduces to Equation 5.2.

There is considerable variance associated with F_{ST} values calculated from segregating sites among the same set of populations. This is because the evolutionary processes that generate population structure (see below) have a large stochastic component, resulting in a wide range of realized values of F_{ST}. Under seemingly restrictive assumptions about the underlying population structure, Lewontin and Krakauer (1973) showed that transformed values of F_{ST} should be χ^2-distributed with $K - 1$ degrees of freedom (where K is again the number of subpopulations). Although more complicated—though not necessarily more realistic—expectations for the variance in F_{ST} have been derived (e.g., Nei and Chakravarti 1977; Weir et al. 2005), the χ^2 approximation seems to fit data collected from natural populations quite well. **FIGURE 5.2** shows the distribution of F_{ST} calculated from across the genome among four human populations. Notice both that there is a large variance in F_{ST} and that it closely resembles a χ^2 distribution with three degrees of freedom (Weir et al. 2005).

Cockerham (1969) pointed out that the method given above for calculating F_{ST} only accounts for the sampling of individuals within populations. If a fixed set of populations is the target of study, then this "fixed"-effect model is completely appropriate. A fixed-effect model is also more appropriate for cases in which there are a limited number of populations—such as in humans. However, if we are instead attempting to relate the value of F_{ST} taken from a small set of populations to a much larger set, then we must also account for the stochasticity associated with the sampling of not just individuals within populations, but also populations within a species. Weir and Cockerham (1984)

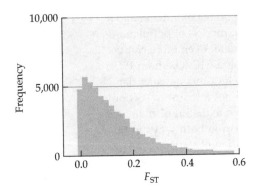

FIGURE 5.2 Distribution of F_{ST} across the human genome. The data come from 599,356 SNPs genotyped on human chromosome 1 in four human populations. (After Weir et al. 2005.)

consequently showed how an F_{ST}-like parameter, which they called θ, can be estimated under this "random"-effect model to better incorporate the multiple levels of sampling using an analysis of variance (ANOVA) approach. Of course the conditions under which the random-effect model is more appropriate are also extremely idealized: all subpopulations must be equally related to one another and have the same population size, and migration rates must be the same among all of them (equivalent to the infinite island model; see Chapter 1). There are likely few species that meet these conditions. In addition, the random-effect model converges on the fixed-effect model when a large number of samples are taken from each of a large number of subpopulations (Weir 1996). Therefore, the fixed-effect estimator (Equation 5.2 or 5.4) is likely to be appropriate for many studies.

One of the first modifications to Wright's original formulation was made by Nei (1973), who introduced a method for calculating F_{ST} when there are more than two alleles at a locus. Nei called his statistic G_{ST} (which is equivalent to the statistic F_{ST} when there are only two alleles) and defined it as:

$$G_{ST} = \frac{H_T - H_S}{H_T} \tag{5.5}$$

where H_T is the expected heterozygosity in the total sampled population (of size n_T individuals) calculated from the frequency of all i alleles using all individuals in all subpopulations:

$$H_T = \frac{n_T}{n_T - 1}(1 - \sum \bar{p}_i^2) \tag{5.6}$$

and H_S is the *average* of the expected heterozygosities calculated from each sampled subpopulation (each with its own sample size n_S) separately:

$$H_S = \overline{\frac{n_S}{n_S - 1}(1 - \sum p_i^2)} \tag{5.7}$$

Note that the quantity $H_T - H_S$ is sometimes referred to as D_{ST} (Nei 1973). These definitions of heterozygosities are highly similar to the ones given in Equation 3.1, but they are now defined with respect to a hierarchical level of population structure. The value of G_{ST} calculated from the populations shown in Figure 5.1 is $(0.526 - 0.356)/0.526 = 0.323$, which is similar to the value obtained using Equation 5.2.

One issue that arises when measuring population differentiation at a locus with more than two alleles is that G_{ST} does not take into account the identity of the alleles. As a consequence, two populations may not share any alleles but will not have $G_{ST} = 1$. To see why this is, note that when H_S is not 0, G_{ST} is constrained to be less than 1 even though differentiation may be complete (i.e., no alleles are shared). This dependence of measures of population differentiation on within-population variation is common to many of the statistics described here and will be discussed further in the next section. To overcome the reliance of G_{ST} on its maximum possible value, Hedrick (2005) proposed a normalized statistic, G'_{ST}, similar to the normalized linkage disequilibrium coefficient D' introduced in Chapter 4. However, G'_{ST} has many issues of its own and cannot be related to the parametric expectations of F_{ST} (Whitlock 2011).

For loci with many alleles—such as microsatellites—we can overcome the limitations of G_{ST} by explicitly taking into account the mutational process, and therefore the mutational distance, among alleles. For microsatellite loci under a stepwise mutational model and assuming an island model, Slatkin (1995) showed that a statistic he referred to as R_{ST} could be used as an estimator of the parameter F_{ST}. This statistic is calculated as:

$$R_{ST} = \frac{V_T - V_S}{V_T} \tag{5.8}$$

where V_T is the variance in repeat number (or allele size when exact repeat numbers are not known) among alleles observed at a single locus across all samples in all subpopulations, and V_S is the average variance in repeat number within each subpopulation. We have previously seen a statistic using the variance in allele size at a microsatellite, θ_V (Equation 3.23); for consistency we will continue to use the notation introduced there, though note that "S" is often used to denote the variances in Equation 5.8 (Slatkin 1995). We can also see a pleasing symmetry between the measurements of population differentiation using allele frequencies (Equations 5.2, 5.4, and 5.5) and those using allele sizes (Equation 5.8): both are based on differences in the variance within and between populations. R_{ST} will generally lead to more accurate inferences for microsatellites than G_{ST}, at least when the stepwise mutation model holds. When it does not, R_{ST} can also be biased (Balloux et al. 2000).

For calculating F_{ST} from full sequence data, we can use the sum of site heterozygosities across a locus (Nei 1982):

$$\gamma_{ST} = \frac{\pi_T - \pi_S}{\pi_T} \tag{5.9}$$

where π_T is calculated according to Equation 3.2, using all samples combined across subpopulations, and π_S is the average of the same calculation for each subpopulation separately (the quantity $\pi_T - \pi_S$ is also sometimes referred to as δ_{ST}). The statistic γ_{ST} is therefore analogous to a multisite version of G_{ST}, such that, as stated in Nei (1982, p. 172), "obviously, π_T, π_S, δ_{ST}, and γ_{ST}, correspond to H_T, H_S, D_{ST}, and G_{ST}" in single-site analyses (obviously). Slatkin (1991) showed that γ_{ST} is an estimator of Wright's parameter F_{ST} under the assumptions of the island model.

Similar statistics measuring population differentiation from full sequence data (usually haplotypes) have been proposed, including corrections for multiple mutations at a single site. One measure was proposed by Lynch and

Crease (1990), who called their statistic N_{ST}. An important distinction between γ_{ST} and N_{ST} is that while the former includes comparisons between sequences from the same subpopulation in calculating the value of π_T, the latter does not (see below for similar measures). This means that N_{ST} is not an estimator of F_{ST}, although N_{ST} and γ_{ST} will converge for large numbers of sampled populations (Lynch and Crease 1990). A further wrinkle was added by Excoffier, Smouse, and Quattro (1992), who introduced an extension of Weir and Cockerham's (1984) ANOVA framework based on haplotypic differences between sequences; their statistic is denoted ϕ_{ST}. Excoffier, Smouse, and Quattro's analysis of molecular variance (AMOVA) approach differs from other methods using full sequence data in that it does not use the sum of site heterozygosities, instead requiring full haplotypes. This is because it takes into account the number of nucleotide differences between all pairs of haplotypes in the dataset and then converts the resulting distance matrix into the hierarchical variance components. For this reason ϕ_{ST} is often applied to mitochondrial DNA (mtDNA) data, though in fact a similar distance matrix can be generated for microsatellite alleles, and ϕ_{ST} can be used with this kind of data as well (Michalakis and Excoffier 1996). Unlike standard analyses of variance, in both Weir and Cockerham's (1984) ANOVA approach and Excoffier, Smouse, and Quattro's (1992) AMOVA approach, the variance components (and therefore θ and ϕ_{ST}) can be negative and do not all have to sum to 1; it is standard in all analyses to set negative values to 0. A discussion of the differences and similarities between these measures of population differentiation can be found in Excoffier (2007).

Alternative measures of population differentiation

F_{ST} and its related statistics are just one way to summarize population differentiation. Alternative methods have been proposed—and are sometimes used—based on private alleles (those alleles only occurring in one population; Slatkin 1985), shared alleles (Bowcock et al. 1994), absolute differences in allele frequency (or their square-root transformed values; Cavalli-Sforza and Edwards 1967), the topology of gene trees when there is no recombination at a locus (Slatkin and Maddison 1989; Hudson, Slatkin, and Maddison 1992), and differences in homozygosity between populations (Nei 1972). This last measure—known as Nei's D—has been an especially popular measure of genetic distance for allozyme and microsatellite data, for which there can be many alleles at a locus. Nei (1972) defined the statistic at a locus for a pair of populations as:

$$D = -\ln\left(\frac{J_{XY}}{\sqrt{J_X J_Y}}\right) \tag{5.10}$$

where

$$\begin{aligned} J_X &= \sum x_i^2 \\ J_Y &= \sum y_i^2 \\ J_{XY} &= \sum x_i y_i \end{aligned} \tag{5.11}$$

and x_i and y_i are the frequencies of the ith allele at a locus in populations X and Y, respectively. In these calculations, J_X represents the total expected homozygosity (assuming random mating) over all alleles in population X, J_Y represents the total expected homozygosity over all alleles in population Y, and J_{XY} represents the total expected homozygosity over all alleles if populations X and Y have exactly the same allele frequencies. If the two populations have the same alleles at the same sample frequencies, then $D = 0$. Increasing differences in allele frequencies between populations lead to larger values of D. Unlike F_{ST}-like statistics, Nei's D is dependent on the mutation rate at a locus. While the independence from mutation rates would appear to make F_{ST} and related statistics more reliable, there are still important caveats in interpreting differences in these measures across loci.

One very important aspect of all the F_{ST}-like statistics described above is that they are strongly influenced by within-subpopulation levels of variation (Charlesworth 1998; Jakobsson, Edge, and Rosenberg 2013). Because of this, we refer to them as *relative* measures of differentiation. In contrast, *absolute* measures of population differentiation are mostly independent of levels of within-population diversity; absolute measures are also known as *genetic distances*. The reliance on within-population levels of variation means that values of F_{ST}-like statistics will differ between types of markers (SNPs, microsatellites, etc.), simply because of differences in the average heterozygosity among markers (e.g., Moyle 2006). The reliance on within-population diversity also means that the same type of markers sampled from parts of the genome with more or less diversity will provide very different views on levels of differentiation. For instance, F_{ST} from regions of reduced recombination—which often have reduced diversity as a result of linked selection—will be higher than F_{ST} from regions of normal recombination for no other reason than that total levels of nucleotide diversity are different (Charlesworth, Nordborg, and Charlesworth 1997; Charlesworth 1998; Noor and Bennett 2009; Cruickshank and Hahn 2014).

To get around this problem, Nei (1973) proposed calculating the average number of pairwise differences between sequences from two populations, *excluding* all comparisons between sequences within populations. I will refer to this statistic as d_{XY}, although it is also referred to in the literature as π_{XY} (Nei and Li 1979), D_{XY} (Nei 1987), and π_B (Charlesworth 1998). This absolute measure of differentiation is independent of the levels of diversity within the two populations being compared (but it is dependent on the mutation rate). It is calculated as (Nei and Li 1979; Nei 1987, eq. 10.20):

$$d_{XY} = \sum_{ij} x_i y_j k_{ij} \tag{5.12}$$

where x_i and y_j are the frequencies of the ith haplotype from population X and the jth haplotype from population Y, respectively, and k_{ij} is the number of nucleotide differences between pairs of haplotypes from each population. This equation is similar to the way in which we calculated the average number of differences among haplotypes within a population (Equation 3.3), although here we are only counting differences between populations. The variance in

d_{XY} is given in Nei (1987, eq. 10.24). Note that d_{XY} can be calculated from un-phased data as well, with x_i and y_j representing allele frequencies and k_{ij} being either 1 or 0 depending on whether or not the alleles differ at a single site. The values are then summed across sites to obtain a locus-wide measure. d_{XY} can also be calculated using a single sequence from each population, as it has the same expectation as the divergence statistic, d (see Chapter 7).

There are a number of relative measures of differentiation based on d_{XY} that are in wide usage. In two populations, X and Y, we refer to the within-population levels of diversity as d_X and d_Y, which are calculated as π in sample X and π in sample Y, respectively. Nei and Li (1979) defined a measure of the "net" nucleotide differences between two populations, d_a (which they called δ), as:

$$d_a = d_{XY} - (d_X + d_Y)/2 \tag{5.13}$$

This statistic (also commonly called D_a) is intended to capture only the differences that have accumulated between populations since they split. It does so by subtracting out the differences that had accumulated before this split (see Figure 7.1), assuming that the level of ancestral variation is equal to the average of the variation found in the two current populations. d_a is therefore a relative measure because its value can be strongly affected by the amount of within-population variation. (Confusingly, Nei [1987] refers to this statistic in different places as d_a, d, and D_m.) We can also represent the number of fixed differences between populations or species as d_f (e.g., Ellegren et al. 2012). The number of fixed differences is also a relative measure because of its reliance on within-population variation.

To see how these absolute and relative measures can give dissimilar results, consider the example given in **FIGURE 5.3**. Figure 5.3A shows one hypothetical genealogical history of two populations for a locus in a region with an average recombination rate (see Chapter 6 for a detailed discussion of gene genealogies). Levels of within-population diversity are proportional to the time to the most recent common ancestor of each population sample and are therefore high relative to the between-population portion of their history. In this case d_X and d_Y are high, and d_a and d_f are commensurately low because they are dependent on within-population variation. Figure 5.3B, on the other hand, shows a hypothetical genealogical history for a locus from a region of low recombination, sampled from exactly the same populations and individuals from which Figure 5.3A is drawn. The time to the most recent common ancestor within populations in this scenario is extremely recent because of linked selection. This means that d_X and d_Y are low, and that d_a and d_f are commensurately high; d_{XY} is exactly the same in the two panels. In this way, d_a, d_f, and d_{XY} can give very different results for loci from the same populations because two are dependent on within-population variation and one is not. As a consequence, Charlesworth (1998, p.538) advised that relative measures of differentiation "are not necessarily appropriate if we wish to compare loci with very different levels of within-population variation."

One final implication of the reliance of many measures of population differentiation on relative levels of diversity is that they may be affected by ascertainment bias. To understand why, recall that a major reason an ascertainment

(A)   (B)

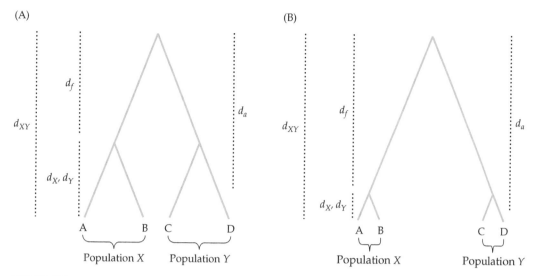

FIGURE 5.3 Demonstrating the differences between d_X, d_Y, d_a, d_f, and d_{XY}. (A) and (B) both show example genealogies relating four sampled chromosomes (A, B, C, and D) from two populations (X and Y). The statistics d_X and d_Y measure the average number of nucleotide differences among samples in populations X and Y, respectively. d_{XY} measures the average number of nucleotide differences between each sample in population X and each sample in population Y, with no comparisons made within a population. d_a represents the difference between d_{XY} and $(d_X + d_Y)/2$ (see Equations 5.12 and 5.13), and d_f represents the total number of fixed differences. The important distinction between (A) and (B) is that there is a difference in d_X, d_Y, d_a, and d_f between the two panels, but no difference in d_{XY}. For simplicity, the genealogies in the two panels have the same height, as do the times to coalescence within populations X and Y in each figure. This does not have to be the case, but it makes it easier to distinguish among the various measures described here.

bias exists is that intermediate-frequency polymorphisms are more likely to be observed in the small "discovery" sample of individuals. This means that segregating sites with population frequencies closer to 0.5 will be overrepresented in the ascertained set of markers. The result of this ascertainment bias on summary statistics is complex (Rosenblum and Novembre 2007; Albrechtsen, Nielsen, and Nielsen 2010) but can affect many inferences about population structure, including the value of F_{ST} (Clark et al. 2005), inferred rates of migration between populations (Wakeley et al. 2001), and the structure of the populations themselves (Foll, Beaumont, and Gaggiotti 2008; Guillot and Foll 2009).

Is there evidence for population differentiation at a locus?

All of the statistics for measuring population differentiation described above provide some quantitative assessment of the degree to which populations differ in allele frequencies. Once these measures are in hand, we can use them to infer something about the structure of the sampled populations and the evolutionary forces that may have contributed to any differences observed.

	Population 1	Population 2
Allele A	8	2
Allele a	2	8

FIGURE 5.4   Allele counts from Figure 5.1. Counts show number of A and a alleles in populations 1 and 2, respectively. A χ^2 test on these data (using a continuity correction because of the small counts in each cell) gives $P = 0.025$.

The simplest question we can ask is whether there is any evidence for population structure—that is, whether or not the samples come from one panmictic population. There are multiple ways to test for evidence of population structure, including testing for deviations from Hardy-Weinberg equilibrium if genotypic data have been collected from individuals (see the earlier discussion of the Wahlund effect). If genotypic data have not been collected, we can still test for structure by comparing allele frequencies between proposed subpopulations. The simplest way to carry out this test is to use a χ^2 test or other test of independence on allele counts (see Goudet et al. 1996). Using the allele counts from Figure 5.1, we can apply such a test by arranging the data as shown in **FIGURE 5.4**. By applying a χ^2 test, we find the probability that these allele counts were drawn from one single population to be $P = 0.025$, and it is therefore likely that the two subpopulations are differentiated.

We can also frame our question in terms of statistics based on these allele counts, such as F_{ST} and its related measures. Testing for population structure is equivalent to testing the null hypothesis that $F_{ST} = 0$. A straightforward way to obtain P-values for such a test is to permute the samples among populations many times, generating a null distribution of values (Hudson, Boos, and Kaplan 1992). The P-value is then based on the position of the observed value of F_{ST} in the simulated distribution; a P-value of 0.01 would therefore require that the observed value be higher than all but 1% of permuted values. For single polymorphic sites there does not appear to be much of an advantage to conducting tests using F_{ST} over allele counts (Hudson, Boos, and Kaplan 1992), but for longer sequences that encompass multiple segregating sites a permutation approach must be used because of the non-independence among sites. Such an approach would calculate γ_{ST} or other sequence-based statistics of differentiation (e.g., Hudson 2000) for the observed data and would then permute the individual haplotypes among subpopulations many times to generate the null distribution.

As I have described them, both the method using allele counts and methods using summary statistics of differentiation are testing the null hypothesis of no structure. It should be obvious then that there is a tight relationship between sample size and effect size, with even very small levels of differentiation detectable when large numbers of chromosomes are sampled. In other words, values as low as $F_{ST} = 0.001$ can be significant given enough data, and simply indicate that there is some low level of population structure. Deciding whether subpopulations should be combined or split therefore depends both on sample size and on the particular questions being addressed (cf. Waples and Gaggiotti 2006). There is no single rule for when to say that two populations are differentiated, and it is likely that most studies will be unaffected by combining subpopulations with a level of differentiation on the order of

$F_{ST} = 0.001$. Conversely, recent studies using hundreds of thousands of markers can detect important patterns of geographic structure, even with an average F_{ST} of 0.004 (Novembre et al. 2008). Later in this chapter I will address several methods for determining population structure with multiple loci using information-theoretic approaches, which allow us to make likelihood statements about the increased fit of the data when more subpopulations are proposed.

Once we have determined that there is significant population structure, how do we interpret our summaries of differentiation? How much is "a lot" or "a little" differentiation? Wright (1978, p. 85) gave the following guidelines: "We will take $F_{ST} = 0.25$ as an arbitrary value above which there is very great differentiation, the range 0.15 to 0.25 as indicating moderately great differentiation. Differentiation is, however, by no means negligible if F_{ST} is as small as 0.05 or even less." Whether these interpretations should be used in all cases—or simply as a rule of thumb—is impossible to know, especially when results depend on the markers being used. Furthermore, the most interesting inferences arising from data on population differentiation will not be limited to the magnitude or meaning of any particular value of F_{ST}, but rather will relate to the evolutionary forces that drive such differentiation.

THE EFFECT OF EVOLUTIONARY PROCESSES ON DIFFERENTIATION

The effects of natural selection on population differentiation

Differences in allele frequencies between populations are driven by the same processes that change allele frequencies within populations: mutation, natural selection, genetic drift, and migration. The effect of mutation will always be quite small and will principally be limited to the introduction of new alleles into individual populations. Recurrent mutation will sometimes introduce alleles identical-by-state into different subpopulations, but the probability of such events is low, depending on θ (Clark 1997). However, selection, drift, and migration—and the interactions among these forces—can have very large effects on population differentiation. Much of the recent work in this field has been focused on teasing apart the patterns generated by these forces.

I will first focus on the overall effects of natural selection on differences between populations, coming back to methods used to identify specific loci under selection in Chapter 10. The effects of natural selection on population differentiation are, as expected, very different for different forms of selection: weak negative selection, balancing selection, and positive selection can produce highly distinctive patterns of differentiation. This means that we can to some degree distinguish among these effects, although it may still be difficult to distinguish each of them from a neutral scenario.

Strong negative selection will of course prevent any variants from reaching appreciable population frequency and will therefore have little effect on the differentiation among segregating polymorphisms. Weak negative selection, however, allows variants to segregate at low frequencies but constrains the range of allele frequencies that are possible. This constraint means that on average F_{ST} will be lower for a weakly deleterious variant than for a typical neutral one that is able to drift to any frequency (**FIGURE 5.5**). In addition, while balancing selection also acts to lower F_{ST} (see the next paragraph), only weak negative selection will cause variants to be at lower frequencies within each population; polymorphisms under balancing selection may be at any frequency.

We expect balanced polymorphisms that are subject to the same (or very similar) selective pressure across sampled populations to be at similar frequencies across populations. As alluded to earlier, in this case selection is again constraining the range of possible allele frequencies, which means that there will be smaller changes in allele frequencies between populations as compared to neutral polymorphisms. All of this means that balancing selection will act to lower F_{ST}. A classic example of such selection is the *ABO* blood-group locus in humans: the *A*, *B*, and *O* alleles are present in almost every human population in a narrow range of allele frequencies (Brues 1954; Chung and Morton 1961). Although the exact form of balancing selection acting on this gene remains unclear, the *ABO* locus shows multiple molecular signatures of balancing selection in addition to very low F_{ST} values (Stajich and Hahn 2005; Calafell et al. 2008).

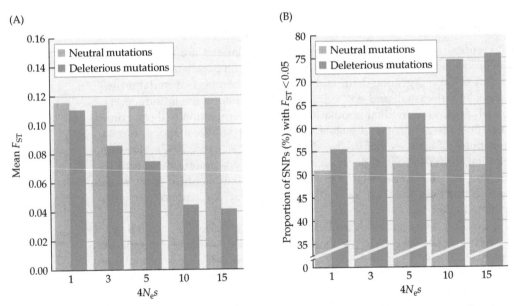

FIGURE 5.5 Effect of negative selection on F_{ST}. The results of simulating neutral and selected polymorphisms shared between two populations are shown, with average F_{ST} set to 0.11 (equivalent to the average differentiation between human populations). With increasing selection against deleterious mutations, (A) mean F_{ST} decreases, and (B) the proportion of all polymorphisms with $F_{ST} < 0.05$ increases. (After Barreiro et al. 2008.)

Any positive selection that acts similarly across all subpopulations within a species is not expected to cause differentiation between subpopulations, except transiently when a new advantageous allele sweeps through each population in succession (but see Bierne 2010). However, positive selection restricted to a subset of populations—referred to as *local adaptation*—can result in very large differences in allele frequencies. (This form of spatially varying selection is sometimes called *balancing selection* because the polymorphism is maintained at the species level. In the context of studying population differentiation I will only refer to processes that maintain polymorphisms within subpopulations as *balancing*.) In the extreme, alleles that are highly deleterious in one environment may be highly advantageous in another: in such cases allele frequencies may range from 0 to 1 between closely related populations. Selected polymorphisms in this scenario can show $F_{ST} = 1$, even if there is no differentiation across most of the genome (e.g., Turner et al. 2010). In less extreme cases the effects of positive selection will be seen as an excess of selected polymorphisms in the upper tail of the distribution of F_{ST} values. **FIGURE 5.6** shows the large excess of nonsynonymous SNPs with $F_{ST} > 0.65$ in the human genome; a similar excess is also seen for nonsynonymous SNPs with $F_{ST} < 0.05$, that is, for deleterious polymorphisms in the lower tail of the distribution (Barreiro et al. 2008). This excess is measured relative to polymorphisms presumed to be under little direct selection (in this example, nongenic SNPs). But because F_{ST} and allele frequency are not completely independent, a proper statistical comparison must compare polymorphisms of similar frequency from both the selected and unselected classes of sites in order to examine the effect of selection on population differentiation (e.g., Barreiro et al. 2008; Langley et al. 2012).

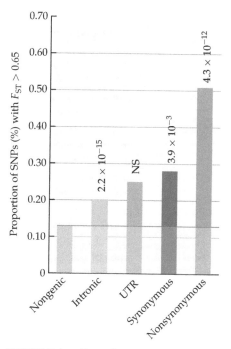

FIGURE 5.6 Effect of positive selection on F_{ST} in the human genome. A total of 851,856 SNPs were genotyped in four human populations, and F_{ST} was calculated according to the method of Excoffier, Smouse, and Quattro (1992). The large excess of nonsynonymous polymorphisms in the upper tail of the F_{ST} distribution is consistent with the action of positive selection (*P*-values indicate excess compared to nongenic SNPs). It is likely that linked selection in coding regions has increased the proportion of synonymous polymorphisms in this tail. (After Barreiro et al. 2008.)

The effects of migration and drift on population differentiation

For neutral polymorphisms, patterns of population differentiation will be determined by migration, drift, and linked selection. Linked selection clearly has an effect on neutral polymorphisms in close proximity to selected ones—see, for instance, patterns of F_{ST} on synonymous polymorphisms in Figure 5.6—and the patterns it generates will be key to finding specific regions that are the targets of spatially varying selection (see Chapter 10).

Linked selection can also have effects on estimates of F_{ST} across large swaths of the genome loosely linked to targets of selection, such that differentiation across the genome will be a mix of loci showing varying effects of this process (Nosil, Funk, and Ortiz-Barrientos 2009). As the outcome of linked selection is largely to reduce or increase local N_e, in the rest of this chapter I will simply consider this mechanism as part of the drift term in determining levels of population differentiation.

Population differentiation is largely determined by the opposing forces of migration and drift. In order to understand the differing effects of migration and drift on population differentiation, it is instructive to explain two opposing models for the history of subpopulations: the aptly named *migration* and *isolation models* (Wakeley 1996a). Consider two questions that are often asked about subpopulations (and sometimes, species): (1) How long ago did they split? (2) How much migration is there between them? It turns out that the model of population history chosen will fundamentally affect the answer to each of these questions.

The *migration model* is essentially Wright's infinite island model (Chapter 1). In this model migration between subpopulations homogenizes allele frequencies, while sampling due to drift results in differences in allele frequencies. In the migration model all subpopulations have reached migration-drift equilibrium, such that there is no longer any historical signal of their split (**FIGURE 5.7A–C**). Under a large number of assumptions that are enumerated below, Wright (1931) showed that the expected value of the statistic F_{ST} for an autosomal locus is:

$$E(F_{ST}) = \frac{1}{1 + 4N_e m} \tag{5.14}$$

where N_e is the size of each subpopulation (which are presumed to be equal) and m is the rate of migration per individual per generation (sometimes the compound parameter $4N_e m$, or just $N_e m$ by itself, is denoted as M). If there is no migration ($m = 0$), then F_{ST} will equal 1 at equilibrium (Figure 5.7C); with increasing migration, F_{ST} will decline toward 0 (Figure 5.7A).

Equation 5.14 appears to offer an easy way to estimate the effective number of migrants moving between populations per generation, $N_e m$: all one has to do is calculate F_{ST} (or F_{ST}-like) statistics from the data. However, a large number of assumptions must hold for this relationship to be true, and violations of any of these assumptions will lead to highly misleading inferences of total migration. The most important of these assumptions—and the most likely to be untrue—are that (1) populations are truly at migration-drift equilibrium and (2) there is no spatial structure among populations (this latter assumption is only relevant when there are more than two populations being considered). The first assumption is violated any time populations share alleles as a result of ancestral polymorphism and not migration; these cases are exactly those to which the isolation model is relevant and are discussed further below. The second assumption is violated when there are not an infinite number of populations each exchanging equal numbers of migrants with one another. For instance, in the stepping-stone model

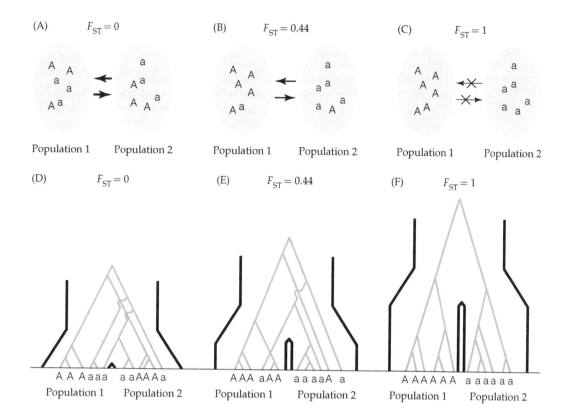

FIGURE 5.7 Differences between the migration and isolation models. (A–C) illustrate the migration model under varying levels of migration between populations. Each oval represents a separate population. Migration rates are highest in (A), medium in (B), and nonexistent in (C) (i.e., there is no migration). All differentiation is a function of migration and effective population size (see Equation 5.14). (D–F) illustrate the isolation model under varying times since population split. Hypothetical genealogical relationships for individual chromosomes in two populations are shown, with the line at the bottom representing the time at which chromosomes are sampled (i.e., the present). The split is most recent in (D), at an intermediate time in the past in (E), and far in the past in (F). All differentiation is a function of the time since the split and effective population size (see Equation 5.15).

(Chapter 1) there is spatial structure, as populations exchange more migrants with their neighboring populations. In this case nearby populations will have more similar allele frequencies and there will be "isolation by distance" (Wright 1943; see Chapter 9). If either of these assumptions—or a number of other assumptions, including that selection has no effect (see Whitlock and McCauley 1999)—are violated, Equation 5.14 cannot be used to accurately estimate migration rates. Several methods have been developed that are able to relax some of these assumptions by allowing for ancestrally shared alleles and asymmetric migration between populations (e.g., Beerli and Felsenstein 1999, 2001; Bahlo and Griffiths 2000). However, these models still assume that populations are at equilibrium, and they do not distinguish between migration and isolation as the processes contributing to population structure.

The *isolation model* describes a population-splitting scenario with no further migration after the split (**FIGURE 5.7D–F**). In this case the two resulting populations will have similar allele frequencies only because they started out with similar allele frequencies: immediately after the split the two populations should have had approximately equal representation of all alleles in the ancestral population (Figure 5.7A). The allele frequencies will diverge over time as a result of drift (and selection)—the so-called lineage sorting process—such that after sufficient time there will be no shared alleles and all chromosomes will be most closely related only to other chromosomes within the same population—that is, there will be reciprocal monophyly (Figure 5.7F). Lineage sorting can take a long time, and the stochastic nature of the process means that individual loci will vary in the time it takes to become reciprocally monophyletic. As a general guideline, Hudson and Coyne (2002) calculated that 95% of sampled autosomal loci will not share any alleles between two lineages after 9 to $12N_e$ generations.

In the isolation model the two most important parameters contributing to population differentiation are the time since the populations split, t, and their effective population sizes, N_e, which are assumed to be equivalent between populations. The expected value of the statistic F_{ST} in the isolation model is then (Wright 1931):

$$E(F_{ST}) = 1 - e^{-t/2N_e} \tag{5.15}$$

where the effects of drift on the variance in allele frequencies at an autosomal locus are simply compounded over t generations. All else being equal, increasing the splitting time increases the amount of drift that has occurred, resulting in larger values of F_{ST} (Figure 5.7D–F). Likewise, increasing the population size decreases the effect of drift, thereby decreasing F_{ST}. Any process that decreases N_e in one or all subpopulations—such as a population contraction—will increase the effects of drift in that subpopulation and lead to increased population differentiation. Likewise, drift will have a greater effect on non-autosomal loci, leading to increased differentiation at these loci (e.g., Keinan et al. 2009).

If we assume that the isolation model holds, then Equation 5.15 can be used to estimate the joint parameter $t/2N_e$. However, again a number of assumptions must be true for this relationship to hold. The main assumption is that no migrants have been exchanged between populations. Because migration homogenizes allele frequencies, any migration will lower F_{ST}, resulting in an underestimation of of $t/2N_e$. This assumption also requires that the population split was instantaneous, with all exchange ceasing immediately at time t generations in the past. These assumptions may be true sometimes, and in these cases parameters of the isolation model—including separate population mutation parameters, θ, for each of the descendant populations and the ancestral population—can be estimated (e.g., Wakeley and Hey 1997; Wang, Wakeley, and Hey 1997; Leman et al. 2005). But in many cases the major assumptions of both the isolation and migration models will be violated, or we will not be able to determine *a priori* which model is a better fit to the data. This is why recent work has turned to a model including both processes, the *isolation-with-migration model*.

Distinguishing between migration and drift: The isolation-with-migration model

One of the most challenging tasks in molecular population genetics is to distinguish between the causes of shared polymorphisms among populations or species. Because both migration and recent splitting events can result in shared polymorphisms (Figure 5.7A–F), both the migration and isolation models can explain a wide range of values of differentiation. The aim of recent research has therefore been to find distinct signatures of each process and to design a statistical framework within which robust inferences can be made about either migration rates, splitting times, or the joint effects of migration and ancestral polymorphism on levels of differentiation. Here I describe some attempts to distinguish between these processes as well as to bring them together in a single model.

Several early methods presented tests meant to discriminate between the isolation and migration models. Wakeley (1996a) showed that the variance in pairwise differences between sequences at a single locus (Equation 4.14) is greater under the migration model than under the isolation model, even though the expected values of pairwise differences are the same. That is, even when the expected value of F_{ST} is the same between the two models, the variances in π will differ. If there is asymmetric migration, the variance in π will be larger in the population accepting the larger number of migrants. The expected difference in variances assumes no recombination within a locus and no selection, two very important restrictions. Because migration increases the variance in values of both π from each of two populations exchanging genes and their total divergence (equivalent to d_{XY} above), a test can be constructed to distinguish isolation from migration. In the test proposed by Wakeley (1996b), the isolation model is the null hypothesis because there is greater statistical power to reject a null model that has the smaller variance than one that has the larger variance. The test appears to have the most power—that is, there is the greatest disparity in variances between the models—when migration is low or the time since divergence is high. As the migration rate increases or time since the split decreases, the variances converge to those expected in a single, randomly mating population; this reduced power in high-migration/recent-split scenarios appears to generally be the case among different methods (see below).

In a different approach, Wakeley and Hey (1997) found that the expected numbers of shared polymorphisms, exclusive polymorphisms in each population (i.e., private alleles), and fixed differences between two populations can be calculated for the isolation model. These summary statistics are informative in comparing migration versus isolation models because individual non-recombining loci can have shared polymorphisms or fixed differences, but not both (they can have neither as well). Gene flow will increase the number of shared polymorphisms while simultaneously decreasing both the number

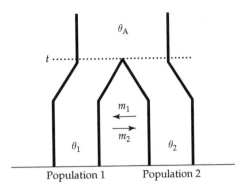

FIGURE 5.8 The isolation-with-migration model for two populations. In the most basic model there are six parameters to be estimated: three population mutation parameters (θ_1, θ_2, θ_A), two migration parameters (m_1, m_2), and time since the populations split (t). The programs *IM*, *IMa*, and *IMa2* scale all parameters by the neutral mutation rate, such that the migration parameter being estimated is m/μ and the time parameter is $t\mu$. (After Hey and Nielsen 2004.)

of fixed differences and the number of exclusive polymorphisms. These four summary statistics can then be used as the basis of tests applied against either simulated values (Wang, Wakeley, and Hey 1997) or expected values (Kliman et al. 2000). Rejection of the null model in either test implies that the pure isolation model can be discarded as a viable explanation for the data because there is too much variation among loci in these summary statistics. All of these tests require a modest number (>10) of phased haplotypes from each of two populations.

The two methods just described test the alternative model of migration after a population split but do not estimate any population parameters. In the isolation-with-migration model involving two populations there are six fundamental parameters (**FIGURE 5.8**): θ-values for the two descendant populations and the single ancestral population, time since divergence, and migration rates in both directions (which are sometimes collapsed to a single migration parameter). To estimate models with a large number of parameters, Bayesian methods often use MCMC sampling of coalescent genealogies to arrive at the most likely parameter values (see Chapter 9). The first method to estimate parameters of the isolation-with-migration model using MCMC could only use phased sequence data from a single locus (Nielsen and Wakeley 2001), but subsequent refinements have allowed multiple loci and more than two descendant populations to be studied (implemented in the programs *IM* [Hey and Nielsen 2004], *IMa* [Hey and Nielsen 2007], and *IMa2* [Hey 2010]). Analyses using MCMC sampling provide a distribution of posterior probabilities for values of each parameter. While the peak of this distribution is generally taken as the estimate of a parameter, the shape of the curve can be used to gauge support for the estimate and to contrast alternative models. Specifically, in attempting to distinguish between the isolation and migration models, "the difference between the probabilities at [the migration parameter] peak and at a migration rate of zero can be used for a likelihood-ratio test of the null hypothesis of zero gene flow" (Pinho and Hey 2010, p.223). A significant result implies that some amount of migration is required to better explain the data and rejects the pure isolation model.

The inference of parameters in the isolation-with-migration model using the *IM* program and its successors offers a powerful way to infer recent evolutionary processes, but these methods are not without problems. Most difficulties arise from the assumptions made by the methods, which will be violated quite often and affect the inferences made in many cases (Becquet and Przeworski 2009; Strasburg and Rieseberg 2010). The main assumptions include selective neutrality at all loci (i.e., no positive or balancing selection), no recombination within genes, no further population structure within sampled descendant populations or the ancestral population, independence between genes, conformity to the chosen mutation model, and no migration from unsampled populations. Strasburg and Rieseberg (2010) found that the *IM* software package was "relatively robust" to moderate violations of any of the above assumptions, although specific violations did result in misleading inferences. One of the most commonly violated assumptions is that there is no intragenic recombination, the effect of which is to increase estimates of θ in each descendant population and to increase the estimated time of divergence between them. To address this limitation, Hey and Nielsen (2004) suggested retaining only non-recombining portions of sampled sequences for analysis, though this approach can create other biases. Violations of the assumptions of *IM* that can (and do) result in large estimated values of the ancestral θ, including unaccounted-for structure in the ancestral population, are also common (Becquet and Przeworski 2009). A final caveat is that these methods appear to overestimate migration rates for very recent splits (e.g., Naduvilezhath, Rose, and Metzler 2011; Hey, Chung, and Sethuraman 2015).

An additional use of the isolation-with-migration model has been to attempt to estimate the timing of migration (e.g., Won and Hey 2004). Especially when considering recent splits between species, it is of considerable interest to distinguish between a model of primary speciation-with-gene-flow and one of secondary contact after divergence. Estimating the timing of migration events should allow researchers to know which model is more likely given the data. However, it has been shown by both simulation (Strasburg and Rieseberg 2011) and theory (Sousa, Grelaud, and Hey 2011) that one cannot infer the timing of any migration events with certainty in this model. There simply does not seem to be enough signal in the data used by the *IM* method to confidently determine when an individual migration event occurred, although it may be possible to contrast models that *a priori* define epochs that have different migration rates (Sousa, Grelaud, and Hey 2011).

Other approaches to estimating parameters in the isolation-with-migration model have also been used, each with their own limitations. Several methods use the same basic approach as *IM* and related programs but require only a very small number of individuals and a very large number of phased (Wang and Hey 2010) or unphased loci (Gronau et al. 2011). Similar methods using summaries of the data allow for intra-locus recombination and can be implemented either as Bayesian (Becquet and Przeworski 2007) or "approximate" Bayesian approaches (see Chapter 9). A quite different approach uses the multi-population allele frequency spectrum—that is, a joint site frequency spectrum for all descendant populations considered—as the input data from which to infer parameters (Wakeley and Hey 1997). Methods using this representation of the data can use hundreds or thousands of individuals and data from thousands of markers, assuming the polymorphisms are independent (see Chapter 9 for more details).

It should be stressed that although these newer model-based methods are being used to distinguish between drift and migration as the causes of population differentiation, for the most part they have not yet been extensively tested. A large body of work has detailed the consequences of violating assumptions for *IM* and its successor programs (e.g., Becquet and Przeworski 2009; Hey 2010; Strasburg and Rieseberg 2010), but the lack of such studies for newer methods should not imply that they are more accurate—only that they have not been tested yet. It is truly difficult to disentangle ancestral polymorphism from migration, even using the most realistic theoretical models and the most advanced computational tools. This difficulty is in part due to the still-unrealistic assumptions that must be made in each model and in part due to the very similar patterns generated by all evolutionary processes when using only part of the available genetic data. Future approaches of this kind may require the use of both improved models and new types (or much larger amounts) of data.

Additional methods for detecting migration between pairs of populations

In Chapter 9 we will discuss a number of additional methods for inferring migration and new models for understanding the patterns generated by migration. But before moving on in this chapter, I want to introduce several approaches that make predictions about patterns of variation in the presence of gene flow without explicit expressions for expected values of individual statistics. These methods use simulations to generate expected values across loci and are therefore relatively sensitive to the accuracy of the parameters used in the simulations. For instance, consider a case in which we use simple summary statistics (e.g., F_{ST} or d_{XY}) to assess whether migration has occurred between two populations. Simulated datasets generated assuming an ancient population split can result in incorrect inferences of migration if the real data came from two populations with a more recent split. Such errors will be a consequence of the fact that the observed statistic values will be much lower than those from the simulation, and a simple interpretation of such an observation would be that migration has occurred. However, in practice most errors will be in the opposite direction, making it harder to detect migration. The reason for this is that divergence times are estimated from the same data as migration, leading to circularity in the tests: if migration is occurring, then sequence divergence among our samples will be lower than expected given the real split time. As a result, simulated data without migration will closely match observed data with migration and an older split, leading to conservative tests for migration.

   One way to attempt to get around these problems with simulations is to try to detect individual loci that show signs of migration between populations, even when introgression across the genome is rare. In this case the hope is that simulations will reflect the non-migration history of the majority of the genome, making outlier loci with signals of migration more easily detected. If outliers are found, this is *prima facie* evidence for migration having occurred at all. This mode of inference can again be carried out with either F_{ST} or d_{XY}, though both present difficulties in correctly interpreting outlier loci. As mentioned above, F_{ST} can be strongly affected by selection, such that loci experiencing balancing selection can appear to have much lower values than neutrally evolving regions (or those affected by hitchhiking). Such a pattern could be mistaken for loci that have been introgressed. Using d_{XY}, loci with exceptionally low neutral mutation rates will have very low divergence levels, again raising the possibility that they will be confused with introgressed loci. In addition, neither F_{ST} nor d_{XY} are sensitive to the presence of low-frequency migrant alleles (Geneva et al. 2015). This means that analyses using these statistics alone may fail to detect recent introgression (e.g., Murray and Hare 2006).

In order to detect even rare introgressed lineages, Joly, McLenachan, and Lockhart (2009) proposed using the minimum sequence distance between any pair of haplotypes from two populations or species. Defining k_{ij} as above (Equation 5.12), the minimum sequence distance, d_{min}, is $\min_{i \in X, j \in Y} \{k_{ij}\}$, the minimum distance among all pairings of haplotypes in the two populations, X and Y (**FIGURE 5.9**). The logic behind this method is that any two sequences that are highly similar to each other—and therefore that represent an ancestor more recent than the population divergence time—can only be explained by introgression. By comparing the observed d_{min} to the expected values under a model without migration, we can obtain positive evidence for introgression. This method has high power when its assumptions are met (Joly, McLenachan, and Lockhart 2009), but, like d_{XY}, it assumes that there is no variation in the mutation rate among loci.

Multiple solutions have been proposed to account for mutation rate variation, especially as the same issue occurs when using d_{XY}. One way to account for this problem is to explicitly include variation in the mutation rate in the simulations, but these rates are rarely known. An alternative that does not require estimates of the mutation rate at each locus is to account for this variation using the *relative node depth* (*RND*) of the two taxa compared to an

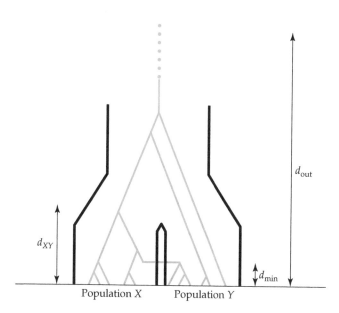

FIGURE 5.9 Additional sequence statistics used to detect migration. A representative genealogy is shown for two populations that have exchanged a migrant lineage. Divergence to an unseen outgroup (d_{out}) is represented by a dotted line. The average distance between all sequences in the two populations, d_{XY}, is only somewhat affected by the migration event. However, the minimum distance between sequences in the two populations, d_{min}, is much lower than expected under a no-migration scenario. (Based on Rosenzweig et al. 2016.)

outgroup (Feder et al. 2005). *RND* is defined as d_{XY} between the two populations divided by the average distance from each to an outgroup:

$$RND = \frac{d_{XY}}{d_{out}} \tag{5.16}$$

where $d_{out} = (d_{XO} + d_{YO})/2$; d_{XO} is the average distance between population X and the outgroup, O; and d_{YO} is the average distance between population Y and the outgroup (Figure 5.9). Low mutation rates are reflected in shortened branch lengths both between X and Y and between each population and the outgroup. Therefore, *RND* is robust to low mutation rates as long as mutation rates have been constant across the tree. This statistic is in fact most often used to find regions that are *not* introgressing when gene flow is widespread (e.g., Nachman and Payseur 2012; Carneiro et al. 2014), which would be manifested as those regions with especially high values of *RND*. When looking for introgressed loci, *RND* gets around the problem of mutation rate variation in a principled way but is still not sensitive to low-frequency migrants.

Two methods have recently been introduced that are sensitive to recent migration while still being robust to variation in mutation rates. Geneva et al. (2015) introduced a statistic, G_{min}, defined as (see Figure 5.9):

$$G_{min} = \frac{d_{min}}{d_{XY}} \tag{5.17}$$

Because a lower mutation rate is expected to affect all lineages at a locus equally, the normalization by the average distance between all sequences in the two populations will account for variable rates of evolution among loci. While G_{min} has power to detect low-frequency migrants, it loses power when the frequency of migrant alleles is higher. This is likely due to the fact that as migrant lineages rise in frequency, d_{XY} gets lower. As a migrant allele approaches fixation, the ratio of d_{min} to d_{XY} moves toward 1, approaching the value expected when there is no migration.

A statistic that is sensitive to both low- and high-frequency migrants—while still being robust to mutation rate variation—combines the best aspects of d_{min}, G_{min}, and *RND*. Rosenzweig et al. (2016) defined their statistic, RND_{min}, as (see Figure 5.9):

$$RND_{min} = \frac{d_{min}}{d_{out}} \tag{5.18}$$

Low mutation rates are reflected in shortened branch lengths to the outgroup, so RND_{min} (like *RND*) is robust to variable mutation rates. Similarly, like both d_{min} and G_{min}, RND_{min} is sensitive to even rare migrant haplotypes. The major advantage of this statistic is that it is powerful even when migrants are at high frequency (Rosenzweig et al. 2016). Arguments similar to those made above can also be used to find individual haplotypes that are *more* distant to all other haplotypes in a single population than expected, as these can represent recently introgressed sequences (e.g., Brandvain et al. 2014).

In addition to the "minimum-distance" methods just described, we can consider the different signals left by recent introgression on patterns of linkage disequilibrium. Machado et al. (2002) pointed out that patterns of LD among shared polymorphisms are expected to differ under the isolation and migration models (**FIGURE 5.10**). In the isolation model shared polymorphisms are presumed to have existed in the ancestral population and therefore are thought to have undergone many rounds of recombination and are not expected to be in strong LD with each other or with exclusive polymorphisms in each descendant population. Conversely, in the migration model shared polymorphisms are thought to be the result of introgression from one population to another and are therefore expected to be in strong LD with one another, at least in the population that receives the migrants. Additionally, this model provides specific predictions about the sign of LD: with migration, the derived alleles at two shared polymorphisms introgressed together will be positively associated, while the derived alleles at a shared polymorphism and an exclusive polymorphism will be negatively associated (Figure 5.10). No particular directionality to the sign of LD is expected under the isolation model. Machado and colleagues (2002) proposed a statistic that takes the difference between two measures of average LD—the first between all pairs of shared polymorphisms and the second between all pairs of sites at which one is a shared and one an exclusive polymorphism—and generated P-values using simulations of the inferred population history under an isolation model. As with many of the other methods described here, this method has the greatest power to detect significant deviations from the isolation model when migration is recent

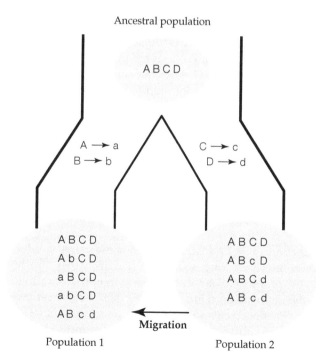

FIGURE 5.10 The effect of migration on patterns of linkage disequilibrium within a recombining locus. Each row of letters represents a haplotype, with the ancestral population monomorphic at all four sites considered. After separation, population 1 has mutations from A→a and B→b, and population 2 has mutations from C→c and D→d. The migration event introduces haplotype ABcd into population 1. Defining the sign of LD based on ancestral and derived states, the result of this introgression is positive LD in population 1 between the sites with shared polymorphisms (C/c and D/d) and negative LD between these shared polymorphisms and the polymorphisms exclusive to population 1 (A/a and B/b). (After Machado et al. 2002.)

(because LD in the introgressed region has not yet been broken down) and when the population split is further in the past (because there will be less "background" LD between shared polymorphisms that have in fact persisted since before the split). The results of this test are also completely dependent on the particular isolation scenario simulated, so that simulations using unrealistic parameters could cause rejection of the null model even when there is no migration.

Although Machado et al. (2002) studied patterns of LD contained within single loci sampled in multiple populations, similar ideas have been used to detect ancient introgression with LD data from only a single population (Wall 2000b; Plagnol and Wall 2006) as well as so-called *migrant tracts* (or *admixture chunks*) among closely related populations using genome-wide datasets (e.g., Falush, Stephens, and Pritchard 2003). The length of such tracts can be used to estimate the timing of inferred migration events or changes in migration rates (e.g., Koopman et al. 2007; Pool and Nielsen 2009; Moorjani et al. 2011), or as an explicit test of the migration and isolation models between populations (Loh et al. 2013). Patterns of LD and the proportion of each genome that is the result of migration between multiple populations have been immensely useful in defining populations in the first place, and in assigning each individual's membership in a population. In the next section I address these basic problems and proposed solutions.

DEFINING POPULATIONS

Identifying populations and the individuals within them

All of the analyses discussed to this point have assumed that we know something about population structure in our species of interest. Specifically, we have assumed that we know how many populations exist and that we can assign samples to these populations. Without having done this, there is no way for us to measure differences in allele frequencies between populations or to ask whether there is evidence for significant population structure, migration, and so on. Such questions all require us first to identify populations and to assign our samples to these populations. However, in many cases these assignments may not be known beforehand or may be inferred based only on the geographic location in which the sample was collected. Furthermore, there may be cryptic population structure that obscures the fact that samples collected from the same location actually belong to distinct populations. In both of these cases we can use statistical tools to help us make probabilistic inferences about the composition of subpopulations within a species.

An important distinction must be made between those cases in which populations are known ahead of time and those in which they are not known. In the former, the problem to be solved is the assignment of individuals to populations—the methods used are therefore called *assignment tests*. All such methods (e.g., Paetkau et al. 1995; Rannala and Mountain 1997; Cornuet et al. 1999) require multilocus genotypes from the individuals to be assigned, as well as population allele frequencies for each of the possible source populations; they also assume HWE at every locus and linkage equilibrium among loci. The optimal assignment of an individual to a source population is then based on the match between an individual's genotype and the allele frequencies in each population, and can be accompanied by a measure of uncertainty in the assignment. The accuracy of assignments is largely determined by the magnitude of the allele frequency differences among populations, the accuracy of population allele frequency estimates, and the inclusion of the correct source populations (Cornuet et al. 1999). Especially when estimates of source population allele frequencies are based on a small number of individuals, having a measure of uncertainty in assignments that takes into account small sample sizes is important. Populations can also be *excluded* as possible sources of individuals because of a mismatch between genotype and population allele frequencies. If the correct source population is unintentionally left out of a study, its absence may be indicated by high exclusion probabilities (or low assignment probabilities) for all sampled populations.

In the case in which populations are not predefined, we must use alternative tools to simultaneously assign individuals to populations and to estimate the number of populations and their allele frequencies at each locus. Furthermore, some individuals in our sample may have mixed ancestry—that is, they may have genetic contributions from multiple populations. We refer to such individuals as *admixed*. We would like to identify these individuals and to estimate the fraction of their ancestry contributed by each source population. Fortunately, there are multiple methods available that can perform all of these calculations, and more. Although each method differs in the details, many use the same underlying principles: namely, minimizing Hardy-Weinberg disequilibrium and linkage disequilibrium. Below I describe the basic approach used by the program *structure* (Pritchard, Stephens, and Donnelly 2000), the most widely used method for inferring the presence of population structure when neither the number of populations nor each individual's membership in these populations are known.

Using the Wahlund effect to identify population structure

We have already seen how the Wahlund effect can explain cases of Hardy-Weinberg disequilibrium when individuals from multiple subpopulations are mistakenly considered to be from the same population. By similar logic, mis-assigning individuals to subpopulations can also cause them to be in Hardy-Weinberg disequilibrium by introducing unlikely genotypes. Consequently, one way to find the best assignment of individuals to populations—or to find the most likely number of populations—is to attempt to minimize the amount of Hardy-Weinberg disequilibrium. Most software packages do this by trying a very large number of different assignments of individuals to populations, often by using MCMC methods. The optimal assignment is then the configuration that results in the least amount of Hardy-Weinberg disequilibrium in each population. These methods all therefore assume that there is Hardy-Weinberg *equilibrium* within each population, with the only exception being cases in which inbreeding is allowed (e.g., Gao, Williamson, and Bustamante 2007).

There is also a second kind of disequilibrium generated by misassignment of individuals to populations: linkage disequilibrium. To see why this is, consider two populations fixed for alternative alleles at two loci (**FIGURE 5.11**). Population 1 has only AB individuals, while population 2 has only ab individuals. Within each population there is no LD, but considered as a single population there is a perfect association between the two sets of alleles (i.e., perfect LD). Wahlund (1928) was actually the first person to point out this problem, but it was later rediscovered by multiple other groups (Cavalli-Sforza and Bodmer 1971; Sinnock and Sing 1972; Nei and Li 1973; Prout 1973). This effect is sometimes called the *multilocus Wahlund effect* (Feldman and Christiansen 1974) or *mixture LD* (Falush, Stephens, and Pritchard 2003).

Fixed differences between populations are not necessary to generate this effect: any differences in allele frequencies can generate LD if individuals are incorrectly grouped into populations. In fact, the magnitude of disequilibrium generated is proportional to the differences in allele frequencies between populations. Consider two loci, each with two alleles: A/a and B/b. If we denote the frequency of the A allele in population 1 as p_{1A} and in population 2 as p_{2A}, and the frequency of the B allele in population 1 as p_{1B} and in population 2 as p_{2B}, then the amount of LD is given by:

$$D = y(1 - y)(p_{1A} - p_{2A})(p_{1B} - p_{2B}) \tag{5.19}$$

where y denotes the fraction of individuals from the total sample that are from population 1, and $1 - y$ denotes the fraction from population 2. One can easily see that this form of sampling will generate linkage disequilibrium even between loci on different chromosomes. As with single-locus disequilibria, the optimal assignment of individuals to populations will be the one that

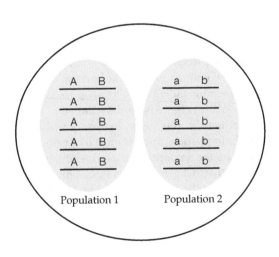

FIGURE 5.11 Generating linkage disequilibrium by sampling multiple populations. Two populations (within yellow ovals) are treated as a single population (white oval). Population 1 is fixed for the AB haplotype, while population 2 is fixed for the ab haplotype. The two loci are in linkage equilibrium within each of population 1 and 2, but when treated as a single population there is perfect LD between them.

minimizes LD among sites. In practice, minimizing both kinds of disequilibria is computationally challenging, especially when one is also attempting to find the most likely number of populations and the possible contributions of each population to the ancestry of each individual. But software packages such as *structure* can do exactly these tasks, as well as identifying the ancestry of individual loci.

Evolutionary inference from the determination of population structure

One of the most important uses of the methods described here is to find the "optimal" number of populations contained within a sample. Because both Hardy-Weinberg disequilibrium and linkage disequilibrium will be further minimized by increasing the number of populations (denoted K), the likelihood of the data will also increase with larger K. As with many statistical problems, we would like to avoid overfitting a model with a large number of parameters, instead choosing the minimum number of parameters that are sufficient to explain the data (cf. Burnham and Anderson 2002). While there are a number of different approaches to inferring the optimal value of K (described below), it should also be remembered that populations can be subjective entities, and the exact partitioning of populations may change based on the questions being asked (Waples and Gaggiotti 2006). Indeed, when population structure is hierarchical, the global "best" K for a dataset may be useful for some questions, while further subdivision of individual populations may be useful for others (e.g., Rosenberg et al. 2002). In the case of populations made up almost entirely of admixed individuals (e.g., African Americans), it is not even clear what the value of K should represent: the number of source populations or the current single population? In a similar simulated example, Pritchard, Stephens, and Donnelly (2000) inferred $K = 2$ for a single hybrid population. As an alternative to choosing a single, optimal value of K, downstream analyses can also be conducted for multiple values of K, with results reported for each value. **FIGURE 5.12** shows the outcome of increasing K for the worldwide sample of humans analyzed in Rosenberg et al. (2002).

There are two main approaches to inferring the optimal number of populations. The first uses the output of programs such as *structure* to conduct a *post hoc* analysis of the likelihood of datasets under each value of K (see Janes et al. 2017 for caveats to this approach). Because *structure* calculates the optimal assignments of individuals for a fixed value of K, independent runs with $K = 1, 2, 3 \ldots n$ must be run up to some maximum number of populations, n. The optimal value of K can then be inferred based on the rate of increase in the likelihood of the data with increasing K (Evanno, Regnaut, and Goudet 2005) or more statistically rigorous model selection methods (e.g., Gao, Bryc, and Bustamante 2011). In the second approach, the number of populations is estimated simultaneously with the assignment process. Software can either assign an upper limit to the number of populations considered within a single run, using the MCMC process to split or merge these populations (e.g., Corander, Waldmann, and Sillanpaa 2003; Corander et al. 2004), or it

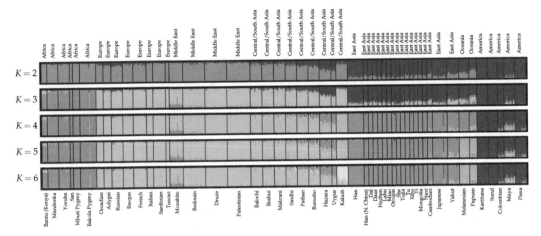

FIGURE 5.12 The structure of human populations. Each individual is represented by a thin vertical line that represents the admixture proportions from each of the K populations. Black lines separate individuals in different predefined populations (named at the bottom), and continental locations of populations are given above. In each row there are K colors used to distinguish admixture proportions for each individual. (From Rosenberg et al. 2002.)

can treat K as a random variable to be estimated given some prior distribution (Dawson and Belkhir 2001; Huelsenbeck and Andolfatto 2007). Some methods can also use the spatial sampling locations of individuals to provide priors on the number of populations and the assignment of individuals to populations (Guillot et al. 2005; François, Ancelet, and Guillot 2006; Hubisz et al. 2009).

In an alternative approach, principal component analysis (PCA) can be used to find hidden structure within a sample (see Chapter 9). While PCA methods do have an underlying relationship to the admixture models described thus far (Engelhardt and Stephens 2010; Lawson et al. 2012), they do not assign individuals to populations. However, Patterson, Price, and Reich (2006) showed that the number of significant eigenvalues in a PCA study is equal to $K - 1$. Therefore, PCA can be used to independently (and much more rapidly) calculate the optimal number of populations in a dataset.

One important issue in inferring the number and identity of populations is the effect of sampling. As with assignment tests that have predefined populations, the accuracy of population identification depends on the number of individuals included, the number of loci used, and the allele frequency differences among populations (Pritchard, Stephens, and Donnelly 2000; Rosenberg et al. 2005). In addition, population identification can be strongly affected by the amount of admixture within individuals and the proportion of admixed individuals in the dataset (Pritchard, Stephens, and Donnelly 2000), as well as by the number of individuals sampled from each "pure" population (Fogelqvist et al. 2010). It is even the case that for a limited range of low

F_{ST} values, significant structure may be identified even when there is none (Orozco-terWengel, Corander, and Schlötterer 2011).

Once an optimal choice of populations has been made and individuals have been assigned to populations, further analyses can identify migrant or admixed individuals. Very recent migrants will have statistically significant assignments to source populations that do not match the populations that overlap their sampling location (Rannala and Mountain 1997). Most such methods will only be able to identify migrants or estimate migration rates in the last few generations (e.g., Wilson and Rannala 2003), as repeated backcrossing will erase the genetic signal of the original source population. However, many individuals identified as very recent migrants may instead be mislabeled or contaminated samples (Rosenberg et al. 2002). The admixture proportion reported by *structure*—and similar methods such as *FRAPPE* (Tang et al. 2005) and *ADMIXTURE* (Alexander, Novembre, and Lange 2009)—reveals the ancestry of individual samples. Identification of admixed individuals is important for many applications, serving as a correction for the confounding effect of population stratification on association studies (e.g., Hoggart et al. 2003; Price et al. 2006), helping to determine the status of endangered populations (e.g., Beaumont et al. 2001), and providing key information in studies of hybrid zones (e.g., Nielsen et al. 2003).

It can be useful to estimate not only a single genome-wide admixture proportion for each individual, but also the ancestry of individual loci (referred to as *global* and *local* ancestry, respectively; Alexander, Novembre, and Lange 2009). Any individual's genome may be divided up into separate chromosomal "chunks" or haplotypes, each with its own ancestral origin, physical length, and population frequency. Although the original version of *structure* required unlinked markers so that "mixture" LD was the only source of disequilibrium (see above), newer versions can take into account the LD generated by the introgression of chunks of linked markers (*admixture LD*) as well as background levels of LD (Falush, Stephens, and Pritchard 2003; see also Patterson et al. 2004; Tang et al. 2006; Sankararaman et al. 2008). As mentioned earlier in the chapter, identification of these admixture chunks can be used to infer features of the migration process (e.g., Koopman et al. 2007; Pool and Nielsen 2009; Loh et al. 2013), as well as to correct for stratification in association studies (e.g., Hoggart et al. 2004), and even to infer recombination rates (Wegmann et al. 2011). Further use of haplotype-specific ancestry, using the methods implemented in the *fineSTRUCTURE* program (Lawson et al. 2012), can reveal fine-scale population structure that is not apparent using unlinked markers.

Finally, one very important caveat to inferences of population structure and admixture must be made. Despite the fact that ancestry assignments in programs like *structure* and *ADMIXTURE* are referred to as *admixture proportions* (a term I have also used here), they do not in fact indicate whether admixture has occurred, and are therefore perhaps misleadingly named. Consider the case in which there is a single large population containing much of the ancestral polymorphism, coupled with several smaller populations that have split off from the larger one and have each individually gone through a bottleneck. In many scenarios, the smaller populations will be identified by software as separate populations, with the single larger population represented as a mixture of ancestry from all the others. In this case the admixture proportions should *not* be taken as evidence for any history of admixture; instead, they should be understood as simply indicating the assignment of shared ancestry with many of the smaller populations. Therefore, *structure* plots (and similar plots) cannot by themselves be used to infer a history of admixture (see Chapter 9 for more discussion).

THE COALESCENT

6

SIMULATING SAMPLES OF DNA SEQUENCES

In order to make accurate inferences about the evolutionary forces affecting molecular variation, we must be able to accurately simulate DNA sequences. Ideally, we would like to be able to simulate populations under a wide variety of models, varying demographic histories, forms and strengths of selection, and any other parameter of interest. The result of such simulations should be a sample of DNA sequences equivalent to one we might have collected from a natural population. This procedure, repeated thousands of times, can then be used either to find the model that best fits our observed data or as a null distribution against which we can test specific hypotheses.

How might we simulate such a population? Naïvely, the most straightforward approach would be to start with a very large number of diploid individuals (equivalent to the size of the population we want to simulate) that are evolving according to our population model of choice. We would apply mutation and recombination to a sequence of appropriate length (possibly equal to the length of the DNA sequence to which we wish to compare the simulations), with mating and any possible natural selection occurring as well. This system would then have to run for many generations, at least until the population reached equilibrium, and finally a small number of chromosomes would need to be sampled from the population to create one simulated dataset. In order to generate thousands of independently simulated samples, this process would have to be run thousands of times. This procedure is often referred to as a *forward* simulation, as it proceeds forward in time from an initially identical population.

It should be obvious from the above description that this approach is very inefficient: we only want to know the properties of a sample, but we are keeping track of the whole population. To get around this inefficiency several mathematical geneticists independently proposed what has come to be known as the *coalescent process* (Kingman 1982a, 1982b, 1982c; Hudson 1983a; Tajima 1983). At its core, the coalescent process gives us a way to generate a sample of DNA sequences according to simple rules for generating a set of relationships among lineages—a *genealogy*. These rules enable us to produce a sample with almost any demographic history, without the need to simulate an entire population.

Because the process starts from the present and runs backward in time, it is referred to as a *backward* simulation. Today there are a number of software packages available for carrying out coalescent simulations (Table 6.1).

Because the coalescent process considers only a sample and not the whole population, it can be a very efficient way to generate a large number of simulated datasets, but it is not without its own limitations, both biological and computational. Biologically, the coalescent is a model of a model: that is, it is an approximation of a Wright-Fisher or Moran population (albeit a very good one), which is itself an approximation of natural populations. A key assumption of the coalescent is that the sample size, n, is much smaller than the effective population size, N_e. Violations of this assumption can result in misleading inferences about population processes (e.g., Wakeley and Takahashi

■ TABLE 6.1 Programs for simulating population samples of DNA sequences

PROGRAM	SOURCE
COALESCENT ("BACKWARD") SIMULATION SOFTWARE	
ms	Hudson 1990; Hudson 2002
GENOME	Liang et al. 2007
SIMCOAL/SIMCOAL 2.0	Excoffier et al. 2000; Laval and Excoffier 2004
CoaSim	Mailund et al. 2005
Recodon	Arenas and Posada 2007
discoal	Kern and Schrider 2016
msprime	Kelleher et al. 2016
WRIGHT-FISHER ("FORWARD") SIMULATION SOFTWARE:	
EASYPOP	Balloux 2001
simuPop	Peng and Kimmel 2005
Nemo	Guillaume and Rougemont 2006
FREGENE	Hoggart et al. 2007; Chadeau-Hyam et al. 2008
ForSim	Lambert et al. 2008
FORWSIM	Padhukasahasram et al. 2008
GENOMEPOP	Carvajal-Rodriguez 2008
SFS_CODE	Hernandez 2008
FFPopSim	Zanini and Neher 2012
fwdpp	Thornton 2014
SLiM/SLiM 2	Messer 2013; Haller and Messer 2017
ARGON	Palamara 2016
APPROXIMATE COALESCENT ("SIDEWAYS") SIMULATION SOFTWARE	
FastCoal	McVean and Cardin 2005; Marjoram and Wall 2006
MaCS	Chen et al. 2009
fastsimcoal/fastsimcoal2	(Excoffier and Foll 2011; Excoffier et al. 2013)

2003). While the standard coalescent can be modified to both more exactly match the Wright-Fisher model and to relax the $n \ll N_e$ assumption, this modification results in the loss of speed of the algorithm (Fu 2006). More importantly from a biological perspective, the coalescent only generates a small, random sample of chromosomes from a population without natural selection. It is not a simulation of a full population and therefore cannot provide results on anything but the properties of a sample. Although methods for simulating the coalescent with selection have been improving (e.g., Spencer and Coop 2004; Teshima and Innan 2009; Ewing and Hermisson 2010; Kern and Schrider 2016), the forms of selection that can be modeled are still limited.

Computationally, the advantages provided by the coalescent were much more important when computer processor speed and memory resources were limited, which has become less and less of an issue. Technological developments have made the efficiencies of the coalescent much less important, except in applications that use Bayesian sampling methodologies (see Chapter 9). Most importantly, however, when simulating very large regions of the genome or regions with very high recombination rates, the coalescent can become very inefficient, to the point that datasets currently being collected cannot be modeled using it. One solution to this problem has been the invention of approximate coalescent methods, which are based on generating related genealogies as one moves along a stretch of DNA rather than one large genealogy for the whole region (Wiuf and Hein 1999); they are therefore *sideways* simulations. These methods are more formally referred to as *sequentially Markovian coalescent (SMC) models* (McVean and Cardin 2005; Marjoram and Wall 2006), and there are multiple programs for carrying them out (Table 6.1). SMC models have become quite useful, but they are even more removed from natural populations: they are a model of a model of a model!

In the future, it may be that forward simulations will be the most widely used of such methods. There are a growing number of fast and flexible forward simulators that can model very large genomic regions under many different forms of natural selection and with many different demographic histories (Table 6.1). Computational advances have made these methods more feasible, and technological advances in DNA sequencing may make them necessary.

While I have thus far only presented the coalescent as an algorithm for generating samples, there are many important uses for this approach (see Hein, Schierup, and Wiuf 2005, and Wakeley 2009 for good overviews). The coalescent provides us with a way to probabilistically model genealogies and therefore provides expectations for a wide array of results in molecular population genetics. Many of these expectations are used in the later chapters of this book. The coalescent also gives us a framework for thinking in terms of genealogies and therefore can give us insight into the underlying relationships between our sampled DNA sequences. This framework can be extremely helpful in gaining an intuitive understanding of population genetic processes, so in the rest of the chapter I will discuss the basics of neutral coalescent genealogies and explain some of the most important aspects of the coalescent. In later chapters we will examine the effects of natural selection and nonequilibrium demographic histories on the genealogy at a locus.

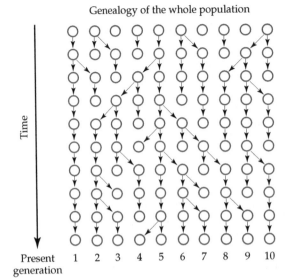

Genealogy of the whole population

Time

Present generation 1 2 3 4 5 6 7 8 9 10

Embedded genealogy of a sample

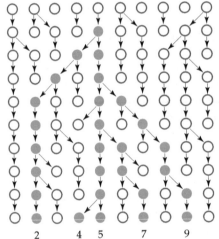

2   4 5   7   9

Genealogy of the sample

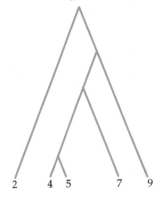

2   4 5   7   9

SIMULATING THE COALESCENT

Coalescent genealogies

Any sample of homologous chromosomes from a population is related in some way: that is, they all share ancestors in the recent past. Some sequences may be more closely related than others, but moving back in time all ultimately share a common ancestor. If we imagine the forward-in-time simulation described above as a model of a Wright-Fisher population, any collection of chromosomes sampled in the most recent generation will have a particular set of relationships at a locus, which we call a *genealogy* (**FIGURE 6.1**; sometimes these are referred to as *gene genealogies*, even when the locus is not a gene). Going backward in time, two sequences that share a common ancestor in a specific generation are said to *coalesce* into a single lineage; going forward in time, this coalescent event represents an individual chromosome giving rise to two daughter chromosomes. For the moment we consider only haploid individuals in a single population with no recombination and no selection, what Hudson (1990) called a "typical garden-variety haploid species." N diploid individuals can be formed by randomly joining $2N$ pairs of haploids. Eventually,

FIGURE 6.1 The connection between Wright-Fisher populations and coalescent genealogies. (A) The ancestor-descendant relationships among 10 haploid chromosomes in a Wright-Fisher population. (B) The genealogy among five of these chromosomes across generations, until the most recent common ancestor of the sample is reached. (C) The resulting genealogy of the sample. (Adapted from Jobling, Hurles, and Tyler-Smith 2004.)

all lineages coalesce to a single sequence—the chromosome that is the most recent common ancestor of our sample.

The coalescent is a method for generating the genealogy relating our sampled chromosomes. The genealogies formed by running the forward-in-time Wright-Fisher simulation over and over can be approximated by a very simple set of rules that put probabilities on the length of time between coalescent events, which can then be turned into a set of relationships represented by a genealogy. We can place mutations on the resulting genealogical tree at random and read off the equivalent of DNA sequences at the bottom. These two steps—generate tree, add mutations—are effectively all that is needed to provide us with a set of sequences approximating one sampled from a Wright-Fisher population. Next I describe these steps in more detail.

A straightforward way to generate coalescent genealogies is based on one simple approximation. If we replace the discrete generations used in the Wright-Fisher model with a continuous approximation, then the time until the next coalescent event is a random variable drawn from an exponential distribution. This exponential distribution is parameterized only by the number of lineages left that can possibly coalesce, which means that times between coalescent events are smaller when there are larger numbers of lineages (for instance, toward the tips of the tree) and largest when only the last two lineages remain. Using the exponential distribution, we can now write down five easy steps for generating genealogies for a sample of size n:

1. Start with $i = n$ chromosomes.

2. Choose a time until the next coalescence from an exponential distribution with parameter $x = i(i - 1)/2$.

3. Choose two chromosomes at random to coalesce.

4. Merge the two lineages that were chosen and set $i \rightarrow i - 1$.

5. If $i > 1$, go to step 2; if not, stop.

As an example, **FIGURE 6.2** shows one possible genealogy generated by following the above steps for $n = 5$. Starting with $i = n = 5$, the process proceeds by choosing two lineages to coalesce, at which point $i = 4$, another coalescent event occurs, $i = 3$, and so on. Each time this process runs a slightly different genealogy will be created, as shown in **FIGURE 6.3**. To generate trees with a non-equilibrium population history we need to adjust the times between coalescent events, but these transformations are relatively straightforward (see Chapter 9), and the resulting trees can be

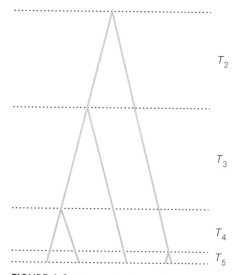

FIGURE 6.2 Example of a coalescent genealogy for a sample size of $n = 5$. The times between coalescent events, T_i, are equivalent to the amount of time that there are i lineages in the genealogy.

FIGURE 6.3 Variation in the height and topology of coalescent genealogies of size *n* = 5. Each tree is a genealogy independently generated from exactly the same population. Simulations were carried out in *discoal* (Kern and Schrider 2016).

interpreted in the same way. Time in this tree is measured in *N* generations for haploid models, 2*N* generations for diploid models, and sometimes even 4*N* generations for convenience.

The construction of a coalescent genealogy is completely independent of the presence or absence of mutations. There is a set of relationships among chromosomes regardless of whether or not there are mutations distinguishing them. This means that we can apply mutations to our genealogy in a second step that is not dependent on the details of how our genealogy was constructed, although the number and placement of mutations can depend on the tree. This independence also allows us to use any kind of mutation (SNP, indel, transposable element, etc.) under any mutation model (infinite sites, infinite alleles, etc.).

There are two methods used for putting mutations on genealogies. The first is referred to as the *fixed-S method* (Hudson 1993) and proceeds by putting a prespecified number of mutations, *S*, on the tree. This method is useful for making direct comparisons between a sample with a certain number of segregating sites and simulated genealogies with the same number of variants. These sorts of comparisons are most common when some aspect of variation other than just the level of polymorphism, such as the allele frequency spectrum, is being investigated. Since the number of segregating sites is prespecified, we only have to place the *S* mutations on branches of the tree at random, with probability proportional to branch length. A second method is used when *S* is not specified initially: in this case we need to generate the number of mutations to place on the tree using the population mutation parameter θ ($=4N\mu$ for autosomal loci in diploids). Using *t* to represent the time on any particular branch of the genealogy, the number of mutations per branch is Poisson-distributed with mean $t * \theta/2$. Here we have combined the generation of mutations with their placement on the tree: longer branches will again have more mutations using this method because more of them will appear over longer periods of time. For computational simplicity we can also sum all of the times in the tree, T_{total}, to first draw mutations from a Poisson distribution with mean $T_{total} * \theta/2$ and then "throw" all of them on the tree at random, as was done in the fixed-S method.

FIGURE 6.4 shows a coalescent genealogy with four chromosomes and six mutations. As stated earlier, these mutations can represent any kind of event under any mutational model; in order to simplify the discussion, however, let us treat them as single nucleotide polymorphisms generated under the infinite sites model. In order to turn these mutations into something that resembles a DNA sequence, we start with a sequence of length $L = 6$ for each of the four chromosomes, setting each position to 0 initially (the ancestral state). Then we propagate each mutation down the tree, such that all descendant lineages receive a 1 if they inherited the mutated version of the allele. The string below each sample in Figure 6.4 shows the resulting sequence, with the position

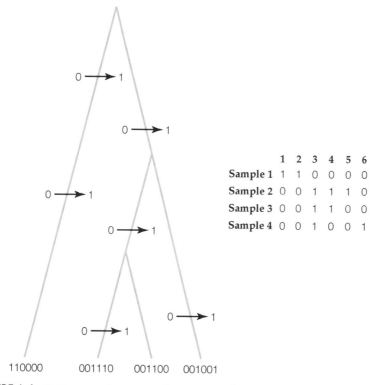

	1	2	3	4	5	6
Sample 1	1	1	0	0	0	0
Sample 2	0	0	1	1	1	0
Sample 3	0	0	1	1	0	0
Sample 4	0	0	1	0	0	1

FIGURE 6.4 Putting mutations on coalescent genealogies. Six mutations have been randomly placed on the coalescent genealogy on the left. Starting from an ancestor represented by the sequence [0, 0, 0, 0, 0, 0], derived alleles are represented by a change from 0 → 1. The sequences at the bottom represent the resulting haplotypes, which are also shown in the alignment to the right. There is no significance to the ordering of mutations in the alignment.

in the string corresponding to the mutations in the tree numbered from left to right and top to bottom. Bearing in mind that we are still dealing with a non-recombining locus, note that nothing is implied by the order of mutations in the alignment, although later in the chapter we will consider the effect of recombination, when order will matter. One can see that mutations that occur on an external branch of the genealogy result in singleton mutations, and that those closer to the root generally occur in a greater number of individuals in the sample. The resulting "sequences" of this process can then be treated like any DNA or protein sequence, with each 0/1 site representing the ancestral/derived alleles.

Genealogies and phylogenies

While coalescent genealogies resemble typical phylogenetic trees in many ways, there are very important differences between the two types of trees. The samples taken from a population are related to one another by some *unknown* genealogy. In fact, the tree describing the relationships among our samples is generally both unknown and nearly unknowable: there is still a genealogical tree even if there are no mutations distinguishing our samples. We are in no way trying to infer the tree from the identity of alleles, as is done in phylogenetics, or when building "gene trees" for mtDNA or Y-chromosome haplotypes within species. Furthermore, as will be discussed in more detail below, in the presence of recombination there can actually be a different genealogy for every nucleotide position in the locus, precluding even the possibility of inferring a single tree.

Instead, the coalescent algorithm presented above generates one *possible* set of relationships between these chromosomes, which may or may not exactly match that of our sample. One of the explicit purposes of the coalescent is to generate many different genealogies, effectively treating the tree as a nuisance parameter that has to be averaged over. The idea that the set of relationships among chromosomes is itself a cause of variance in our dataset highlights the two sources of stochasticity in the coalescent process and, by extension, in any molecular dataset: the variance introduced by sampling sets of relationships (i.e., genealogies) and the variance introduced by sampling mutations from a group of chromosomes with any given set of relationships. These sources of variation are obvious in retrospect when considering the two steps we used to carry out coalescent simulations—generate trees, add mutations—each of which had an important element of stochasticity. Often these two sources of randomness are referred to as the *evolutionary variance* and the *sampling variance*, respectively. When using explicit hypothesis tests it will be very important to know which sources of variation we must take into account and which can be ignored (see Chapters 7 and 8).

UNDERSTANDING THE COALESCENT

Important genealogical quantities

Thus far the discussion has not touched on the mathematical basis for the coalescent, nor on how exactly different aspects of the coalescent genealogy affect measures of molecular diversity. While a deep theoretical understanding of the coalescent is beyond the purview of this book, below I discuss four mathematical descriptions of genealogies that should at least shed light on the connection between the coalescent and population genetic processes.

PROBABILITY OF A COALESCENT EVENT In any given generation, it is an evident requirement that every chromosome had an ancestor in the previous generation (Figure 6.1). If there is no natural selection, each chromosome in the current generation can pick an ancestor at random; alternatively, we can say that each chromosome in the previous generation has an equal probability of being a parent to a given chromosome in the current generation. In a Wright-Fisher population of constant size $2N$ this means that there are $2N$ possible ancestors from which to choose, each with probability $1/2N$ of being chosen. Recall that a coalescent event occurs when two individuals choose the same ancestor, which for any particular pair of chromosomes has a probability of $1/2N$: that is, conditional on one of the pair picking an ancestral sequence, the second has a $1/2N$ probability of picking the same sequence. If we have a sample of size n, then there are $n(n - 1)/2$ possible pairs of sequences that can choose the same ancestor in the preceding generation. In general, in any generation in the past, for i lineages there are $i(i - 1)/2$ possible pairs that could choose the same ancestor with probability $1/2N$. So the probability of any coalescent event occurring in a generation—that is, the probability of going from i to $i - 1$ lineages—is the product of these two terms:

$$P(i \rightarrow i - 1) = \frac{i(i - 1)}{4N} \tag{6.1}$$

This probability assumes that at most two chromosomes can choose the same ancestor in any generation, an assumption closely tied to the requirement that $n << N$. We will see how the probability of a single coalescent event leads to further insight in the next section, but for now note that Equation 6.1 implies that larger numbers of sample lineages are more likely to have a coalescent event in a given generation.

TIME BETWEEN COALESCENT EVENTS The probability of a coalescent event among i lineages can directly tell us about the expected time for which i lineages remain. With probability $1 - P(i \rightarrow i - 1)$ there are *no* coalescent events in any particular generation with i lineages remaining. If no coalescent event occurred then there is again a probability of $P(i \rightarrow i - 1)$ of a coalescent event and $1 - P(i \rightarrow i - 1)$ of no coalescent events in the generation prior, and so on. The expected mean waiting time until a coalescence occurs (i.e., the time it takes to go from i to $i - 1$ lineages)—referred to here as T_i (Figure 6.2)—is therefore approximated by an exponential distribution with a mean that is the inverse of Equation 6.1:

$$E(T_i) = \frac{4N}{i(i - 1)} \tag{6.2}$$

Just as the probability of a coalescent event was higher for larger numbers of lineages, Equation 6.2 shows that the time between coalescent events is smaller. This means that, for instance, the time to go from 10 to 9 lineages will be much smaller than the time to go from 5 to 4 lineages (e.g., Figure 6.3). On average, times between coalescent events will be smallest toward the tips of the tree and longest while waiting for the last two lineages to coalesce to the *most recent common ancestor (MRCA)* of the entire sample.

HEIGHT OF THE COALESCENT GENEALOGY We may be interested in the expected amount of time before every lineage in our sample coalesces to a single lineage, the MRCA. Since we know that the expected waiting times to go from $i \to i - 1, i - 1 \to i - 2, i - 2 \to i - 3, \dots, 2 \to 1$, are $T_i, T_{i-1}, T_{i-2}, \dots, T_2,$ we can determine the average tree height by summing these times:

$$E(T_{\text{MRCA}}) = 4N\left(1 - \frac{1}{n}\right) \tag{6.3}$$

For a sample of two chromosomes, Equation 6.3 says that the tree height—the time to the most recent common ancestor—is on average $2N$ generations. As the sample size gets larger, the height of the tree approaches $4N$.

TOTAL TREE LENGTH The total length of the tree is the sum of all branches in the tree, and this value will therefore determine how many mutations are observed in the sample. We can derive the expected total tree length, T_{total}, by noting that Equation 6.2 gives the expected amount of time that there were i lineages, such that the total branch length among all lineages during this period is $i * E(T_i)$ (see Figure 6.2). Summing the total amount of time there were i lineages, $i - 1$ lineages, $i - 2$ lineages, and so on, gives us:

$$E(T_{\text{total}}) = \sum_{i=2}^{n} i\frac{4N}{i(i - 1)} = 4N\sum_{i=1}^{n-1}\frac{1}{i} \tag{6.4}$$

For $n = 2$, the expected total tree length is $4N$ (twice the value of T_{MRCA} because there are only two branches), and the tree gets larger as the sample size gets larger. But note that each new chromosome added to the sample adds less and less length to the tree on average: for $n = 5$ the expected length is $4N * 2.08$, while for $n = 10$ it is $4N * 2.83$, and for $n = 20$ it is only $4N * 3.55$. Most of the tree length (and height) is contributed by only two chromosomes, and adding more individuals is likely to only add more coalescences that occur toward the tips of the tree.

The coalescent and measures of polymorphism

We can now relate the above genealogical measures to actual quantities collected from samples of chromosomes. We can also informally prove that several of the statistics for measuring diversity in a sample (Chapter 3) are in fact estimators of the population mutation parameter θ under a Wright-Fisher population at equilibrium.

Recall that we earlier defined the statistic π as the average number of nucleotide differences between any two sequences. From Equation 6.3 we can see that the average time to the MRCA of any two randomly sampled sequences is $2N$ generations. Assuming mutations occur with probability μ

per generation and that populations are measured in effective sizes (N_e), we expect there to be $2 * 2N_e * \mu = 4N_e\mu$ differences between two sequences. This demonstrates that π is expected to be an estimator of θ under our neutral-equilibrium assumptions.

While the statistic π is based on the average number of pairwise differences between sequences, we define the statistic θ_W based on the number of segregating sites, S, so that:

$$\theta_W = \frac{S}{a} \tag{6.5}$$

where a is equal to:

$$a = \sum_{i=1}^{n-1} \frac{1}{i} \tag{6.6}$$

The term in Equation 6.6 should be familiar—it is used in defining the total length of the coalescent tree, T_{total} (Equation 6.4). It is obvious that collecting more sequences will allow us to find more segregating sites, but the θ_W statistic penalizes us less and less for each additional sample. As seen above, each additional sequence also adds less and less to the total length of the coalescent tree, which of course results in fewer new polymorphisms added. We can see this pattern explicitly by rearranging Equation 6.5 to give $S = \theta_W a$. Again assuming (1) that mutations occur with probability μ per generation and uniformly randomly across the tree and (2) an infinite sites model, the expected number of segregating sites in a sample is then:

$$E(S) = \mu * 4N \sum_{i=1}^{n-1} \frac{1}{i} = 4N\mu * a = \theta a \tag{6.7}$$

using the relationships given in Equations 6.4 and 6.6. This helps us to see that θ_W is expected to be an estimator of θ under our neutral-equilibrium assumptions.

The structure of coalescent genealogies gives us a way to graphically represent the frequency of alleles and the allele frequency spectrum. We previously defined the number of segregating sites with the derived allele present in i chromosomes, and the ancestral allele present in $n - i$ chromosomes, as S_i. We can see from the structure of the tree that for any particular derived allele present on i sequences, the mutation must have occurred on a branch with exactly i descendants (Figure 6.4). For instance, an allele found three times in a sample must have arisen on a branch with exactly three descendants. Here we are assuming an infinite sites model, since under a finite sites model the same site may have either reverted to the ancestral state on a descendant lineage or mutated to the same state on a completely different branch of the tree. Such events will complicate the correspondence between single mutations and their frequency.

Singleton mutations—those present on only a single chromosome—occur on the branches leading directly to the present-day samples. These are sometimes referred to as *external branches*, in contrast to all branches leading to two or more descendants, which are called *internal branches* (Fu and Li 1993a). Not all allele frequencies will be possible with every genealogy, as there are not necessarily internal branches with 2, 3, ... $n − 1$ descendants in every tree (there are always n branches with one descendant). The exact topology of the genealogy will determine what allele frequencies are possible and therefore the possible allele frequency spectra. However, averaging over all genealogies, we can predict the expected number of segregating sites at any frequency (i.e., the allele frequency spectrum) with:

$$E(S_i) = \frac{\theta}{i} \tag{6.8}$$

where i goes from 1 to $n − 1$ (Fu 1995; Griffiths and Tavaré 1998). This implies that there are θ singletons, $\theta/2$ doubletons, $\theta/3$ tripletons, and so on, and that singletons will be the most common type of site. Equation 6.8 also shows why the statistic θ_e ($=S_1$; Chapter 3) is an estimator of θ in a Wright-Fisher population at equilibrium, and in general that iS_i is an estimator of θ for all i between 1 and $n − 1$.

Using θ_W as an estimator of θ, for any sample we can write the expected number of neutral alleles at every frequency as:

$$E(S_i) = \frac{S}{i * a} \tag{6.9}$$

where S represents the total number of segregating sites in the sample and a is defined as in Equation 6.6. **FIGURE 6.5A** shows an example of the expected neutral allele frequency spectrum at a locus with $S = 28$ and $n = 11$.

All of the above discussion has referred to what we called the *unfolded* frequency spectrum, for which we were able to assign ancestral and derived alleles (Chapter 3). For the *folded* allele frequency spectrum we do not know whether an allele present on i chromosomes is due to a new mutation that occurred on a branch with i descendants or an ancestral allele that remains after a mutation occurred on a branch with $n − i$ descendants. Regardless, we can also write down the expected number of segregating sites at frequencies 1 to $n/2$ (rounding down for odd-sized samples):

$$E(S_i) = \frac{n * \theta}{i(n − i)} \tag{6.10}$$

and we can again substitute θ_W as an estimator of θ to give:

$$E(S_i) = \frac{n * S}{i(n − i)a} \tag{6.11}$$

FIGURE 6.5B shows the expected folded allele frequency spectrum at a locus with $S = 28$ and $n = 11$, the same as in Figure 6.5A, but with no assignment of derived and ancestral alleles.

The effects of ascertainment bias on the allele frequency spectrum

The shape of the allele frequency spectrum is an important measure in molecular population genetics. Knowing its expected shape in an equilibrium population when there is no effect of selection means that we are able to detect

(A)

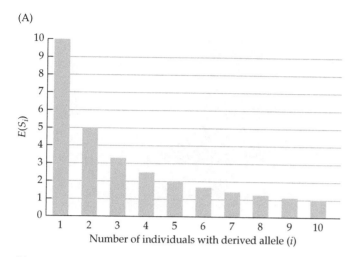

FIGURE 6.5 The expected number of segregating sites according to frequency in a sample. Given a locus with $S = 28$ and $n = 11$, the expected number of polymorphisms at frequency i $[=E(S_i)]$ is shown for (A) the unfolded spectrum and (B) the folded spectrum. The unfolded spectrum shows the expected frequency of derived alleles, while the folded spectrum shows the expected frequency of minor alleles.

(B)

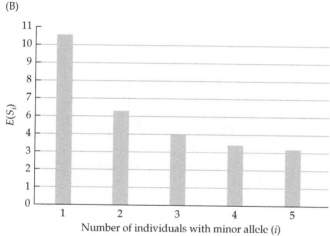

deviations from this expectation, whether they are due to nonequilibrium demographic histories (see Chapter 9) or linked selection (see Chapter 8). However, as mentioned in Chapter 2, if we genotype a set of variants in a large sample that was previously ascertained from a smaller sample, we introduce a bias into our allele frequency spectrum. An important consequence of the small size of the initial discovery sample is that we are more likely to discover intermediate-frequency variants and less likely to discover any particular low-frequency variants (i.e., those whose minor allele frequency is low). This tendency means that the allele frequency spectrum inferred from pre-ascertained polymorphisms can be highly biased compared to the theoretical expectation (**FIGURE 6.6**).

As was discussed with regard to pre-ascertained polymorphisms and measures of both linkage disequilibrium (Chapter 4) and F_{ST} (Chapter 5), there is nothing inaccurate about the sample frequencies of the individual segregating sites. These estimates are expected to be the correct frequencies and may

FIGURE 6.6 The effect of ascertainment bias on the allele frequency spectrum. Given a discovery sample of size k, the expected folded allele frequency spectrum for a second, independent sample of size $n = 41$ is shown. The unbiased sample represents the expected frequency spectrum when there is no prior discovery process. (Based on Marth et al. 2004.)

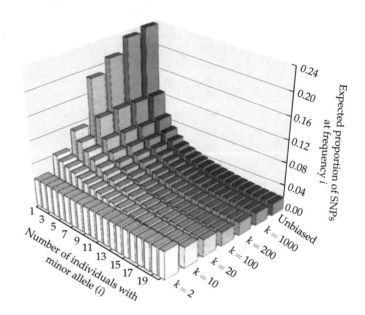

even be the same between the discovery sample and the second, often larger, sample if they are drawn from the same population. However, the expected proportion of polymorphisms at each frequency in the genotyped sample will not match the expected values if there was no pre-ascertainment. In fact, a slight bias will be introduced even if the discovery sample is larger than the genotyped sample, largely because low-frequency alleles unique to the second sample will not be represented. If the ascertainment scheme is known, we can correct for it to recover the underlying (unbiased) allele frequency spectrum (e.g., Nielsen and Signorovitch 2003; Nielsen, Hubisz, and Clark 2004; Clark et al. 2005). However, no correction is perfect, and caution must be used when interpreting even these updated spectra (Clark et al. 2005).

EXTENDING THE COALESCENT

The coalescent with recombination

The above discussion has focused on the properties of a locus large enough (or with a low enough mutation rate) to meet the assumptions of the infinite sites model, but small enough that it will have no intra-locus recombination. These assumptions are clearly unrealistic for most datasets, so we must now consider the effect of recombination on coalescent genealogies.

As stated earlier, each individual nucleotide in a sequence has its own genealogy describing the relationships among chromosomes in the sample. If there is no recombination between a nucleotide and the nucleotide immediately adjacent to it, then they both have the same genealogy (**FIGURE 6.7A**). If there is free recombination—as between two nucleotides on different chromosomes—then the two sites will have completely independent genealogies. The more interesting cases occur when sites are partially linked: when there are a limited

(A)

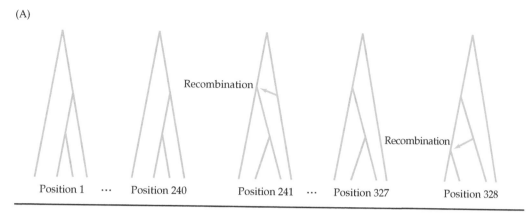

(B)

Ancestral recombination graph

FIGURE 6.7   The effect of recombination on coalescent genealogies. (A) Genealogies across a hypothetical locus. From positions 1 to 240 there are no recombination events, and therefore there is only a single genealogy. A recombination event occurs between positions 240 and 241, resulting in a new tree. Again, there is no recombination between positions 241 and 327. A second recombination event occurs between positions 327 and 328, and a new tree is created moving to the right. (B) The ancestral recombination graph summarizing the three unique genealogies represented in (A). Relative to the tree at position 1, new co-alescent events introduced by recombination are indicated with dashed lines.

number of recombination events separating them. In terms of the coalescent genealogy, each recombination event between two sites causes a rearrangement in the tree, resulting in a change in location of the coalescent event pre-dating the recombination event (Figure 6.7A). Note that both trees still retain all of the properties described above for single trees—recombination does not change the average topology of a tree (Kaplan and Hudson 1985). The two neighboring genealogies that result from a single recombination event will no longer be identical, but will be highly similar; many properties, such as their heights and total lengths, will be highly correlated (Griffiths 1981; Hudson 1983b; Kaplan and Hudson 1985). This means that many measures of variation, such as π and θ_W, will also be correlated among linked loci. More and more recombination events will lead to more and more changes to the genealogy, until eventually two genealogies separated by many recombination events will look nothing alike and will in fact be completely independent.

Considering a single locus with a limited number of recombination events, R, we can now imagine it as a set of $R + 1$ non-recombining segments, each with its own genealogy (as in Figure 6.7A). We refer to the individual genealogies for each segment as the *marginal genealogies* for the locus. Given all of the marginal genealogies, we can construct an *ancestral recombination graph* (ARG) that represents the combined history of all the individually non-recombining segments (**FIGURE 6.7B**). Coalescent events are again represented in the ARG as lineages that bifurcate, while recombination events cause lineages to join together (looking forward in time). Although this description of ARGs makes them seem limited to simple graphical representations of a collection of marginal trees, these graphs have many important applications in population genetics (e.g., Song and Hein 2005). This discussion of ARGs should also make it clear that for an autosomal locus with recombination, there is no single tree that describes the relationships among the samples. Instead, there is a *series* of trees, one for each segment with no recombination events in its history. This implies that it is not possible to infer a single, bifurcating gene tree for a recombining region, although such an inference will still be possible using Y-chromosome or mtDNA loci.

As was the case with the relationship between coalescent genealogies and summaries of nucleotide variation presented in the previous section, consideration of the effect of recombination on a genealogy can lead to a deeper understanding of statistics that summarize the history of recombination events in a sample, as well as of the variances associated with multiple measures of nucleotide polymorphism. In Chapter 4 we considered two measures of gametic linkage disequilibrium, D' and r^2. Recalling that "complete" LD means that $D' = 1$ and "perfect" LD means that $r^2 = 1$, a coalescent genealogy gives us an intuitive graphical understanding of what both of these values mean. Because two polymorphisms can have $D' = 1$ as long as there is no (detectable) recombination between them, in the coalescent context this value just means that the two segregating sites share the same genealogy. The mutations do not necessarily have to have occurred on the same branch of the genealogy, and as a result there can be three haplotypes in the sample. In contrast, two polymorphisms have $r^2 = 1$ only when they are perfectly correlated, and there are therefore only two haplotypes in the sample. This means that the sites must share the same topology *and* that the mutations must have occurred on the same branch or on either side of the root node.

We have also seen the population recombination parameter, $\rho = 4N_ec$ and several estimators of it in Chapter 4. Just as the population mutation parameter, θ, is proportional to the number of polymorphic sites in a sample, ρ is proportional to the number of recombination events in a sample, R. As discussed in Chapter 4, estimating the true number of recombination events in a sample can be difficult because most such events are not detectable. However, the structure of the coalescent genealogy still allows us to see the relationship between ρ and R. Using c to represent the recombination rate per nucleotide per generation, we can again imagine that recombination events occur uniformly randomly across space and time—that is, recombination can occur anywhere within a locus with equal probability, and we can "throw" recombination events on our coalescent genealogy in the same manner we did with mutations (though recombination changes the underlying shape of the genealogy). Given the expectation for total tree length in Equation 6.4, the expected number of recombination events in our sample is thus (Hudson and Kaplan 1985):

$$E(R) = c * 4N \sum_{i=1}^{n-1} \frac{1}{i} = 4Nc * a = \rho a \qquad (6.12)$$

which is exactly analogous to the expectation for the number of segregating sites given in Equation 6.7, with θ replaced by ρ. There are three important differences between the relationships given in Equations 6.7 and 6.12, however. First, as mentioned a number of times, most recombination events are not detectable, and therefore using the number of detectable events, such as R_m, will not yield a good estimator of ρ. Because most mutation events are detectable under the infinite sites model—and adjustments can be made for finite-site scenarios (Chapter 3)—S is a good estimator of θ, at least under the equilibrium assumptions of the Wright-Fisher model. Second, unlike the case of simulating mutations alone, we must now explicitly choose a location along the sequence at which individual recombination events and mutations occur. The ordering of recombination events is key to generating the marginal genealogies at a locus. Third, the exact tree onto which we are throwing recombination events may not be clear, as putting a recombination event on a tree necessarily changes the topology. One way to think about this situation is to consider the fixed genealogy at the first nucleotide to the left of a region being studied. We can then imagine each recombination event in our sample being put on this fixed tree, with new genealogies being generated to the right of this site, within our region of interest. This method is not the way in which standard backward coalescent algorithms simulate recombination events (see Hudson 1983b, 1990), but it may help to give a more intuitive understanding of the process.

Finally, recall that the variance in estimators of the population mutation parameter, such as π and θ_W, go down as recombination rates go up (Equations 3.4 and 3.7). The coalescent provides a genealogical perspective on this association. Within a locus with no recombination, there is a single genealogy describing the relationships among chromosomes. The height and total length of this genealogy are themselves drawn from distributions with large variances, and the fact that we have a single tree means that there is effectively a single draw from these distributions. Within a locus with recombination, higher recombination means that there are more and more marginal genealogies and these individual genealogies are increasingly independent from one another. In effect, this means that we have drawn a large number of independent samples from the distributions that indirectly determine levels of polymorphism (tree height and length); the variance is therefore much reduced. In the extreme case of free recombination among a very large number of sites, there is essentially no effect of evolutionary variance because we are averaging over a very large number of independent trees. The only variance is introduced by sampling mutations on the tree, and for θ_W it is therefore equivalent to the variance of the Poisson process by which we chose the number of mutations (the mean is equal to the variance for Poisson distributions). The take-home message from these results is that accurate estimates of recombination are key to producing accurate confidence intervals around statistics of interest.

The coalescent and reference genomes

The reference genome for a species generally comes from sequencing a single individual, whether inbred or outbred (Chapter 2). Because there is nothing necessarily special about this individual—it does not represent the ancestral state for the species, nor is it a "consensus" individual—the reference genome will contain both ancestral and derived states for independently segregating polymorphisms, just as any chromosome sampled from nature would. In the context of short-read sequencing, the reference genome is used as the backbone upon which all reads are mapped and therefore often takes a more preeminent role in population samples than perhaps it deserves. For instance, conservative methods for calling polymorphisms from short-read datasets sometimes default to the reference base when there is not high confidence in alternative alleles. This can have important consequences for estimating the derived allele frequency spectrum (note that defaulting to an N is therefore better). One interesting (and helpful) use of coalescent genealogies can thus be to ask: *What proportion of segregating sites in a sample will have the derived allele in the reference genome?*

If the sample is of size $n = 2$, the answer should be obvious: there are expected to be on average θ differences between any two chromosomes, with the mutations occurring equally along the lineages leading to each chromosome from the ancestor. Therefore, the proportion of derived alleles on either of the two branches (arbitrarily assigning one to be the reference) will be $1/2$. For samples larger than 2, slightly more sophisticated arguments are needed. There are actually two ways of answering this question, each of which is just a restatement of the central problem. To find the proportion of segregating sites with the derived allele on any arbitrarily chosen chromosome, we can find either (1) the expected average frequency of derived alleles in a sample, or (2) the expected proportion of mutations that have occurred in the history of a single chromosome relative to the entire sample.

The average frequency of derived alleles in a sample gives the expected proportion of derived alleles on any chromosome, averaging across all segregating sites. To see this, imagine choosing a single chromosome at random from nature. For any given polymorphism with derived allele frequency p, the probability of choosing a chromosome that contains the derived allele will also be p, so the probability of sampling a derived allele at frequency $p = 0.8$ will be 0.8, the probability of sampling an allele at frequency $p = 0.01$ will be 0.01, and so on. Therefore, averaging across many segregating sites, the probability of sampling the derived allele will simply be the average derived allele frequency; this measure will be equivalent to the fraction of sites at which the derived allele has been sampled. From Equation 6.9, we have the expected number of segregating sites with derived alleles found on i chromosomes, S_i. We now want to know—across all segregating sites—the average number of chromosomes in a sample of size n that contain the derived allele at each site, which we denote as m. Letting $q(m)$ be the distribution of m in a sample (i.e., the derived allele frequency spectrum), Griffiths and Tavaré (1998) showed that the expected value of m in a Wright-Fisher population at equilibrium is:

$$E(m) = \sum_{i=1}^{n-1} i \frac{1}{i*a} = \frac{1}{a}(n-1) = \frac{n-1}{\sum_{i=1}^{n-1}\frac{1}{i}} \tag{6.13}$$

such that the average frequency of derived alleles in the sample is m/n. For the case of $n = 2$, Equation 6.13 again gives $m = 1$, for an average derived allele frequency of $1/2$. For a sample of size $n = 11$ (as in Figure 6.5), $m = 3.59$ and the average derived allele frequency is 0.326.

To get a more genealogical intuition for this result, consider the alternative way of phrasing the question: What is the expected proportion of mutations that have occurred in the history of a single chromosome, out of the whole sample? Because we are conditioning on seeing segregating sites, we only need to trace the history of the chromosome back to the MRCA of the sample; all mutations that occurred before this point are by definition fixed. Recognizing that derived mutations on any single chromosome must have occurred since the MRCA, we need to know the time back to this ancestor. Fortunately, Equation 6.3 gives the expected time back to the MRCA from any of the tips of the genealogy. To get the fraction of all mutations that occurred on only this branch, we divide Equation 6.3 by Equation 6.4, the total time in the tree (**FIGURE 6.8**). If mutations are again thrown down at random across branches and we assume the infinite sites mutation model, this relationship gives the expected proportion of sites that have the derived allele on any single chromosome, b:

$$E(b) = \frac{4N\left(1 - \frac{1}{n}\right)}{4N\sum_{i=1}^{n-1}\frac{1}{i}} = \frac{1 - \frac{1}{n}}{\sum_{i=1}^{n-1}\frac{1}{i}} \tag{6.14}$$

which is equivalent to m/n, the average frequency of derived alleles, as defined above for Equation 6.13.

These relationships show that the average derived allele will be low in frequency, with a range of frequencies of approximately 0.20 to 0.50 for low- to moderate-sized samples. As could be guessed from a glance at the expected derived allele frequency distribution (Figure 6.5), the minor allele at a segregating site is most often the derived allele. In fact, Watterson and Guess (1977) have shown that the probability that an allele represents the ancestral state is simply equal to its population frequency, p; the same holds for the frequency of alleles in a sample (Griffiths and Tavaré 1998).

With respect to a reference genome, the above results show that a substantial fraction of polymorphic sites in a sample (0.20–0.50) will have the derived allele in the reference genome. As mentioned above and in Chapter 2, this

FIGURE 6.8   The proportion of all mutations in the sample that have occurred in the history of a single chromosome. If mutations occur uniformly randomly across the tree, this proportion is equivalent to the amount of time since the MRCA for a single chromosome, divided by the total time in the tree. For an arbitrary sampled chromosome denoted with a *, the thicker black line traces its ancestral lineages. The proportion of total time in the tree represented by this lineage is given by Equation 6.14.

finding implies that base-calling in a sample of individuals may often default to both derived and ancestral reference alleles. If base-calling is too conservative, this process will result in a skewed derived allele frequency spectrum, as polymorphisms with the derived state in the reference will be pushed to higher apparent sample frequencies. Therefore, when base-calling defaults to the reference allele conservative criteria for identifying variant sites may be worse. Fortunately, we can correct for such biases, either by lowering the quality thresholds used to call polymorphic sites, taking the sampling bias inherent in the reference genome into account, assigning low-quality bases to be Ns rather than the reference allele, or iteratively identifying varying sites and re-calling genotypes based on the updated probability that a site differs from the reference (e.g., DePristo et al. 2011).

DIRECT
SELECTION

The consequences of natural selection for molecular variation can be usefully divided by separately considering the effects of selection on mutations that are themselves advantageous, deleterious, or neutral (direct selection) and the effects on mutations closely linked to those under selection (linked selection). The expectations for patterns of polymorphism are often different in these two cases, and therefore different methods will be best at detecting one type of selection or the other. A deeper understanding of these methods will require at least a passing familiarity with the theory and assumptions underlying them; in this chapter I therefore consider these assumptions and their implications for inferring the effects of direct selection. The methods discussed here have been used successfully in a large number of studies to examine the effects of natural selection and are a key tool in understanding DNA sequence evolution.

THE ACCUMULATION OF SEQUENCE DIVERGENCE

The rate of nucleotide substitution

John Maynard Smith called $k = \mu$ one of the most important equations in all of evolutionary genetics (Maynard Smith 2002). I could not agree more: it is both an amazingly elegant result (first applied to molecular data by Kimura 1968, but also found in Wright 1938) and the apotheosis of the neutral theory (discussed in Chapter 1). But it is not only an elegant result—it is also one of the building blocks of many of the tests of natural selection covered in the rest of this book. Below we begin to explore its role in molecular evolution.

The variable k is defined as the substitution rate of new alleles—that is, the rate at which alleles are fixed over long periods of time (also sometimes denoted as ρ). The substitution rate is commonly measured per generation or per million years. The value of k determines how quickly two sequences are expected to diverge over time. Defining d as the genetic distance between two orthologous sequences, the contribution of the rate of substitution to the expected amount of divergence can be seen in the following equation:

$$E(d) = k2t + \theta_{\text{Anc}} \tag{7.1}$$

where t is the time since the species split (Gillespie and Langley 1979; also see Li 1977). For simplicity, assume that we are calculating all values per site so that we can ignore the length of the sequence. Within a locus, k therefore represents the average substitution rate across sites. Recall that we use $2t$ because substitutions can occur on both branches of the phylogenetic tree and we are considering only the pairwise divergence between sequences. We add the average amount of nucleotide variation expected between two sequences in the ancestor (θ_{Anc}) because at the time of speciation these differences have already accumulated along the two lineages (**FIGURE 7.1**). This expected distance is the same as that used for the statistic d_{XY} (Chapter 5); however, we only have to sample a single sequence from each species to calculate d.

In cases in which divergence levels are much greater than the expected levels of polymorphism in the ancestral species, we may simply write:

$$E(d) \approx k2t \tag{7.2}$$

This equation demonstrates an obvious relationship: species differ because new alleles arise and are fixed, and the rate at which this process happens and the time elapsed since the lineages split both contribute to the total amount of divergence. Nevertheless, this equation does not tell us what affects the rate of substitution.

There are two quantities that determine the rate of substitution. The first is the probability of fixation of any mutation, which we denote u. This probability is a function of the frequency at which mutations are found, p, but for now we are only interested in the probability of fixation of new

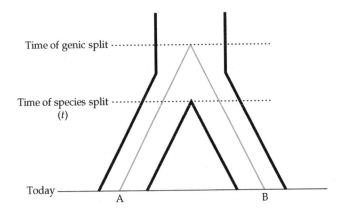

FIGURE 7.1 Distinction between the time since the species split and the time since the genes being compared split. The history of a single ancestral species splitting into two descendant species is circumscribed by the darker lines. The history of a single orthologous sequence sampled in each of the two descendant species (A and B) is denoted by the thinner, blue line. Prior to the species' split, two sequences within the ancestral population are expected to have on average θ_{Anc} differences. If the ancestral population met the assumptions of the Wright-Fisher model, the time of genic split is expected to have been $2N_e$ generations prior to the time of the species split, t.

mutations, which always begin at frequency $1/2N$ for autosomal loci. The probability of a mutation fixing when it has no effect on fitness—when it is neutral with respect to alternative alleles—is simply equal to its current frequency. So for a mutation at $p = 0.4$, $u = 0.4$, and there is a 40% chance it will eventually go to frequency 1 (and conversely a 60% chance it will be lost). For a new, neutral mutation of frequency $1/2N$, its probability of fixing is therefore simply $u_0 = 1/2N$.

For new, advantageous mutations ($s > 0$) and large effective population sizes, the probability of fixation is (Haldane 1927; Fisher 1930; Wright 1931):

$$u_a \approx 2s_a \tag{7.3}$$

(In order to indicate the selective advantage in these papers, Haldane used the symbol k, Fisher used the symbol a, and Wright used the now-standard s.) Here s_a is the selective advantage of the new allele in a heterozygote and $2s_a$ is the advantage in a homozygote. For new, deleterious mutations ($s < 0$) that do not have large effects, the probability of fixation is (Kimura 1957):

$$u_d \approx \frac{2s_d}{1 - e^{(-4Ns_d)}} \tag{7.4}$$

Here s_d is the selective *dis*advantage of the new allele in a heterozygote and $2s_d$ is the disadvantage in a homozygote. **FIGURE 7.2** shows the probability of a new selected mutation fixing relative to a neutral mutation such that these probabilities are equal at $Ns = 0$ (i.e., their ratio is 1). It can be seen that slightly advantageous mutations are not that much more likely to fix than neutral mutations, and also that even slightly deleterious mutations have some probability of fixing. Even though these slightly deleterious mutations—often referred to as *nearly neutral*—must have extremely small selective effects for Ns to be close to 0, they are expected to make a contribution to sequence divergence in smaller populations (see below).

The second quantity contributing to the rate of substitution is the total number of mutations that arise and can possibly be fixed. If the probability of a mutation at a nucleotide in each generation is ν, then in a population of N diploid individuals there will be $2N\nu$ new mutations per generation (at a single site) in the total population at autosomal loci. Of these mutations, $2N\nu f_0$ ($=2N\mu$) will be neutral, with f_0 representing the fraction of mutations that are neutral. The remaining mutations are composed of the fraction that are advantageous, f_a, and the fraction that are deleterious, f_d (such that

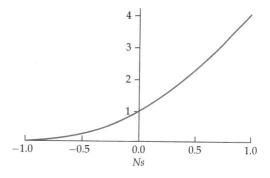

FIGURE 7.2 Probability of fixation, relative to a neutral allele, of new, selected mutations. The probability of fixation of neutral mutations is $1/2N$, with advantageous ($Ns > 0$) and deleterious ($Ns < 0$) mutations having increased and decreased probabilities of fixation, respectively. Values are calculated from Equation 7.4, which can be used for positive or negative selection coefficients.

$f_0 + f_a + f_d = 1$). There are therefore also $2N\nu f_a$ new advantageous mutations and $2N\nu f_d$ new deleterious mutations every generation.

We can now consider the two determinants of the substitution rate together. If there are no advantageous mutations, and if deleterious mutations do not fix at appreciable levels, then the substitution rate is a function of only the total number of neutral mutations that arise and the probability that each of them fixes. We can write this relationship as:

$$k = \left(2N\nu f_0\right)\left(\frac{1}{2N}\right) = \nu f_0 \tag{7.5}$$

or, substituting the symbol μ for the total rate at which neutral mutations arise, νf_0, the rate of substitution of neutral mutations is:

$$k = \mu \tag{7.6}$$

In words, this relationship says that *when considering only neutral mutations, the substitution rate is equal to the mutation rate, regardless of population size.* This independence from population size may at first seem unintuitive, but it can be understood by considering the fact that while more mutations arise in large populations, each of them has a smaller chance of eventually going to fixation. Likewise, it is more likely that any single new mutation will fix in a small population, but there are fewer mutations overall. Thus population size exactly cancels out when considering neutral mutations. This result will be used again and again in our quest to uncover cases in which natural selection is acting.

Let us finally consider the rate of substitution for alleles that are not neutral. Of the total mutation rate, ν, we have already defined f_a and f_d as the fractions of all mutations that are advantageous and deleterious, respectively, and u_a and u_d as the probabilities of fixation for new mutations of each type. We can now write the rate of substitution for advantageous mutations as:

$$k = (2N\nu f_a)(2s_a) = 4N\nu f_a s_a \tag{7.7}$$

To obtain the rate of substitution for deleterious mutations we use Equation 7.4 to find:

$$k = \left(2N\nu f_d\right)\left(\frac{2s_d}{1 - e^{\left(-4Ns_d\right)}}\right) = \frac{4N\nu f_d s_d}{1 - e^{\left(-4Ns_d\right)}} \tag{7.8}$$

From Equations 7.7 and 7.8 we can see a key difference between neutral mutations and selected mutations: N, the population size, plays an important role in the rate of substitution of selected mutations. All things being equal, more advantageous mutations will fix in larger populations than in smaller populations, while more deleterious mutations will fix in smaller populations relative to larger populations. We next consider how to determine the contribution of each of these types of mutation to divergence between species.

Estimating sequence divergence

One of the most straightforward methods for quantifying the effects of natural selection does not use population-level data at all. While this characteristic might seem to make it inappropriate in a text about population genetics, it is a standard test in every population geneticist's toolbox. And because substitutions between species are a long-term consequence of polymorphism within species (Kimura and Ohta 1971), these methods are a good starting point for understanding more powerful methods that do use polymorphism data.

Earlier we defined d as the genetic distance between two orthologous sequences. To be specific, we generally calculate d by taking a single sequence from each species and counting the number of positions that differ between them, divided by the total number of aligned nucleotides. As with calculating π between individuals in the same species (Chapter 3), we only count the number of nucleotide differences at positions without a gap. If we are considering a multiple-sequence alignment, this process often results in a position in the alignment with a gap in any sequence being ignored in all species. For this reason, pairwise divergence values calculated from alignments generated for each pair of species can differ slightly from divergence values calculated from a single multiple-sequence alignment including all species. As an example, **FIGURE 7.3** shows an alignment of sequences from different species that is 15 nucleotides long. Using the entire multiple-species alignment we would calculate the divergence between sequences 1 and 2 as 0.273 (3 differences/11 sites without gaps); however, if we were to realign only sequences 1 and 2 and recalculate divergence, we would get $d = 0.23$. Clearly, this difference can lead to biased estimates of divergence if both closely and distantly related sequences are included in the same alignments, as only those regions less prone to insertions and deletions will be included in analyses.

There are two other important issues to be considered when calculating d. First, because it is being calculated from only a single sequence per species (or per locus, because d can also be calculated between paralogs), this statistic by itself does not represent only fixed differences. That is, any derived, polymorphic allele present in the sequence chosen to calculate d will be counted toward interspecific divergence. Earlier we also considered the contribution of ancestrally polymorphic alleles, but these are in fact fixed between current species and therefore do not conflate polymorphism and divergence. This measure of divergence includes polymorphic sites because any chromosome sampled from nature is expected to contain on average θ derived alleles that are still polymorphic in the population. You can see where this expectation comes from by considering that the time to the most recent common ancestor of a population is expected to be $4N$ generations (Equation 6.3). This implies that there are $4N\mu$ mutations that have occurred—and are still polymorphic—on the lineage leading to any sequence (**FIGURE 7.4**). Assuming levels of diversity are the same in the two species being compared, this means that approximately 2π differences between orthologous sequences are not actually fixed differences between the species. These

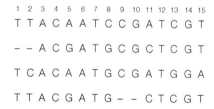

	1	2	3	4	5	6	7	8	9	10	11	12	13	14	15
	T	T	A	C	A	A	T	C	C	G	A	T	C	G	T
	–	–	A	C	G	A	T	G	C	G	C	T	C	G	T
	T	C	A	C	A	A	T	G	C	G	A	T	G	G	A
	T	T	A	C	G	A	T	G	–	–	C	T	C	G	T

FIGURE 7.3 Example alignment of four sequences, each from a different species.

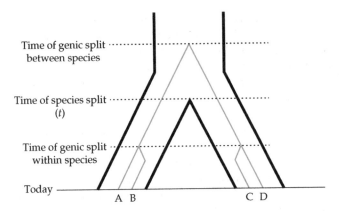

FIGURE 7.4 Comparisons of single sequences from two different species include both fixed and polymorphic differences. The history of a single ancestral species splitting into two descendant species is circumscribed by the darker lines. The history of a single orthologous sequence sampled in each of the two descendant species is denoted by the thinner, blue line; assume sequence A is sampled from species 1 and sequence C is sampled from species 2. For clarity, the genealogies of two maximally diverged samples from each species (A and B in species 1; C and D in species 2) are shown. The time to the most recent common ancestor is not necessarily the same for the samples within each species but is assumed to be the same for simplicity. All changes that arose from the time of genic split between species to the time of genic split within species are fixed differences. Mutations that have arisen on the single branch leading to A and C since the genic split within species are polymorphic. The number of such differences in a single sequence is expected to be π for both A and C, such that the total number of differences, d, between A and C includes 2π $(= \pi_1 + \pi_2)$ total polymorphic sites.

calculations imply that we can correctly count the amount of fixed divergence between two species as $d^* = d - 2\pi$. This approach is similar to the logic used to calculate the statistic d_a seen in Chapter 5, but in this case we are attempting to remove current polymorphism rather than ancestral polymorphism. If there are not equal levels of diversity we can subtract different values of π, one from each species, such that $d^* = d - (\pi_1) - (\pi_2)$. In most cases this correction is ignored because the level of divergence is much larger than levels of polymorphism (i.e., $d >> \pi$).

A second issue arises in cases in which the species being compared are too distant. It was recognized by early practitioners of molecular evolution that simply counting the proportion of sites differing between two species would undercount the true number of differences when multiple mutations have arisen and fixed at the same site. In the most extreme case, when every site has changed multiple times, we would still have 25% of the bases matching at random (assuming no gaps) and $d = 0.75$. The simplest way to account for the effect of multiple substitutions is to carry out a *Jukes-Cantor correction* (Jukes and Cantor 1969). Defining a as the observed proportion of sites that differ between two sequences (also referred to as p or the *p-distance*), the Jukes-Cantor corrected value of d is:

$$d = -\frac{3}{4} \ln \left(1 - \frac{4}{3} a \right) \tag{7.9}$$

Kimura (1980) made an adjustment to account for the fact that transition mutations (between purines or between pyrimidines) occur more often than transversion mutations (between purines and pyrimidines). The Kimura two-parameter correction is:

$$d = -\frac{1}{2} \ln \left(1 - 2a - b \right) - \frac{1}{4} \ln \left(1 - 2b \right) \tag{7.10}$$

where a is now the proportion of sites that differ by a transition and b is the proportion that differ by a transversion. There are a number of further elaborations that can be made, such as allowing different parameters for transitions involving purines versus pyrimidines (e.g., Hasegawa, Kishino, and Yano 1985; Tamura and Nei 1993).

   None of these methods ensures accurate measures of divergence when distances are very large, because in this case even the corrected estimates have large variances associated with them. Values of a greater than 0.75 are undefined in both Equations 7.9 and 7.10, and, in general, values of d larger than 0.50 will be less accurate than is ideal. Values of d much greater than 1 will essentially be guesses as to the true level of divergence.

DETECTING SELECTION USING DIVERGENCE
d_N, d_S, and d_N/d_S

In addition to measuring the overall divergence between two sequences, in coding regions we can also separately measure divergence that is due to nonsynonymous and synonymous changes. In particular, we define d_N as the number of nonsynonymous differences per nonsynonymous site (also referred to as d_R, K_A, or K_N) and d_S as the number of synonymous differences per synonymous site (also referred to as K_S). Methods for calculating the number of nonsynonymous and synonymous differences between two sequences are described in detail in **BOX 7.1** and **BOX 7.2**. Here we focus on how the quantities d_N and d_S are used to make inferences about natural selection.

■ **BOX 7.1**
Calculating d_N and d_S: Counting Methods

The simplest methods for calculating d_N and d_S first find the number of nonsynonymous and synonymous *differences* between sequences in an alignment (denoted N_d and S_d, respectively) and

then count the number of nonsynonymous and synonymous *sites* in each sequence (denoted N_c and S_c, respectively). In the end we will take the number of nonsynonymous differences per

(Continued)

■ BOX 7.1 *(continued)*

nonsynonymous site to be $d_N = N_d/N_c$, and the number of synonymous differences per synonymous site to be $d_S = S_d/S_c$. Because values for N_c and S_c will differ slightly for each sequence in the alignment, we usually use their average when calculating d_N and d_S.

There are a number of different ways of counting nonsynonymous and synonymous sites, and the use of different methods can often give different answers. One simple method assumes that the rate of transitions and transversions is equal, and that there is no codon usage bias (Nei and Gojobori 1986). For example, the codon TTA codes for the amino acid leucine. If we consider only mutations at the third position, we see:

TTG Leucine

TTA Leucine → TTC Phenylalanine

TTT Phenylalanine

Using the Nei and Gojobori method for counting sites, the third position would be 1/3 synonymous and 2/3 nonsynonymous, as only one of the three possible mutations is to a synonymous codon and two are to nonsynonymous codons; we refer to such codons as *two-fold degenerate*. The first two positions in the codon would both be 3/3 nonsynonymous and 0/3 synonymous, because all mutations at both of these positions are nonsynonymous. The total counts for this codon would therefore be $N_c = 8/3$ and $S_c = 1/3$.

Other methods take into account transition/transversion bias when calculating the effective number of sites (Li 1993; Pamilo and Bianchi 1993; Comeron 1995; Ina 1995). This bias can have important consequences for counting sites. For example, strong transition/transversion bias will cause two-fold degenerate codons to be more than 1/3 synonymous at the third position, as the genetic code allows all transitions in the third position of two-fold degenerate codons to be synonymous. You can see from the above example that if mutations from A→G occur much more often than either A→C or A→T mutations, then in a sense the third position is more than 1/3 synonymous. Weighting these sites

appropriately—using the known or estimated ratio of transitions to transversions—corrects for this bias. Still other methods take into account unequal codon usage within a sequence, effectively correcting for the fact that some synonymous codons are rarely used (Yang and Nielsen 2000).

There are also a number of different approaches to counting nonsynonymous and synonymous differences between sequences. All of these methods give the same result when codons differ at only one nucleotide. For instance, in the following alignment of a single codon between two species:

GTT
GTA

there is only one difference, and it is synonymous; hence $N_d = 0$ and $S_d = 1$. Complications only arise when there is more than one difference in a single codon between the two sequences; we refer to such cases as *complex codons*. The reason these cases are more complicated is that there are now two possible pathways that evolution might have taken, depending on which of the two mutations came first. In some instances a nucleotide difference might be either synonymous or nonsynonymous, as seen in the following example:

TTT
GTA

The two possible pathways are:

Pathway 1   TTT (Phe) ↔ GTT (Val) ↔ GTA (Val)

Pathway 2   TTT (Phe) ↔ TTA (Leu) ↔ GTA (Val)

In pathway 1 the change from T→A in the third position is a synonymous change, and there is a total of one nonsynonymous and one synonymous difference ($N_d = 1$, $S_d = 1$). In pathway 2 the T→A change is nonsynonymous and there is a total of two nonsynonymous substitutions ($N_d = 2$, $S_d = 0$). The Nei and Gojobori (1986) method simply averages the different evolutionary paths, giving each equivalent weight; doing this gives $N_d = 1.5$ and $S_d = 0.5$. Other methods give more weight to pathways that include synonymous substitutions (Li, Wu, and Luo 1985) or to pathways that include common nonsynonymous changes (Miyata and Yasunaga 1980) and then take the weighted average among pathways to find N_d and S_d. Unless sequences are highly diverged, all of these methods for counting nonsynonymous and synonymous differences should give similar results.

Finally, after summing N_c, S_c, N_d, and S_d across all codons in the alignment, it is recommended to correct both d_N and d_S for multiple possible nucleotide substitutions at a single site. Equations 7.9 and 7.10 describe two common methods for carrying out such corrections. More details on all of these methods can be found in Nei and Kumar (2000) and Yang (2006).

■ BOX 7.2
Calculating d_N and d_S: Likelihood Methods

As an alternative to the "counting" methods for calculating d_N and d_S outlined in Box 7.1, several likelihood methods have been proposed—and are commonly used—for this purpose (Goldman and Yang 1994; Muse and Gaut 1994). These methods are based on Markov chain models of substitutions between

codons, specifying an instantaneous rate matrix, Q, for independent transitions between the 61 non-termination codons. For one of the most popular likelihood models (Yang 1998), the rate matrix consists of separate transition rates, q_{ij}, between each pair of codons i and j ($i \neq j$), where:

$$q_{ij} = \begin{cases} 0, & \text{if } i \text{ and } j \text{ differ at more than one position} \\ \pi_j, & \text{for a synonymous transversion} \\ \kappa\pi_j, & \text{for a synonymous transition} \\ \omega\pi_j, & \text{for a nonsynonymous transversion} \\ \omega\kappa\pi_j, & \text{for a nonsynonymous transition} \end{cases}$$

Here the equilibrium frequency of codon j is denoted π_j, the transition/transversion bias is denoted κ, and the relative probability of observing nonsynonymous substitutions relative to synonymous substitutions is denoted by ω. In this context, we can think of ω as the ratio of nonsynonymous to synonymous substitutions, or d_N/d_S; this method is explicitly *not* separately estimating d_N and d_S. In fact, while methods implementing this model can allow ω to vary among codons, they assume that the synonymous substitution rate is constant across a gene—therefore, differences among codons are inferred to be due only to differences in selection on nonsynonymous

changes. However, many of the assumptions of the simple model can be relaxed: more than one change between codons can be allowed (e.g., Whelan and Goldman 2004), correlated rates of change among codons can be allowed (e.g., Siepel and Haussler 2004a), and synonymous substitution rate variation among codons can be allowed (e.g., Pond and Muse 2005).

When used with a multiple-sequence alignment and a prespecified phylogenetic topology, parameters of the model described above can be estimated via maximum likelihood methods. The most commonly used software packages for doing this are *PAML* (Yang 2007) and *HyPhy*

(Continued)

■ BOX 7.2 (continued)

(Pond, Frost, and Muse 2005); here I describe the methods available in *PAML*. To calculate the average strength of natural selection, ω, across all codons, the simplest model is denoted M0. This model places no restrictions on the values taken by ω, and separate ω parameters can be estimated for every branch of the tree if this is needed. However, unless positive selection acts on many sites across the whole gene, M0 is very unlikely to have $\omega > 1$. Instead, models that allow ω to vary among codons (referred to as *site models*) are most powerful for detecting positive selection.

The most straightforward way to test whether there are *any* codons in the alignment that have evidence for positive selection is to conduct a likelihood ratio test between two nested site models, one with and one without a parameter allowing $\omega > 1$. Model M1a has two ω parameters: one for codons with $\omega < 1$ (ω_0) and one for codons with $\omega = 1$ (ω_1; there are also parameters for the proportion of sites assigned to each ω parameter, p_0 and p_1). Model M2a has three ω parameters: one for codons with $\omega < 1$ (ω_0), one for codons with $\omega = 1$ (ω_1), and one for codons with $\omega > 1$ (ω_2), along with their corresponding parameters for the proportion of sites assigned to each ω (p_0, p_1, and p_2).

Because model M1a has two free parameters and model M2a has four free parameters—and model M1a is nested within M2a—a likelihood ratio test with two degrees of freedom can be used to test whether the model allowing sites to evolve under positive selection (M2a) provides a significantly better fit to the data. This test can be used with only two taxa (i.e., two sequences), although power improves when more taxa are added (Anisimova, Bielawski, and Yang 2001). Note, however, that this method assumes that the underlying phylogenetic tree comes from sequences of different species, and not within species; *PAML* requires that there be a single tree for all sites in the alignment, and sampling sequences from within a species that recombine would violate this assumption and could lead to erroneous inferences of positive selection (Anisimova, Nielsen, and Yang 2003). Further refinements to site models allow for tests of positive selection on a subset of codons on specific branches of the phylogenetic tree (referred to as *branch-site models*), which in turn allow for the identification of specific lineages undergoing adaptive evolution at a subset of sites (Zhang, Nielsen, and Yang 2005). More details on these methods and their uses can be found in Yang (2006; 2007).

It should be obvious from the discussion of substitution rates that natural selection has a profound effect on the number of nonsynonymous mutations that are fixed. If we use the expectations for the rate of substitution of neutral, advantageous, and deleterious mutations—presented in Equations 7.6, 7.7, and 7.8, respectively—and multiply all rates by $2t$ to convert them to distances, then we can write down the total expected nonsynonymous divergence as:

$$E\left(d_{N}\right) = \left(2N\nu f_0\right)\left(\frac{1}{2N}\right)2t + \left(2N\nu f_a\right)\left(2s_a\right)2t + \left(2N\nu f_d\right)\left(\frac{2s_d}{1 - e^{(-4Ns_d)}}\right)2t \tag{7.11}$$

which reduces to:

$$E\left(d_{N}\right) = \nu 2t\left(f_0 + f_a 4Ns_a + f_d\frac{4Ns_d}{1 - e^{(-4Ns_d)}}\right) \tag{7.12}$$

Because the total nonsynonymous divergence in a region is due to all three types of mutations, our expression for d_N includes all three terms. This equation says that a higher underlying mutation rate, ν, and longer divergence times, t, will increase the amount of divergence; that the proportion of advantageous mutations fixed will be a function of the frequency at which they arise and their average selective effect; and that deleterious mutations can also contribute to divergence if selection is weak enough.

Although it is not true that all synonymous mutations are selectively equivalent, we will make this assumption for now, coming back to the effect of selection on synonymous mutations later. Assuming all synonymous changes are neutral, that is, $f_0 = 1$ and $f_a = f_d = 0$ for these sites, then the total expected amount of synonymous divergence between two sequences is:

$$E\left(d_S\right) = \nu 2t \tag{7.13}$$

This expression mirrors Equation 7.2 because, for neutral mutations, the substitution rate is simply equal to the mutation rate.

In order to ask about the effects of natural selection on sequence evolution, we cannot just compare values of d_N among genes, even if they are all from the same pair of species. This is because the underlying mutation rate varies throughout the genome and along chromosomes, leading to different values of d_N at loci where selection may be acting in exactly the same way. Kimura (1977) informally suggested comparing the ratio of nonsynonymous to synonymous divergence at a gene, d_N/d_S, in order to control for differences in mutation rates among loci (see also Miyata and Yasunaga 1980). Because both ν and t will be approximately the same for nonsynonymous and synonymous sites in the same gene, dividing Equation 7.12 by 7.13 gives:

$$\frac{E\left(d_N\right)}{E\left(d_S\right)} = f_0 + f_a 4Ns_a + f_d \frac{4Ns_d}{1 - e^{\left(-4Ns_d\right)}} \tag{7.14}$$

This expression shows us that, relative to synonymous divergence, the level of nonsynonymous divergence is again due to the fractions of mutations that are neutral, advantageous, and deleterious. Note that to avoid sub-subscripting, I am not using different notations for the fraction of neutral mutations at synonymous versus nonsynonymous sites; the f_0 in Equation 7.14 represents only the nonsynonymous mutations.

Interpreting d_N/d_S

Although d_N/d_S is determined by the combined effects of neutral, advantageous, and deleterious mutations, it can tell us a lot about the general impact of natural selection on sequence evolution. An example of the range of d_N/d_S values observed in nature is given in **FIGURE 7.5**, which shows values from 11,492 orthologs between *Arabidopsis thaliana* and *A. lyrata* (Yang and Gaut 2011). As a general observation across many taxa, the mean value of d_N/d_S calculated between species is 0.15 to 0.25, though values for individual genes vary from 0 to >2 (Rat Genome Sequencing Project Consortium 2004; Mikkelsen et al. 2005; *Drosophila* 12 Genomes Consortium 2007; Yang and Gaut 2011). This range indicates that *at least* 75 to 85% of nonsynonymous mutations are deleterious and do not fix. The reason this represents a minimum estimate of the fraction of strongly deleterious mutations is that

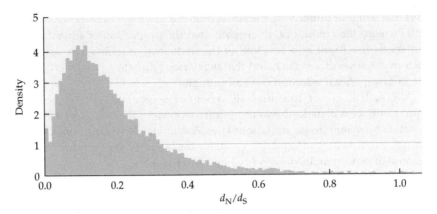

FIGURE 7.5 Distribution of values of d_N/d_S from 11,492 orthologs between *Arabidopsis thaliana* and *A. lyrata*. (After Yang and Gaut 2011.)

even a small number of advantageous mutations will contribute dispropor-tionately to divergence (see Equation 7.14). Even if the fraction of advanta-geous mutations, f_a, is as low as 1% and there were no neutral mutations, sufficiently strong selection and large population sizes could push d_N/d_S far above 0.01. For example, if 99% of nonsynonymous mutations were strongly deleterious and 1% were advantageous with an average $Ns = 10$, d_N/d_S would equal 0.4. A simple estimate of the fraction of deleterious mutations, $1 - d_N/d_S$, would underestimate this fraction by about 50% (0.6 vs. 0.99). Only if there are no advantageous mutations and deleterious mu-tations do not fix at appreciable rates does $d_N/d_S = f_0$ (and therefore $1 - d_N/d_S = f_d$). While estimates of selective constraint based on this assumption are commonly used, they are almost certainly underestimating the fraction of all mutations that are deleterious.

The value of d_N/d_S at a single gene does not tell us what fraction of the nonsynonymous mutations that do fix are neutral, advantageous, or slightly deleterious. Multiple combinations of values for the frequencies of each type of mutation, their average selective effect, and population size can give the same value of this statistic. One method that has been used to estimate the fraction of mutations that are slightly deleterious is to compare the average value of d_N/d_S between pairs of taxa that are known to differ in population size: the prediction is that d_N/d_S will increase in the species with smaller pop-ulation sizes, in proportion to the fraction of slightly deleterious mutations. Comparisons of multiple species suggest that 15 to 30% of all mutations are slightly deleterious (Eyre-Walker et al. 2002). For instance, the average value of d_N/d_S between mouse and rat is 0.14 and between human and chimpanzee is 0.20 (Mikkelsen et al. 2005). While there are no *a priori* reasons to believe that there are overall more and/or stronger advantageous mutations in primates, it is known that population sizes are much smaller in this clade, which would lead to higher values of d_N/d_S.

There is only one generally accepted method for demonstrating—using d_N/d_S—that there have been advantageous mutations (i.e., positive selection) in the history of a gene: the value of this statistic must be greater than 1 (Hill and Hastie 1987; Hughes and Nei 1988). To see why this is, consider a gene with no advantageous mutations. The maximum value of d_N/d_S in this case is 1, which will only occur if all possible nonsynonymous mutations are selectively equivalent. The only way to obtain $d_N/d_S > 1$ (with statistical significance) is for there to be some fraction of advantageous mutations (Equation 7.14). We can also see that this is a very stringent requirement, as more and more advantageous mutations are needed to get $d_N/d_S > 1$ when the fraction of deleterious mutations grows. Indeed, only those genes in which there are repeated adaptive fixations of amino acid substitutions will ever have the chance to raise the value of d_N/d_S over 1. These genes are usually proteins associated with immune and reproductive functions, as they are often involved in "arms race" dynamics between hosts and pathogens or between males and females.

Significance testing for $d_N/d_S > 1$ has been accomplished in a number of different ways. The likelihood method outlined in Box 7.2 allows for a likelihood ratio test between models with and without positive selection. The simpler counting methods outlined in Box 7.1 are not based on any model and therefore do not allow for such a straightforward testing procedure. These latter methods, however, have given rise to a number of different approaches. Comparing only pairs of sequences, Hughes and Nei (1988) calculated the standard error on d_N and d_S, testing the hypothesis that $d_N - d_S > 0$ using a t-test for significance. Messier and Stewart (1997) carried out a similar analysis on individual branches of the primate tree for the lysozyme gene, reconstructing ancestral states at each node and calculating the number of nonsynonymous and synonymous differences and nonsynonymous and synonymous sites between parent and daughter nodes; again, testing proceeded via a t-test. Both of these methods present problems. First, the t-test assumes that d_N and d_S are approximately normally distributed, and unless there are a large number of both types of substitutions this is unlikely to be the case. Second, the approaches using reconstructed ancestral states fail to take into account uncertainty in these reconstructions, leading to uncertainty in d_N and d_S. One solution to the first problem is to use a small-sample test to compare counts of nonsynonymous and synonymous substitutions. Zhang, Kumar, and Nei (1997) arrived at this solution by using a 2×2 test of independence on the counts of nonsynonymous and synonymous substitutions versus the number of nonsynonymous and synonymous sites (i.e., N_d, S_d, N_c, and S_c from Box 7.1); testing proceeded via Fisher's exact test. This method is preferred to the large-sample approximations, although the choice of method for counting nonsynonymous and synonymous sites can have a large effect on the outcome (see Box 7.1). Likelihood methods (Box 7.2) can be used to average across reconstructed ancestral states, effectively taking uncertainty into account.

One further complexity is that different parts of proteins are likely to be under different selective forces—some may be under strong positive selection and some under strong negative selection. While the average of all regions may have $d_N/d_S < 1$, subsets of functional domains may have $d_N/d_S > 1$ (as in the original papers of Hill and Hastie [1987] and Hughes and Nei [1988]). It can therefore be useful to examine this statistic in different domains if such domains are identified ahead of time; *post hoc* sliding window analyses may be highly misleading (Schmid and Yang 2008). Likelihood methods can also be used to find individual codons with $d_N/d_S > 1$ in a statistically sound way as long as the assumptions of the test are met (Box 7.2).

It is very possible that many or most proteins with $d_N/d_S < 1$ have some history of positive selection, but this statistic does not allow us to determine whether this is true. Any amount of amino acid change could be due to advantageous mutation, and therefore any gene with $0 < d_N/d_S \leq 1$ could have contributions from neutral, advantageous, and slightly deleterious mutations. Because of the unknown contributions of each type of mutation, our inferences about selection must be commensurately cautious. Here are some general guidelines for interpreting d_N/d_S:

- $d_N/d_S \ll 1$ The vast majority of nonsynonymous mutations are deleterious, and negative (purifying) selection is predominant.
- $d_N/d_S < 1$ The majority of nonsynonymous mutations are deleterious, but there may be some unknown fraction of advantageous mutations.
- $d_N/d_S = 1$ This situation can occur in two cases: First, there is no selection and all nonsynonymous mutations are neutral. Second, there is simply a large number of neutral and advantageous mutations (as well as deleterious mutations).
- $d_N/d_S > 1$ There are many advantageous nonsynonymous mutations and positive selection is predominant, but there are still many deleterious mutations.

Although this distinction has already been made in Chapter 1, it should be repeated here that observing $d_N/d_S = 1$ for a gene *does not* indicate that it is evolving "neutrally." As detailed above, there are multiple combinations of selective forces that will result in $d_N/d_S = 1$, only one of which is the absence of selection. Even more importantly, the absence of selective constraint is not the same as neutral evolution. Many genes with $d_N/d_S < 1$ may be evolving neutrally—that is, they may have not fixed any adaptive substitutions.

Several caveats with regard to interpreting d_N/d_S must also be mentioned. First, small values of d_S can bias the estimate of d_N/d_S upwards. While such biases should never lead to values of d_N/d_S that are significantly greater than 1, comparisons between species pairs that have very different divergence times can be misleading because of differences in d_S (Wolf et al. 2009). Second, we have assumed that all synonymous mutations are neutral. This is clearly not the case in every species: while there are many reasons for the nonrandom usage of synonymous codons, natural selection is an important factor (Ikemura 1981; Akashi 1994; Chamary, Parmley, and Hurst 2006; Lawrie et al. 2013). If the fraction of all synonymous mutations that are neutral is not equal to 1, then the value of d_S should be determined by an expression such as the one presented in Equation 7.12 rather than 7.13. Exactly how this will affect d_N/d_S is not clear: if negative selection predominates, then d_S will be smaller than is expected under the case of no constraint, and consequently d_N/d_S will be higher. Whether this will lead to cases with $d_N/d_S > 1$ in the absence of advantageous mutations—which would effectively require no constraint on nonsynonymous mutations but strong constraint on synonymous mutations—is not known (but is unlikely). On the other hand, if there are many advantageous synonymous substitutions (e.g., Resch et al. 2007), then the value of d_N/d_S will be lower than expected, and therefore the strength of negative selection will be inferred to be greater than it truly is.

In general, d_N/d_S is a relatively robust statistic with few assumptions about how samples were collected, the demographic history of populations, or any of the other problems introduced by truly population genetic methods. On the other hand, only repeated bouts of positive selection will provide evidence of advantageous mutation, and therefore only more powerful methods will be highly informative about adaptive evolution.

DETECTING SELECTION USING POLYMORPHISM

The effect of selection on the frequency of polymorphism

In the previous section we considered how selection opposes or favors the eventual fixation of selected alleles and how we can make inferences about evolutionary processes from data on divergence between species. We will now discuss the effects of selection on the frequency of polymorphisms within populations, initially without considering fixed differences. A number of different methods can take advantage of just this sort of data, although they are generally less powerful than those methods that combine both polymorphism and divergence. Later in the chapter we will see one such method, the McDonald-Kreitman test.

On timescales shorter than those required for mutations to fix, selection will change the mean frequency of alleles in a population. As would be expected from the different probabilities of fixation discussed above, advantageous alleles will be at higher mean frequency relative to neutral alleles, and deleterious alleles will be at relatively lower frequencies. For new mutations, the density of polymorphisms found at frequency q is given by (Wright 1969, p. 381):

$$f(q) = \frac{2\nu}{q(1-q)} \frac{1 - e^{(-4Ns)(1-q)}}{1 - e^{(-4Ns)}} \tag{7.15}$$

where ν is again the total rate of mutation. Advantageous mutations have $s > 0$, while deleterious mutations have $s < 0$.

Examples of frequency spectra for advantageous, neutral, and deleterious alleles are shown in **FIGURE 7.6**. As can be seen from the distribution, advantageous alleles are shifted toward higher frequencies, while deleterious alleles are shifted toward lower frequencies. Balancing selection will maintain alleles at intermediate frequencies, with the exact form of balancing selection greatly affecting the particular probability density of allele frequencies. Distributions

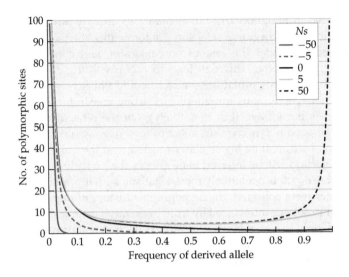

FIGURE 7.6 Allele frequency spectra for advantageous (*Ns* > 0), neutral (*Ns* = 0), and deleterious (*Ns* < 0) mutations. (After Fay, Wyckoff, and Wu 2001.)

for variants with relatively strong selection—for example, |*Ns*| > 1—will be highly skewed for both deleterious and advantageous alleles. Most strongly deleterious mutations are immediately removed from the population, and most strongly advantageous mutations fix very rapidly (or suffer rapid stochastic loss); both spend little time in a polymorphic state. These tendencies imply that, at a locus with only very deleterious mutations and neutral mutations, the allele frequency spectrum observed in a moderately sized sample will much more closely resemble a neutral one.

A simple test of the neutrality of nonsynonymous polymorphisms was proposed by Sawyer, Dykhuizen, and Hartl (1987) and is based on the differences in allele frequency spectra between slightly deleterious and neutral variants. Sawyer and colleagues counted the number of nonsynonymous polymorphisms that were singletons—those that are present only on one chromosome—and the number that were present on multiple chromosomes ("multitons"). All 12 of the nonsynonymous polymorphisms in their sample (of the gene *gnd* in *Escherichia coli*) were singletons. They then compared these two numbers (12 singletons vs. 0 multitons) to the configuration of synonymous polymorphisms: 34 synonymous polymorphisms were singletons and 32 were multitons (**FIGURE 7.7**). Comparison of these values in a 2 × 2 contingency table was highly significant

	Multiton	Singleton
Nonsynonymous	0	12
Synonymous	32	34

FIGURE 7.7 A test for the presence of weakly deleterious nonsynonymous polymorphisms (*P* = 0.001; Fisher's exact test). Multitons are polymorphisms present more than once in the sample; singletons are polymorphisms present only once. Data come from the *gnd* gene in *Escherichia coli*. (After Sawyer, Dykhuizen, and Hartl 1987).

and indicated an excess of singleton nonsynonymous variants. This finding is exactly what one would expect if nonsynonymous polymorphisms were on average slightly deleterious.

In addition to simply showing that allele frequencies differ as a result of selection acting on polymorphisms, one can use Equation 7.15 with only a few assumptions (such as free recombination among sites) to find the average selection coefficient associated with a set of polymorphisms (e.g., Sawyer and Hartl 1992). That is, we can take the observed allele frequency spectrum for a large set of polymorphisms to estimate the selection parameters of Equation 7.15 (or analogous equations). It can also be useful to treat the selection parameter as a distribution itself, in order to estimate not just the mean effect of selection but also the relative contributions of advantageous and deleterious mutations. The expectations for the allele frequency spectrum with direct selection assume an equilibrium population, such that any nonequilibrium population demography or even linked selection will confound inferences because both can affect the frequency spectrum (see Chapters 8 and 10). Because of these confounding factors, commonly used methods for inferring selection parameters at nonsynonymous sites also use synonymous sites or intronic sites to simultaneously control for nonequilibrium processes (e.g., Loewe et al. 2006; Keightley and Eyre-Walker 2007; Boyko et al. 2008). **FIGURE 7.8** shows the unfolded allele frequency spectra for nonsynonymous, synonymous, and noncoding polymorphisms sampled from 40 human chromosomes across 301

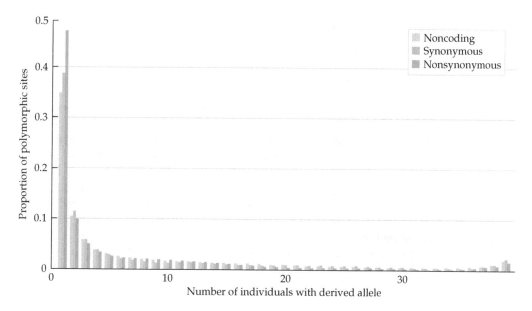

FIGURE 7.8 Allele frequency spectra for noncoding, synonymous, and nonsynonymous polymorphisms in 301 human genes from 40 chromosomes. There is a large excess of low-frequency nonsynonymous polymorphisms. The slight excess of high-frequency polymorphisms of all types is ascribed to ancestral state mis-specification. (After Williamson et al. 2005; copyright 2005, National Academy of Sciences USA.)

genes (Williamson et al. 2005). In this case all three classes of mutations differ from the neutral-equilibrium expectations, but the excess of deleterious nonsynonymous polymorphisms can be inferred by a comparison among classes.

Accurate inference of selection parameters using these methods requires data from a large number of sites and from a more modest number of chromosomes, approximately $n = 10$ (Keightley and Eyre-Walker 2010). Some methods make use of fixed differences (i.e., the frequency class with $p = 1.0$) and some do not, though not all methods using fixed differences take advantage of the full frequency spectrum of polymorphism (e.g., Bustamante et al. 2002). The use of fixed differences increases the power to detect highly advantageous alleles that would otherwise not be detected as polymorphisms. However, some of the methods using fixed differences are prone to giving highly skewed values of Ns, possibly because of assumptions that must be made about the shape of the distribution of selection intensities. Nonetheless, we discuss several such methods at the end of this chapter, after introducing the McDonald-Kreitman test.

Direct selection has a profound influence on the average frequency of polymorphisms. Although estimating average selection coefficients (or, more accurately, the product of the selection coefficient and population size) usually requires data from multiple loci, one can test whether these values are different than 0 or whether selection coefficients differ between classes of polymorphisms (e.g., nonsynonymous vs. synonymous). While this information does not tell us about the history of a particular gene, it does provide a global view of the effects of selection. In the following section I will introduce additional statistical tests based on the expectations described here that will allow us to learn the history of specific loci.

π_N/π_S

By analogy with the logic of the comparison of d_N and d_S laid out earlier in the chapter, within a species we can compare the average number of nonsynonymous differences per nonsynonymous site (referred to as π_N or π_A) to the average number of synonymous differences per synonymous site (π_S). The calculations of π_N and π_S are straightforward, combining the methods for calculating π (Chapter 3) with the methods for calculating nonsynonymous and synonymous changes and sites described in Box 7.1. Only sequences from within a single population or species are needed. In most cases the calculation will often be easier than between-species comparisons, as the limited number of polymorphisms will also limit the number of complex codons that must be accounted for.

The interpretation of the ratio π_N/π_S is also much the same as for d_N/d_S, with an important exception. Values of π_N/π_S below 1 are again evidence for the predominance of purifying selection, and the vast majority of all coding loci show $\pi_N/\pi_S < 1$. Some slightly deleterious polymorphisms may be sampled, but because they are kept at low frequency they contribute relatively little to the average heterozygosity. The main difference in interpreting π_N/π_S comes when values are greater than 1: since positive selection will rapidly fix advantageous mutations, these adaptive changes will rarely be found in studies of polymorphism. Even if one or two such mutations are sampled on their way to fixation, the increase in π_N is unlikely to result in values of π_N/π_S much above 1. Instead, we find $\pi_N/\pi_S > 1$ in genes under multiallelic balancing selection—that is, in cases in which a large number of amino acid polymorphisms are maintained by some form of balancing selection, such as heterozygote advantage or negative frequency-dependent selection. There are few genes that appear to be under such selection, with the major histocompatibility complex (MHC) genes and other immune system–related or sex and reproduction–related genes as the most prominent exceptions. Hughes and Nei (1988) were the first to calculate π_N/π_S for human and mouse MHC loci and found values greater than 1; confusingly, they referred to their statistics as d_N and d_S.

As with the requirement that d_N/d_S be greater than 1 for strong evidence of positive selection, $\pi_N/\pi_S > 1$ is a very strict criterion for detecting balancing selection. Single sites under very strong selection will never contribute enough to values of π_N to push π_N/π_S greater than 1, so only rare cases of multiallelic selection will be detectable using this method. Even in these cases, evidence for balancing selection may be limited to known functional regions (as in Hughes and Nei 1988). Still, comparisons of π_N and π_S are a standard method in the population geneticist's toolbox, and we will see next how slight modifications of these statistics can be combined with between-species divergence to create a truly powerful test for natural selection.

DETECTING SELECTION USING POLYMORPHISM AND DIVERGENCE

The McDonald-Kreitman test

We have seen how a comparison of d_N and d_S allows us to detect the action of positive selection in a theoretically and statistically rigorous way. But because approximately two-thirds of all possible coding mutations lead to changes in amino acids, d_N/d_S will only be greater than 1 when more than two-thirds of all differences between species are nonsynonymous. The selective conditions under which this can occur are likely limited to those in which there are a very large number of recurrent adaptive fixations of nonsynonymous differences. In this section I introduce a test first suggested by McDonald and Kreitman (1991) that is similar to d_N/d_S but gets around the very strict conditions for detecting positive selection in a principled way. This test combines both polymorphism and divergence in order to test predictions of the neutral model. The new test will still not be able to detect adaptive evolution on only one or a few fixed differences, no matter the strength of selection—to do so, we will have to wait for methods introduced in the next chapter. But the McDonald-Kreitman (MK) test offers one of the most powerful, robust methods we have for detecting the action of natural selection. I first explain how the test is conducted, followed by the reasoning underlying it and the interpretation of results.

Experimental design for the McDonald-Kreitman test

The four quantities needed to carry out the MK test for a single gene are the number of nonsynonymous polymorphisms (P_N), the number of synonymous polymorphisms (P_S), the number of nonsynonymous fixed differences (D_N),

	Fixed	Polymorphic
Nonsynonymous	D_N	P_N
Synonymous	D_S	P_S

FIGURE 7.9 The McDonald-Kreitman test. The four counts needed are the number of nonsynonymous fixed differences (D_N), the number of synonymous fixed differences (D_S), the number of nonsynonymous polymorphisms (P_N), and the number of synonymous polymorphisms (P_S).

and the number of synonymous fixed differences (D_S). These values are placed in a 2 × 2 contingency table (**FIGURE 7.9**), and standard tests of independence—usually Fisher's exact test when counts are low, but either χ^2- or G-tests for large-sample approximations—can be used to calculate significance.

The general experimental design used to obtain these four quantities involves sampling multiple chromosomes from a single species, as well as either a single chromosome or multiple chromosomes from a second species. When multiple chromosomes are sampled from more than one species, either the counts of polymorphisms can be combined (as in the original McDonald and Kreitman study of the *Adh* gene in *Drosophila melanogaster*, *D. simulans*, and *D. yakuba*), or separate tests can be run separately for each species, especially if ratios of P_N/P_S differ significantly among the species. In general, the MK test is fairly robust to the sampling of chromosomes from true "populations": sampling one chromosome from each of multiple subpopulations within a species will not lead to incorrect inferences of positive selection, though it can result in other incorrect conclusions (see below).

All values used in the MK test are *counts*—there is no need to calculate the number of nonsynonymous or synonymous sites. While there still may be some issues involving complex codons when both nonsynonymous and synonymous changes have occurred in the same codon, these can be resolved using the same methods outlined in Box 7.1 (although only integer-valued numbers are used in the MK test). There are few issues of ascertainment bias with respect to sample sizes when using the standard MK test because allele frequencies are not used, though obviously all mutations must be ascertained without respect to whether they are nonsynonymous or synonymous (genotyping platforms are often enriched for nonsynonymous variants). In fact, one really only needs to sample two chromosomes from within a single species and one chromosome from an outgroup to be able to conduct the test, as long as there are enough polymorphisms and fixed differences within the sample. How many are enough? It is impossible to get a *P*-value of less than 0.05 in a 2 × 2 test unless the sum of the counts in the rows and the sum of the counts in the columns are both 4 or greater (i.e., there are at least four polymorphisms and four fixed differences, regardless of the type of mutation). So sampling should proceed—by collecting more chromosomes either from within a population or from a more distant outgroup, depending on what is needed—to increase statistical power. Otherwise it will not be possible to get a statistically significant result.

One important difference between calculations of divergence in the MK test and d_N/d_S is that in the McDonald-Kreitman test we are explicitly counting *fixed* differences for the values of D_N and D_S in the sample. A fixed difference in this case means that the identity of the allele found in the population sample is different than the nucleotide found in the outgroup. This is a slightly different meaning of the term *divergence* than is used in calculations of d_N and d_S. If there are two different alleles at a site in the population sample and they both differ from the outgroup, then we count them as one polymorphism and one fixed difference. While differences found to be fixed in a small number of chromosomes may not be fixed species-wide, as long as the same criterion is used for calling nonsynonymous and synonymous fixed differences there is no problem with this definition. In fact, fixed differences in the MK test do not necessarily have to come from comparisons of only two species. In cases in which chromosomes from three species are sampled, regardless of whether polymorphism data are collected from all of them, we can polarize changes along single lineages to use as fixed differences. It is best to compare polarized changes along a lineage to the polymorphism data collected from the tip of the same lineage, as we expect the neutral mutation rates to be the most similar in such comparisons. In addition, the MK test can be carried out by comparing fixed differences between gene duplicates, even if they are in the same species. This type of comparison comes with additional caveats, however, and should be considered carefully (see below).

Expectations for levels of polymorphism and divergence

The McDonald-Kreitman test explicitly relies on expectations based on the neutral theory of molecular evolution. The 2×2 table outlined above tests whether the ratio of nonsynonymous to synonymous changes is the same for polymorphism and divergence. The neutral expectations for the four values used in the test are:

$$
\begin{aligned}
E(P_N) &= 4N_e\mu_{nonsyn} * a * L_{nonsyn} \\
E(P_S) &= 4N_e\mu_{syn} * a * L_{syn} \\
E(D_N) &= 2t\mu_{nonsyn} * L_{nonsyn} \\
E(D_S) &= 2t\mu_{syn} * L_{syn}
\end{aligned}
\tag{7.16}
$$

where μ_{nonsyn} is the neutral mutation rate across nonsynonymous sites, μ_{syn} is the neutral mutation rate across synonymous sites, a is Watterson's correction for sample size (Equation 3.6), and L_{nonsyn} and L_{syn} represent the number of nonsynonymous and synonymous sites, respectively. Recall that we require two different neutral mutation rates because the fraction of all mutations that are neutral will differ for nonsynonymous and synonymous sites at any particular gene. Earlier we assumed that μ_{nonsyn} is equal to some fraction, f_0, of the total mutation rate, ν, and that μ_{syn} is simply equal to the total mutation rate because all synonymous mutations are neutral. Regardless of what these fractions are, and whether all synonymous mutations are truly selectively equivalent, if we assume the same neutral mutation rate over time within both types of sites then the ratio of both P_N/P_S and D_N/D_S is equal to μ_{nonsyn}/μ_{syn}. Alternatively, we can view the MK test as a comparison of the ratios of polymorphism to divergence for nonsynonymous and synonymous mutations, with no change in interpretation. In this case both P_N/D_N and P_S/D_S are equal to $4N_e/2t$, at least within a single locus.

In either comparison, the expectation built into the 2×2 table is that the ratios should be equivalent. This expectation is derived under the assumption that all of the polymorphisms and fixed differences observed are neutral: there have been no fixations of advantageous mutations, and there are no balanced polymorphisms or deleterious polymorphisms found in the sample. If this assumption is met (and often when it is not), the McDonald-Kreitman test will not be significant. The method is therefore explicitly testing whether there have been positively selected fixations, or whether there are balanced and/or slightly deleterious polymorphisms. The MK test will *not* be significant when there is strong purifying selection—the expectation of the ratios under strong purifying selection is exactly the same as under no selection. Unlike the comparison of d_N and d_S, detecting purifying selection is not an object of the method. If we included the number of nonsynonymous and synonymous sites as two additional cells and made the table a 3×2 table we would be able to make inferences about whether there is any signature of purifying selection, but this approach is not generally used. In addition, the MK test will not be significant when there is an excess of both polymorphic and fixed nonsynonymous differences (i.e., both $\pi_N/\pi_S > 1$ and $d_N/d_S > 1$). In these cases the ratios in the table will be the same, and, again, because we are not using the number of sites in our test, we will not be able to detect such departures from neutrality.

Interpreting the McDonald-Kreitman test

FIGURE 7.10 shows the values of P_N, D_N, P_S, and D_S collected in the original study of the *Adh* gene by McDonald and Kreitman (1991). The ratio of nonsynonymous to synonymous polymorphism is 2:42, while that for divergence is 7:17. The McDonald-Kreitman test on these data gives $P = 0.007$ and is generally interpreted to show evidence for adaptive natural selection. This interpretation comes from the apparent excess of nonsynonymous fixed differences in the comparison. Of course this is a 2×2 test of independence, and therefore it is possible that the significant result is due to an excess or deficit of counts in any of the cells. However, the interpretation of positive selection is a biologically motivated one, as large changes in the number of synonymous variants—without concomitant changes in the number of nonsynonymous variants—are not expected to occur. In the next section we discuss several situations in which this interpretation may not be correct because assumptions of the method have been violated.

	Fixed	Polymorphic
Nonsynonymous	7	2
Synonymous	17	42

FIGURE 7.10 A McDonald-Kreitman table with evidence for positive selection ($P = 0.007$; Fisher's exact test). Data are from the *Adh* gene with polymorphism and fixed differences from *Drosophila melanogaster, D. simulans,* and *D. yakuba.* (After McDonald and Kreitman 1991.)

A significant excess of nonsynonymous fixed differences at a gene is expected to occur when there is recurrent fixation of amino acid substitutions, as with $d_N/d_S > 1$. Note, however, that the number of fixations needed for the MK test to be significant is much lower than in the d_N/d_S comparison (indeed, d_N/d_S for *Adh* between the same three *Drosophila* species is 0.114). The MK test is still unable to detect positive selection on only one or a few substitutions, no matter how strong selection may be, but the power is greatly increased over d_N/d_S. In addition, as is the case for both d_N/d_S and the MK test, we do not know which particular amino acids have been selected for.

The MK test gets its real power by using the polymorphism data (P_N and P_S) as a proxy for the ratio of nonsynonymous to synonymous neutral mutations without the contribution of positively selected variants. Although it is obvious how positive selection will contribute to the number of fixed differences, advantageous mutations spend so little time as polymorphisms as they speed through a population that they are not expected to contribute much to P_N. Smith and Eyre-Walker (2002) provide a striking example of the dichotomy in the contribution of adaptive mutations to polymorphism and divergence: if mutations with an average advantage of $N_e s = 25$ occur at 1% the rate of neutral mutations, they will contribute 50% of all fixed differences but only 2% of all polymorphisms. Therefore, assuming that neutral mutation rates at nonsynonymous and synonymous sites stay constant over time, any excess nonsynonymous fixed differences that are due to positive selection will show up as an increased ratio of D_N/D_S in the 2 × 2 table, and the MK test will be significant (assuming enough variant sites were sampled).

In addition to being significant because of an excess of nonsynonymous fixed differences, the MK test can also be significant when there is an excess of nonsynonymous polymorphisms. The assumption that all of the nonsynonymous polymorphisms in the sample are neutral is often violated. In fact, even the original *Adh* data of McDonald and Kreitman include a well-known balanced amino acid polymorphism (Hudson, Kreitman, and Aguadé 1987). Although balancing selection on a single site is expected to raise polymorphism levels at linked sites (see next chapter), this increase should equally affect linked neutral nonsynonymous and synonymous polymorphisms; only the single balanced polymorphism will contribute to an excess change in the ratio of P_N/P_S. In general, balancing selection appears to be a limited explanation for significant excesses of nonsynonymous polymorphism. Only when there is strong multiallelic balancing selection will there be an excess of multiple amino acid polymorphisms, and in these cases there may also be an excess of nonsynonymous fixed differences, leading to a nonsignificant MK test.

The more common explanation for a significant excess of amino acid polymorphism is that these variants are slightly deleterious. **FIGURE 7.11** shows an example of a McDonald-Kreitman table in which there is a significant excess of nonsynonymous polymorphism (Nachman, Boyer, and Aquadro 1994). If these variants are slightly deleterious they will contribute to levels of nonsynonymous polymorphism but will not be fixed. Note that this is only true of *weakly* deleterious variants—strongly deleterious mutations will not rise

	Fixed	Polymorphic
Nonsynonymous	2	11
Synonymous	23	13

FIGURE 7.11 A McDonald-Kreitman table with evidence for weakly deleterious polymorphisms ($P = 0.004$; Fisher's exact test). Data are from the *ND3* gene with polymorphism data from *Mus domesticus* and fixed differences relative to *M. spretus*. (After Nachman, Boyer, and Aquadro 1994.)

to appreciable frequency and are therefore unlikely to be sampled; in these cases the MK test will not be significant (Akashi 1999).

However, any inference that there are slightly deleterious polymorphisms in a sample can be strongly influenced by the way in which chromosomes are sampled. Even though I stated earlier that nonrandom sampling of chromosomes within a species is unlikely to result in an apparent excess of nonsynonymous fixed differences in the MK framework, the same is not true of an excess of nonsynonymous polymorphisms. The conclusion that unusually high numbers of amino acid polymorphisms are due to weakly deleterious polymorphisms rests on the assumption that the sample comes from a single population. If, for instance, single individuals are each sampled from a different subpopulation, it is possible that local adaptation in each subpopulation will have led to differences at a number of amino acids. These differences could be misinterpreted as an excess of weakly deleterious nonsynonymous polymorphism rather than an excess of locally adaptive polymorphisms. While this sort of sampling scheme may seem unusual, it is a regular occurrence in species with low levels of variation within subpopulations, such as in the selfing plant *Arabidopsis thaliana*. The fact that these slightly deleterious polymorphisms are expected to be at lower frequency in the population (see Equation 7.15) does not help to distinguish between deleterious and locally adaptive variants: because only a single accession is sampled from each locale, under either scenario the nonsynonymous variants may be low in frequency in the sample but high in frequency within a subpopulation.

Methodological extensions of the MK test

The presence of non-neutral polymorphisms—especially slightly deleterious ones—can act to mask patterns of positive selection. To see why this is the case, consider a hypothetical locus that has both an excess of fixed adaptive nonsynonymous substitutions and an excess of slightly deleterious nonsynonymous polymorphisms (**FIGURE 7.12A**). If reduced to a comparison of just the numbers of polymorphic and fixed sites, the MK test would not be significant (**FIGURE 7.12B**; $P = 0.093$). This is because the MK test assumes that all polymorphisms are neutral: the excess deleterious nonsynonymous polymorphisms increase the ratio of nonsynonymous to synonymous segregating sites, making it appear as though the high number of nonsynonymous fixed differences is expected given neutral evolution.

How do we deal with the confounding presence of slightly deleterious alleles? Templeton (1996) suggested an extension of the MK test that separates singleton polymorphisms (of both types) from all other polymorphisms. The resulting 3 × 2 table (**FIGURE 7.12C**) attempts to disentangle the effects of slightly

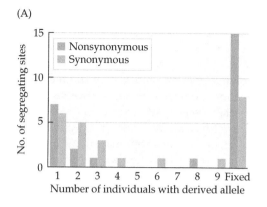

(B)

	Fixed	Polymorphic
Nonsynonymous	15	11
Synonymous	8	17

(C)

	Fixed	Multiton	Singleton
Nonsynonymous	15	4	7
Synonymous	8	11	6

(D)

	Fixed	Multiton
Nonsynonymous	15	4
Synonymous	8	11

FIGURE 7.12 A hypothetical example of a gene with an excess of both low-frequency and fixed nonsynonymous differences. (A) The hypothetical allele frequency spectrum of nonsynonymous and synonymous changes for a gene with $n = 10$ sampled chromosomes. Fixed alleles are included as a frequency bin. (B) The standard McDonald-Kreitman table for this gene ($P = 0.09$; Fisher's exact test). (C) The 3×2 table proposed by Templeton (1996), separating singleton polymorphisms from all others ($P = 0.07$; Fisher's exact test). (D) The 2×2 table proposed by Fay, Wyckoff, and Wu (2001), including only polymorphisms found on more than one chromosome ($P = 0.04$; Fisher's exact test).

deleterious polymorphisms, which are expected to be overrepresented at low frequencies, from those of neutral polymorphisms present at higher frequencies. While there are neutral nonsynonymous polymorphisms at low frequency, under neutrality there should also be a proportional number of synonymous polymorphisms at the same frequency. This table is essentially a combination of the test proposed by Sawyer, Dykhuizen, and Hartl (1987; Figure 7.7) and the McDonald-Kreitman test (Figure 7.9). Akashi (1999) proposed a further extension to this approach, comparing the full frequency spectrum of both nonsynonymous and synonymous alleles (not just singletons and multitons) and including fixed differences as one of the frequency classes (as in Figure 7.12A). He showed that this approach is more powerful than both the standard MK test and Templeton's extension to the test, although a large number of polymorphisms may be needed to adequately populate each frequency class.

Testing for significance in a 3×2 table or using the full frequency spectrum offers more power to detect an excess of adaptive fixed differences, but the results can also be difficult to interpret when the ratio of nonsynonymous to synonymous differences varies among frequency classes. Fay, Wyckoff, and Wu (2001) conducted a much simpler comparison by simply removing all polymorphisms at frequencies <15% in their sample. They showed that by removing low-frequency polymorphisms—and simply conducting a 2 × 2 test using common polymorphisms and fixed differences (**FIGURE 7.12D**; $P = 0.045$)—they were able to more easily detect patterns of adaptive natural selection. Although the 15% cutoff used by Fay, Wyckoff, and Wu is seemingly arbitrary, it turns out that this value is a near-optimal solution to the problem of segregating deleterious amino acids, though many instances of positive selection will still be missed (Charlesworth and Eyre-Walker 2008). In most moderately sized samples simply removing singletons from the counts of polymorphism in the MK test will have a similar effect.

Assumptions of the MK test

Before considering further uses of the McDonald-Kreitman test, it is worth reviewing the major assumptions being made in making inferences from this test. Overall, we can say with some confidence that inferences of positive selection using the MK test are robust to most assumptions, including how chromosomes are sampled, the recent population history of the sample, the presence of non-neutral polymorphisms, recombination among sites, and even the selective equivalency of all synonymous mutations. The most important assumption may simply be that there is no bias in detecting nonsynonymous versus synonymous differences. However, there are several instances in which the assumptions of the test can be violated so as to cause a rejection of the null hypothesis, especially with regard to several novel uses of the MK test discussed in the next section.

Possibly the most unrealistic assumption of the MK test is that the neutral mutation rate is constant over time for both nonsynonymous and synonymous changes. For nonsynonymous changes this means that the fraction of all mutations that are neutral does not change over time (i.e., that selective constraint is constant), even though this value may be quite sensitive to changes in effective population size, N_e. To be explicit, this assumption comes from the fact that expectations for both P_N/P_S and D_N/D_S in the 2 × 2 test of independence are equal to μ_{nonsyn}/μ_{syn}; if these rates change over time, then this assumption no longer holds. In most cases we expect that even when constraint does change at a single gene over time, this change does not have a directionality, and there is therefore little effect on the MK test (Fay and Wu 2001; **FIGURE 7.13**).

At the whole population level, however, consistent changes in the effective population size over time may result in wholesale differences in levels of constraint. For instance, if population sizes for a species of interest were on average smaller in the distant past (when most fixed differences accumulate) than they are in the recent history of a sample (when most polymorphisms accumulate), then we would expect the neutral mutation rate to be higher for fixed differences relative to polymorphisms because some fraction of currently slightly deleterious mutations would have been effectively neutral. This

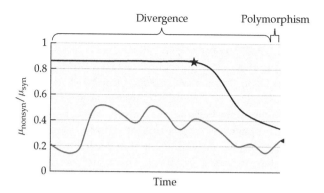

FIGURE 7.13 Hypothetical change in the ratio of nonsynonymous to synonymous mutations that become fixed or are polymorphic. For a single-copy gene (red line) the strength of selection may vary over time, with the arrow denoting the $\mu_{\text{nonsyn}}/\mu_{\text{syn}}$ ratio found within the polymorphism data. For a newly duplicated gene (black line) the strength of selection is initially relaxed and then changes when a new function is acquired (denoted by the star). In these cases the comparison of polymorphism to divergence can be positively misleading. (Modified from Fay and Wu 2001.)

situation could cause a rejection of the null hypothesis that is not due to positive selection. Conversely, population sizes may have crashed in the recent past, increasing the proportion of mutations that are neutral and making it seem as though there were a large number of segregating deleterious polymorphisms. Lower recent population sizes may also result in smaller numbers of segregating polymorphisms, lowering the overall statistical power of the test (e.g., Parsch, Zhang, and Baines 2009). In general, the magnitude of population size change must be quite large to have an effect on most inferences of positive selection (Eyre-Walker 2002). In order to rule out the possibility that a recent increase in population size was responsible for the patterns of adaptive natural selection they observed in *D. melanogaster*, Fay, Wyckoff, and Wu (2002) tested several ancillary predictions of the increased population size model. They compared P_N/P_S ratios in African flies to non-African flies to see if the ratio was lower outside of Africa (the native range of this species) and found no difference. They also asked whether there were more rare nonsynonymous variants in non-African samples—as might be expected if the average selective effect against deleterious polymorphisms had increased—and found no difference. Finally, they reasoned that an overall increase in population size would affect all genes in the genome, causing an across-the-board increase in the ratio of D_N/D_S; instead, they found heterogeneous effects among genes, with a substantial divide between a set of neutrally evolving genes and a set of adaptively evolving genes. Together, these results indicate that population size changes were not responsible for generating misleading patterns of positive selection.

There is one scenario in which the MK test—and the assumption of constant neutral mutation rates—will be consistently violated such that positive selection will be inferred when none has necessarily occurred. This scenario is that of recent gene duplicates (as discussed in Jones and Begun 2005; Thornton and Long 2005; Arguello et al. 2006). A common pattern in the evolution of gene duplicates is that selection is relaxed on the newly duplicated gene (reviewed in Conant and Wolfe 2008; Hahn 2009). If there is little constraint early in the history of a duplicate, then many nonsynonymous fixed differences will accumulate because they are neutral. If selection then gets stronger in the recent past—as when duplicates find new functions—then the ratio of nonsynonymous to synonymous polymorphism will provide a much different picture than that of divergence. This outcome can lead to a rejection of the neutral hypothesis and an interpretation of positive selection in the history of the gene duplicate simply because there is an excess of fixed nonsynonymous differences relative to nonsynonymous polymorphisms. But positive selection on the coding sequence has not necessarily occurred, as an environmental change—or even a single advantageous regulatory change—may have been responsible for the new function and the consequent change in selective pressure.

The second major assumption made by the MK test is that the average genealogical history of nonsynonymous and synonymous differences in the sample is the same. In a non-recombining region this is obviously true: there is only one (possibly unknown) genealogy relating all the samples and therefore all the sites. With recombination, however, different parts of the sequence have different genealogies and may therefore have longer or shorter times to their most recent common ancestor (and therefore more or fewer segregating sites). If there is an association between time to the most recent common ancestor and either nonsynonymous or synonymous changes in a region, there may be a difference in the number of segregating sites of each type simply because of the evolutionary variance among regions (see Chapter 6 for a discussion of evolutionary variance). To be explicit, this assumption comes from the fact that expectations for both P_N/D_N and P_S/D_S in the 2 × 2 test of independence are equal to $4N_e/2t$. If the expected time to the MRCA of the sample (represented here by the value of N_e) differs among regions, this assumption does not hold. But because nonsynonymous and synonymous changes within a locus are usually interdigitated among one another, the *average* genealogy is expected to be the same. At a single gene, even if nonsynonymous and synonymous changes are spatially separated, it is still unlikely that there are large differences in the genealogies of adjacent regions. In both of these cases the only source of variance needed to be accounted for in the 2 × 2 table is sampling variance among chromosomes and mutations.

However, there are a growing number of instances in which the classes of mutations being compared (e.g., nonsynonymous and synonymous) are spatially separated in the genome (see next section). In these cases it is no longer appropriate to apply the 2 × 2 test of independence, as the variance in the expected number of segregating sites among regions is inflated relative to a non-recombining locus. This inflated variance leads to an increase in false rejections of the neutral model (Andolfatto 2008). Although such applications do violate the assumptions of the MK test, there are several solutions to the problem. Andolfatto (2008) proposed carrying out coalescent simulations with recombination to establish more accurate significance levels for hypothesis testing. Such simulations can be used to simulate any amount of recombination between loci and can therefore be used in a wide variety of circumstances. Alternatively, testing can explicitly take both the evolutionary and sampling variance into account. The HKA test (see Chapter 8) is similar in conception to the MK test but is intended for use at multiple unlinked loci; the statistical framework of this test therefore incorporates the expected increase in variance. The details of this test will be laid out in the next chapter, but for now it is sufficient to say that a test taking into account the evolutionary variance among loci must be used if different classes of sites do not have similar genealogies.

Biological extensions of the MK test

The key insight of the MK test is that by comparing different classes of sites—those under selection versus those putatively under no selection—we can largely control for the fact that populations are unlikely to meet all of the equilibrium assumptions of the Wright-Fisher model. In other words, comparison to a putatively functionless class of polymorphisms provides an internal control for nonequilibrium histories that are due to demography or linked selection. Given this insight, it is natural to ask whether such comparisons can be made between other kinds of polymorphisms, especially if they can naturally be divided into selected and non-selected classes.

In fact, several different extensions of the MK test have been made to test for selection in different types of mutations. Akashi (1995) compared polymorphism and divergence among preferred and unpreferred synonymous substitutions in *Drosophila*. In this and other organisms it has often been found that one of the multiple synonymous codons coding for the same amino acid is favored over the others, either for translational accuracy or for translational efficiency (reviewed in Plotkin and Kudla 2011). The favored codons are called *preferred* in these cases, while the unfavored codons are call *unpreferred*. Rather than comparing nonsynonymous and synonymous changes in the MK table, Akashi used preferred and unpreferred synonymous changes and was able to demonstrate that many unpreferred polymorphisms are weakly deleterious. Because the synonymous changes were again interdigitated throughout single genes, this extension of the MK test did not have the problems associated with increased evolutionary variance.

A further extension of the MK test examines polymorphism and divergence in regulatory sequences. This application was first introduced by Ludwig and Kreitman (1995) and Jenkins, Ortori, and Brookfield (1995) but has since been applied to a growing number of datasets (e.g., Crawford, Segal, and Barnett 1999; Kohn, Fang, and Wu 2004; Andolfatto 2005; MacDonald and Long 2005; Holloway et al. 2007; Jeong et al. 2008); a test analogous to d_N/d_S can also be used in regulatory regions (e.g., Hahn et al. 2004). These tests compare binding sites in noncoding regions either to interspersed sites that do not bind transcription factors or to synonymous sites in nearby regions. In the case of comparisons among polymorphisms and fixed differences in unlinked (or only loosely linked) regions, one again must be cautious in accounting for the different histories among loci (Andolfatto 2008).

Almost as important as accounting for evolutionary variance are several important differences between the evolution of nonsynonymous changes in coding regions and regulatory changes in noncoding regions (Hahn 2007). First, because functional regulatory sequences are often only characterized in a single species, assignment of functional homology to a set of homologous nucleotides in interspecies comparisons is not always warranted. Experimentally verified binding sites present in a well-studied focal species may be absent in other species; conversely, sequences with no known function in the focal species may actually be binding sites in the other species used in a comparison. Both types of errors will lead to misclassification of nucleotides, some error in estimates of the strength and direction of selection, and violations of the assumption that the neutral mutation rate is constant. A second caveat is that the genetic code of binding sites is unknown at present. We have little, if any, information on the effect of changes in binding sites on binding affinity. Although the classification of any change within a binding site as selected is therefore largely a hypothesis, it may be just as good as considering any amino acid change in a protein to be functionally relevant. The third major caveat in applying the MK test to regulatory sequences concerns the manner in which positive selection acts. While repeated substitution of amino acids in a protein seems like good evidence for positive selection, it is harder to imagine how exactly this might work in a regulatory region. This skepticism follows from some important features of regulatory sequences: binding sites are often not restricted to specific positions, binding sites arise through point mutation quite often, and multiple changes in a binding site often result in the complete loss of binding affinity (reviewed in Wray et al. 2003). None of these reasons preclude natural selection from acting in this manner; rather, they simply suggest that instances in which repeated substitutions due to directional selection are detectable will be rare.

Given all of these caveats, there are in fact a few examples of positive selection being detected through use of the MK test on noncoding regions. Crawford, Segal, and Barnett (1999) studied variation in the regulatory region of the *Ldh-B* locus between two subspecies of the killifish, *Fundulus heteroclitus*. They found an excess of fixed differences in nucleotides identified as being responsible for transcription factor binding relative to substitutions in interspersed sites, as well as differences in expression driven by the *Ldh-B* regulatory sequence between species. These results are consistent with repeated positive selection leading to the fixation of regulatory mutations.

Summarizing selection across the genome

Given the results of McDonald-Kreitman tests from more than one gene, we would like to be able to compare results among genes or to summarize results across a single genome. The *P*-values taken from a test of the 2 × 2 contingency table are not ideal for comparisons among genes. One reason this is the case is that genes with evidence of positive selection and genes with evidence for segregating deleterious polymorphisms can have the same *P*-value. A second reason is that *P*-values are highly dependent on the counts in each cell—and therefore longer or more rapidly evolving genes will have more extreme P-values regardless of the strength of selection. Simply summing contingency tables across genes to increase counts is also not a good solution, as it can lead to misleading results (Shapiro et al. 2007; Stoletzki and Eyre-Walker 2011).

In order to provide a simple summary statistic that gives an interpretable value to the MK test's 2×2 contingency table, Rand and Kann (1996) proposed the neutrality index (NI):

$$\text{NI} = \frac{\left(\dfrac{P_N}{D_N}\right)}{\left(\dfrac{P_S}{D_S}\right)} \tag{7.17}$$

Under neutrality, the expected value of the neutrality index is 1, as the ratio of P_N/D_N and P_S/D_S should be equal. Values greater than 1 represent an excess of nonsynonymous polymorphisms, and values less than 1 represent an excess of nonsynonymous fixed differences. The neutrality index, or variants of it (see below), has been widely used to compare patterns of molecular evolution across organisms and across genes. As stated in the original paper, "the index is intended to provide a qualitative indicator of direction and degree" of deviations from neutrality (Rand and Kann 1996, p. 737). A P-value must still be calculated to infer statistically significant deviations from NI = 1, but the index allows a quick comparison of evolutionary forces among genes.

There are several problems in interpreting NI besides the general issues of bias and increased variance in a statistic that is a ratio of ratios. One issue arises when D_N, D_S, or P_S has a value of 0. In these cases NI is undefined and no value can be calculated for such genes. The common solution to this problem is to add a "pseudocount" of 1 to each cell in the 2×2 table, which ensures that NI is defined, although hypothesis testing should occur without the pseudocounts. A second issue arises because NI values are not symmetric around 1, in the sense that NI = 2 and NI = 0.5 represent equivalent deviations from the neutral expectation and yet are very different in magnitude (Stoletzki and Eyre-Walker 2011). A solution to this problem is to take the log (or negative log) of NI values, which produces symmetric deviations (e.g., Tachida 2000; Presgraves 2005). The negative log also has the pleasing aesthetic side effect of giving positive values to genes showing evidence for substitutions with positive selection coefficients and negative values to genes showing evidence for polymorphisms with negative selection coefficients (Li, Costello, et al. 2008). **FIGURE 7.14** shows negative log(NI) values plotted for thousands of genes in yeast, flies, and humans and immediately demonstrates the large number of genes showing evidence for positive selection in flies.

A related statistic is α, intended to represent the proportion of nonsynonymous substitutions fixed by positive selection (Smith and Eyre-Walker 2002). The value of α is given by:

$$\alpha = 1 - \frac{D_S P_N}{D_N P_S} \tag{7.18}$$

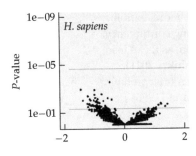
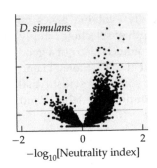
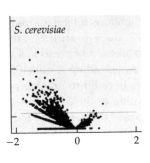

FIGURE 7.14 Neutrality indices for humans, flies, and yeast. Each point represents the neutrality index (NI) for a specific gene and the *P*-value associated with the McDonald-Kreitman test for that gene. For plotting purposes NI is defined as $(P_N + 1/D_N + 1)/(P_S + 1/D_S + 1)$, and the negative $\log_{10}$ of these values is shown. The lower horizontal line in each graph represents $P = 0.05$; the upper horizontal line represents the Bonferroni-corrected $P = 0.05/n$, where n is the number of tests conducted. (From Li, Costello, et al. 2008.)

which is the same as $1 - \text{NI}$ (Smith and Eyre-Walker 2002). In fact, α is not a proportion because it can take negative values. It may be better to think of α as a summary statistic, as it is not clear how to interpret α when it is negative (i.e., when there is an excess of nonsynonymous polymorphism). Positive values do represent the excess of amino acid substitutions that is due to adaptive natural selection, however. Rather than calculating a separate value for every gene in a genome, α is most commonly used to summarize data across genes. This can be done by averaging counts of D_N, D_S, P_N, and P_S across genes in an unbiased manner. Unfortunately, the original papers outlining methods for this summarization described procedures that violated either Simpson's paradox (that combining two datasets can lead to results not present in either dataset) or Jensen's inequality (that the average of ratios is not equal to the ratio of averages). While maximum likelihood methods do exist for calculating α (Bierne and Eyre-Walker 2004; Welch 2006), recent work has provided a simple unbiased method for calculating the average α from a large number of genes (Stoletzki and Eyre-Walker 2011):

$$\bar{\alpha} = 1 - \frac{\sum_{i}^{m} D_{S_i} P_{N_i} / \left(P_{S_i} + D_{S_i}\right)}{\sum_{i}^{m} P_{S_i} D_{N_i} / \left(P_{S_i} + D_{S_i}\right)} \tag{7.19}$$

where the subscript indicates the *i*th gene in the set out of *m* genes total. Calculation of α for nonsynonymous differences across the genomes of *Drosophila simulans* and *D. melanogaster* found $\alpha = 0.54$, a huge proportion of positively selected substitutions (54%; Begun et al. 2007).

Finally, by combining data on the allele frequency spectrum with data on fixed differences, several methods are better able to estimate the proportion of nonsynonymous substitutions fixed by positive selection (e.g., Boyko et al. 2008; Eyre-Walker and Keightley 2009). These methods use the reasoning introduced by Templeton (1996) and Fay, Wyckoff, and Wu (2001) to first account for the presence of slightly deleterious nonsynonymous polymorphisms before estimating this proportion. While the methods differ slightly—Eyre-Walker and Keightley use the folded allele frequency spectrum while Boyko et al. use the unfolded spectrum—both attempt to first estimate the proportion and frequency of nonsynonymous polymorphisms by comparison to the spectrum of synonymous polymorphisms. By fitting a demographic model to the synonymous dataset and assuming all nonsynonymous polymorphisms are either neutral or deleterious, these methods allow for an estimate of the fraction of nonsynonymous fixed differences that are due to adaptive natural selection. This and similar approaches can also be used to calculate the value of the joint parameter $\gamma = 2N_e s$ (also confusingly denoted α), which represents the average effect of selection on nonsynonymous variants (Keightley and Eyre-Walker 2010). Simulations show that these methods can provide an accurate estimate of the average value of α if data from a large number of genes are collected, even if the sample size is small; much larger sample sizes are needed to accurately estimate parameters at a single locus (Keightley and Eyre-Walker 2010). As sequencing becomes cheaper, even single-locus studies of the frequency of deleterious and advantageous mutations will become possible.

LINKED SELECTION

8

In the previous chapter we discussed the effects of natural selection on individual mutations that affect organismal fitness. If there are enough of these mutations of a single type—such as advantageous nonsynonymous changes—then we can often make strong inferences about the dominant mode of selection. However, none of the methods previously described are useful for detecting selection on a single advantageous mutation, no matter how large an effect it has, or on a single balanced polymorphism. In order to identify selective variants of this kind we must instead consider the effects they have on linked neutral variation: polymorphisms that themselves have no influence on fitness but are affected by selection occurring nearby. While methods that use linked neutral variation are often still limited to detecting strong, recent selection, they offer perhaps the best window into the recent selective history of the genome. I examine the effects of linked selection on three distinct aspects of variation and for each discuss tests designed to detect the action of selection on specific facets of this variation.

DETECTING SELECTION USING THE AMOUNT OF POLYMORPHISM

Selection has relatively straightforward effects on levels of linked neutral diversity: it can either raise them or lower them. Understanding the general conditions under which levels of neutral variation will go up or down will help us gain intuition into how tests for selection work and exactly which modes of selection can be detected.

The effects of positive selection on levels of linked neutral variation

Positive selection acting to fix a newly arisen advantageous mutation will lower levels of diversity. This scenario has been termed a *selective sweep* (Begun and Aquadro 1991; Berry, Ajioka, and Kreitman 1991; Harrison 1991) or *hitchhiking* (Maynard Smith and Haigh 1974), with each phrase stressing a slightly different outcome of the process (see also Kojima and Schaffer 1967). The phrase *selective sweep* evokes the removal of variation that occurs as the chromosome containing the adaptive allele becomes the predominant one in the population (**FIGURE 8.1**).

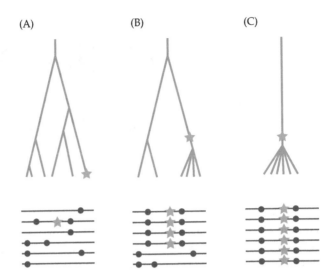

FIGURE 8.1 Phases of a selective sweep. For each phase an example genealogy and an alignment with derived alleles (denoted as circles) is shown. (A) An advantageous mutation (shown as a star) arises on a neutral genealogy. (B) The advantageous allele rises in frequency but is not yet fixed in the population. (C) The advantageous allele has fixed, with all chromosomes coalescing soon after it appears. The branch extending from the root of the sample is shown for clarity.

All completely linked variation is swept aside, with the only remaining variation arising during the transit to fixation. The phrase *hitchhiking* (or "hitchhiking," in Maynard Smith and Haigh's original spelling) stresses the fact that neutral polymorphisms that happen to reside on the lucky chromosome upon which the advantageous mutation arises also rise in frequency (Figure 8.1). These hitchhikers will also fix in the population, unless recombination farther away from the selected locus unhitches them from the advantageous allele.

To get a slightly more quantitative feeling for how a selective sweep reduces polymorphism at a locus, consider the time it takes a mutation to fix in a population. This period of time determines how much variation there will be: in the most extreme example all chromosomes descend from a single lineage containing a highly advantageous mutation that arose in the previous generation. In such a population there would be almost no variation, as all polymorphisms would have had to arise in a single generation. The longer a mutation takes to fix, the more time there is for associated polymorphisms to accumulate. For instance, conditional on its fixation, a neutral mutation takes approximately $4N_e$ generations to fix (Kimura and Ohta 1969), though there is a large variance associated with this expectation (Kimura 1970). There is an obvious connection between the time to fixation of a neutral mutation and the amount of time in a neutral genealogy until all lineages coalesce to their most recent common ancestor, which also approaches $4N_e$ with large samples (Equation 6.3). The amount of variation at a neutral locus—and the shape and height of the genealogy—is therefore intimately connected to the time it takes neutral variants to fix. Now imagine a new additive advantageous mutation arising once in a diploid population. The time it takes this mutation to fix is (Nei 1973, p. 383):

$$T_{Fix} \approx \frac{2 \ln \left(2 N_e \right)}{s} \tag{8.1}$$

If the effective size of a population is $N_e = 1,000,000$ with a selection coefficient of $s = 0.0001$ (a relatively small advantage), this equation implies that it will take the advantageous allele approximately $0.29 N_e$ generations to fix. While this is still a long time (approaching 290,000 generations), it is quite rapid in comparison to the neutral expectation. As a result of the rapid rise to fixation, the genealogy at such a locus will be much shorter and will have a shape different from the one expected under neutrality (Figure 8.1C). These two aspects of the genealogy will form the basis for multiple methods that can be used to detect the action of selection.

Three very important factors interact to determine differences in the magnitude of reduction in diversity at or near a selected locus: the strength of selection (s), the rate of recombination (c), and the time since the sweep ended (**FIGURE 8.2**). Selective sweeps generate a valley (or *trough*) in levels of polymorphism surrounding the location of the advantageous allele (Figure 8.2); the depth and width of the valley are determined by these factors. The strength of selection, together with the population size, determines how quickly the advantageous allele fixes (Equation 8.1). The more quickly it fixes, the lower the height and total branch length of the resulting genealogy, and therefore the lower the level of diversity around the selected site at the end of the sweep (and the deeper the resulting valley). With weaker selection there will be more time since the common ancestor of all lineages in the population and therefore a taller genealogy with greater total branch lengths, resulting in more variation. The sweeping allele will also remove any variation linked to it, except when recombination allows nearby neutral polymorphisms (often referred to as *neutral loci*) to escape this effect. The recombination rate in the region around the advantageous substitution will therefore affect the width of the swept region. Similarly, the time to fixation affects the width of this region, as a longer transit time also means that there is more time for recombination off the sweeping haplotype (Figure 8.2). The size of the region with lower levels of variation will therefore be proportional to s/c (Kaplan, Hudson, and Langley 1989; Barton 1998). From this relationship we can see that selection and recombination counteract each other to determine the width of the diversity valley. Finally, once the sweep has ended levels of diversity will begin to return to normal, reaching equilibrium close to $4N_e$ generations later. However, long before a locus has reached equilibrium, levels of diversity will have risen to within the normal bounds of variation seen at loci evolving neutrally (Figure 8.2, middle sweep). As we will discuss later in the chapter, this means that the power to detect selection diminishes rapidly after the sweep has ended (Simonsen, Churchill, and Aquadro 1995; Fu 1997; Przeworski 2002).

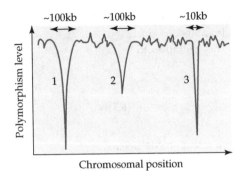

FIGURE 8.2 Effects of selection intensity and time since fixation on the reduction in diversity as a result of selective sweeps. Sweeps 1 and 2 had a similar selective advantage, but sweep 2 ended longer ago than sweep 1. Sweeps 1 and 3 ended at the same time, but sweep 3 had a smaller selective advantage than sweep 1. (Adapted from Macpherson et al. 2007.)

Not all sweeps observed in nature have fixed the advantageous allele, either because this allele is sampled on its way to fixation or because a change in conditions has altered the selection coefficient associated with the relevant allele. These *partial sweeps* often have their own unique patterns of variation at linked neutral polymorphisms (Figure 8.1B) and will be covered later in this chapter. Additionally, not all sweeps are the result of a single advantageous mutation arising once and fixing. The scenario in which the adaptive allele has only a single origin is often labeled a *hard sweep*, to distinguish it from so-called *soft sweeps* (Hermisson and Pennings 2005). Soft sweeps occur when multiple copies of the advantageous allele are fixed, either because selection acts on standing variation that was previously neutral or deleterious, or because mutations to the same advantageous state occur on different backgrounds before the original copy fixes. Selection from standing variation is obviously a common occurrence in nature, and there is also good evidence that adaptive alleles can arise multiple times in the same population (e.g., Karasov, Messer, and Petrov 2010). In either scenario, the general effects of soft sweeps on levels of variation are much reduced relative to hard sweeps (Hermisson and Pennings 2005; Pennings and Hermisson 2006a). Variation around the selected locus is not nearly as low in soft sweeps as hard sweeps, as the advantageous allele is associated with multiple different haplotypes. Therefore, even if a single selected allele is fixed in the population, the surrounding sites can be polymorphic. This makes soft sweeps much harder to detect than hard sweeps (Przeworski, Coop, and Wall 2005; Pennings and Hermisson 2006b), though there may be unique signatures associated with each type of fixation (Garud et al. 2015; Schrider and Kern 2016).

The effects of balancing selection and strongly deleterious alleles on levels of linked neutral variation

We have just seen how positive selection on a newly arising advantageous mutation can reduce levels of linked neutral variation. Conversely, balancing selection acting to maintain two or more alleles at intermediate frequency in a population will *increase* levels of linked neutral variation. Balancing selection acts to counter the effects of drift, resulting in a much slower rate of loss of alleles than would otherwise be expected. This effect extends to linked variation, such that these neutral polymorphisms are also not lost. The genealogy at a locus under balancing selection is much higher than a neutral genealogy and often has two or more long internal branches (**FIGURE 8.3**). As with sweeps, these two aspects of the genealogy can be used to detect the action of balancing selection.

The effect of recombination on signatures of balancing selection is again to narrow the window of diversity that is affected. As nearby neutral polymorphisms recombine away from the balanced alleles, they can be lost. Given that one of the main features of balanced polymorphisms is their greater-than-expected age, the effects of recombination can be large: only a small window of excess diversity around the selected site may remain (**FIGURE 8.4**). This means that balancing selection can be difficult to detect and may be commensurately more difficult for older balanced polymorphisms (e.g., Leffler et al. 2013).

Finally, deleterious alleles can also have an effect by reducing linked neutral variation. This process—dubbed *background selection* (Charlesworth, Morgan, and Charlesworth 1993; Charlesworth, Charlesworth, and Morgan 1995)—occurs because deleterious variants are removed from the population, removing linked neutral polymorphisms with them. The main effect of background selection is to reduce the height of the local genealogy proportional to the fraction of chromosomes carrying a deleterious allele (Equation 3.19), though the shape largely remains the same as in the neutral case. Background selection can sometimes change the shape of the genealogy (Charlesworth 2012), but the area of parameter space over which this occurs is relatively small (Hudson and Kaplan 1994).

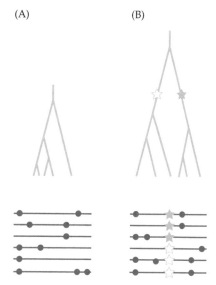

FIGURE 8.3 Effects of balancing selection. (A) A typical neutral genealogy. (B) A genealogy under balancing selection. The site of the balanced polymorphism is denoted by the stars, with the two allele classes associated with one or the other star. Note, however, that one of the balanced alleles represents the ancestral state and likely did not arise on the internal branch of the genealogy. It is shown simply for clarity.

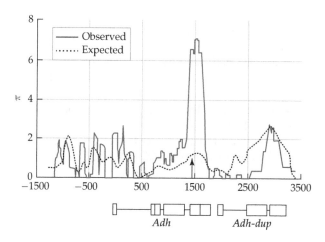

FIGURE 8.4 Balancing selection at *Adh* in *D. melanogaster*. The solid line shows a 100-bp average of π across the region, while the dotted line shows the value expected under a neutral model. The site of the suspected balanced polymorphism is marked with an arrow. (From Kreitman and Hudson 1991.)

Recombination interacts with background selection in much the same way it does with the other forms of selection discussed here: higher recombination limits the distance over which there is an effect, with low recombination regions showing the broadest reductions in linked neutral variation. On nonrecombining chromosomes such as the Y, the effects of background selection may be quite strong (e.g., Wilson Sayres, Lohmueller, and Nielsen 2014).

The most underappreciated result in molecular population genetics

Before moving on to consider statistical methods for detecting linked selection, we need to discuss one of the most important and most underappreciated results in modern molecular population genetics. Thus far we have examined the effects of selection on levels of linked neutral *polymorphism*. Now we want to ask about the effects of selection on levels of neutral *divergence* at linked sites. The naïve expectation may be that hitchhiking will increase levels of divergence and that background selection will decrease levels of divergence (and maybe that balancing selection will have some unpredictable effect). But simple calculations can be used to show that none of these processes will change the rate of neutral substitution. Therefore, linked selection should have no effect on levels of divergence, other than any effect it has on levels of ancestral polymorphism that contribute to divergence (Gillespie and Langley 1979). This is a key result with many important implications for tests of selection.

Although this result was understood by many within the field, it was not explicitly spelled out until a paper by Birky and Walsh (1988). Here I present the mathematical logic used to show that hitchhiking has no effect on neutral substitution rates. Consider first the probability that a derived neutral allele will fix in the population, thus contributing to divergence. This probability is equal to p, its current frequency in the population. Low-frequency alleles are less likely to fix relative to high-frequency alleles, and new mutations have probability $1/2N_e$ of fixing (Chapter 7). If we next imagine that an unconditionally advantageous mutation that will definitely go to fixation enters the population, we can see that any neutral alleles completely linked to it will also go to fixation. But what is the probability that the advantageous mutation will arise on a chromosome carrying an arbitrary derived neutral allele? This probability is also p, the neutral allele's current frequency. So hitchhiking does not change the overall probability of fixation for a neutral allele, and consequently levels of divergence (as measured by d, d_{XY}, or similar statistics) do not increase. The importance of this result cannot be overstated, as it means that we can use divergence as a control dataset to understand the effect of selection on polymorphism.

Although linked selection will not affect the rate of substitution of neutral alleles, it will have a substantial influence on advantageous and deleterious alleles. This relationship was dubbed the *Hill-Robertson effect* by Felsenstein (1974), after a paper by Hill and Robertson (1966). It is also commonly known as *Hill-Robertson interference*. The effect of selection is to lower the probability of fixation of linked advantageous alleles and to increase the probability of fixation of linked deleterious alleles. There are multiple ways to think about why this effect occurs. An abstract way to think about it is to say that selection at a linked site lowers the effective population size at the site of interest because selection and (some) recombination increase the variance in offspring number. Since the product of N_e and s is what determines fixation probabilities for both advantageous and deleterious alleles (Figure 7.2), lowering N_e causes both of these values to approach the probability of fixation for a neutral allele. A more direct way to think about it is to see that the fixation probability of any chromosome is proportional to the total of the selection coefficients associated with it. If an advantageous allele is linked to either deleterious alleles or other advantageous alleles that occur on multiple backgrounds, the probability that it will fix goes down. Inversely, linkage to advantageous alleles actually raises the probability of fixation for deleterious alleles, as long as they are not too deleterious. As recombination rates go up the consequences of the Hill-Robertson effect go down, so those regions of the genome with the highest rates of recombination are expected to be the least affected (Comeron, Williford, and Kliman 2008).

Detecting selection via levels of polymorphism: The HKA test

We have seen how selection at linked sites can either raise or lower levels of neutral variation. In the rest of the chapter we will discuss specific ways to test for selection at a locus and to determine what kind of selection might be acting. One very important caveat to everything covered here is that nonequilibrium demographic histories can have profound effects on the outcome of our tests. Sometimes demography can cause us to miss selection, and sometimes it can cause us to incorrectly infer selection where none has occurred. In Chapter 9 we will cover some of the ways that nonequilibrium demography can mimic selection, returning to an in-depth discussion of the interaction between selection and demography in Chapter 10. But understanding exactly which signals each of our tests can detect will be important in seeing how they can be misled.

Under a neutral model levels of polymorphism should be proportional to levels of divergence: the amount of polymorphism in a population is expected to be $4N_e\mu$, while divergence between species is expected to be $2t\mu$. Although the mutation rate and the neutral mutation rate can change along a chromosome, at any particular locus we expect a correlation between polymorphism and divergence because of the shared parameter, μ. Arnheim and Taylor (1969) were the first to use this expectation to test the neutral hypothesis, though there was little data available to them. With the ability to collect DNA sequences from multiple individuals that emerged in the early 1980s came more sophisticated methods for testing this prediction. The most commonly used such method was introduced by Hudson, Kreitman, and Aguadé (1987) and is consequently referred to as the *HKA test*.

As originally applied, the HKA test addresses a simple question: If one observes a locus with exceptionally low levels of polymorphism, is this due to low mutation rate at the locus or a recent selective sweep? Similarly, if one observes a locus with high levels of polymorphism, is this due to a high mutation rate or balancing selection? If there was a recent selective sweep (or balancing selection), then the level of divergence at the locus of interest will not be exceptionally low (or high)—instead, it will be selection that has affected levels of polymorphism. A key point is that levels of linked neutral divergence are unaffected by selection (see above), allowing us to use divergence as a proxy for the mutation rate. Therefore, a straightforward test of neutrality should test for the proportionality of polymorphism and divergence across loci. The HKA test does this in a statistically sound manner.

Experimental design for the HKA test

The HKA test is normally carried out using two measures from each of two or more loci. For each locus we need variation from one species (S) and divergence between this species and another closely related species (d). The variation data can come from as few as two homologous chromosomes (as long as variable sites are observed), and both polymorphism and divergence can use any type of change: that is, it does not matter whether they are synonymous, nonsynonymous, or noncoding. There is no requirement that the HKA test be carried out on coding regions, as information about the type of change observed is not used. However, it is somewhat common to see researchers use only synonymous sites when coding regions are analyzed, to increase the probability that the variation they see is not affected by direct selection.

To ask whether polymorphism is proportional to divergence across loci, we calculate a χ^2 statistic using the observed and expected values:

$$\chi^2 = \sum_{i}^{m} \left(S_i - E(S_i) \right)^2 / Var(S_i) + \sum_{i}^{m} \left(d_i - E(d_i) \right)^2 / Var(d_i) \tag{8.2}$$

where S_i is the observed number of segregating sites at the ith locus (out of m loci total), d_i is the observed number of differences between randomly chosen chromosomes in the two species, and $E(\bullet)$ and $Var(\bullet)$ denote the expectation and variance of different quantities. We have previously seen equations for most of these quantities: $E(S_i)$ is found by rearranging Equation 3.5, $Var(S_i)$ is given by Equation 3.7, and $E(d_i)$ is given by Equation 7.1 (with all loci sharing the same divergence time, t). The only statistic not given previously is:

$$Var\left(d_i \right) = E\left(d_i \right) + \left(\theta_{Anc_i} \right)^2 \tag{8.3}$$

where θ_{Anc_i} is the value of θ in the ancestral population at the ith locus (Gillespie and Langley 1979). Remember that for all of these values we must be sure we are using the formulas appropriate to the chromosome on which each locus is found (e.g., X, autosome, etc.). The expectations and variances can be estimated if we first assume that the ancestral population size is the same as the current population size and then solve for the unknown parameters: θ_{Anc_i} at each locus and the species split time, t (which appears in the expectation of d_i). The resulting statistic should be χ^2-distributed with $m - 1$ degrees of freedom. Often simulations are used to generate the null distribution rather than any particular parametric distribution.

There are two important aspects of this experimental and statistical procedure that should be noted before discussing how to interpret the HKA test. The first is the use of d rather than some other measure of divergence. As pointed out in the original paper, d_{XY} has the same expectation as d (see Equation 7.1 and related discussion), but the variance term for d_{XY} is slightly more complicated (Nei 1987, eq. 10.24). There is no other reason not to use d_{XY} as the measure of divergence, and d is used only because of the simpler variance term. For very recent split times the use of d (or d_{XY}) leads to reduced power in the HKA test because of a large amount of segregating ancestral polymorphism. To avoid this problem, Ford and Aquadro (1996) proposed a modified HKA test that uses the number of fixed differences between species. Use of this method is expected to increase power to detect selection when species are very closely related (i.e., when d is small relative to θ).

Second, the reader may wonder why we cannot apply a simple test of independence using the observed values of polymorphism and divergence. For instance, given a study that has sequenced two loci, why not simply use a 2 × 2 table as in the McDonald-Kreitman test? In fact, this is exactly what Kreitman and Aguadé (1986) did using data from the *Adh* locus and its flanking 5' region. However, a 2 × 2 table (or any test of independence) assumes that the only source of variation in S is a Poisson mutational process. There is a second source of variation when sampling different loci: the variation contributed by the genealogical process, or what we previously referred to as *evolutionary variance*. Loci can differ in the number of polymorphisms they contain not just because of the variation in the mutational process, but also because of variation in their total height and length that is due to the stochasticity of the genealogical process (Chapter 6). Without accounting for this source of variance we may falsely reject the null hypothesis even when it is true, as the variation in S will be greater than is expected under the model. Another way to see this is to recall that in conducting the McDonald-Kreitman test we assume that N_e is the same for synonymous and nonsynonymous polymorphisms; when these sites are interdigitated among each other at the same locus, this is indeed likely to be the case. But if the two classes of sites are separated along the chromosome we can no longer assume that they have the same value of N_e—this difference in N_e is exactly what is reflected in the variation in the height of genealogies. As a consequence, we must use a goodness-of-fit method like the HKA test.

Interpreting the HKA test

In the pre-genomic era the HKA test was usually carried out with a very small number of loci: one of the loci represented the gene or region of most interest, while the other loci were designated as "neutral" regions that would provide information about background levels of polymorphism and divergence in the absence of linked selection. Often the neutral loci also did not reject any other test of neutrality that had been applied to the data. Rejection of the null hypothesis of the HKA test with a significant χ^2 was taken to indicate that the focal locus was the target of selection, and *post hoc* comparisons of diversity among loci indicated whether selection was positive or balancing. If diversity was low at the focal locus this was taken as evidence of recent positive selection, and if diversity was high at this locus it was viewed as evidence of balancing selection.

However, as with all goodness-of-fit tests, identification of the locus under selection—and the type of selection acting—must be made via some external information. Given only measures of polymorphism and divergence at a set of loci, it is possible that those initially designated as "neutral" are in fact targets of selection. Any one of the values of polymorphism or divergence at any locus could be an outlier, or they could all (or all but one) be affected by linked selection. In their original paper, Hudson, Kreitman, and Aguadé (1987) argued that the 5' region was much more likely to represent variation unaffected by linked selection, and given the similar levels of divergence the high level of polymorphism at *Adh* was interpreted as evidence for balancing selection (Figure 8.4). This overall problem is similar to that in the McDonald-Kreitman test, as part of which a researcher must still make decisions about which of the four input values is driving the statistical result.

With larger and larger population genetic datasets it has become more difficult to identify the specific targets of selection. Often we do not have *a priori* candidates for selection, so the HKA test is asking whether there is an overall fit between the data and a neutral model. There are several approaches we can take to identify the most extreme candidates for selection, however. If the overall HKA test is significant, the contribution of each individual locus to the result can be determined by removing one locus at a time and repeating the test (e.g., Moore and Purugganan 2003). Sometimes the overall test is not significant, but pairwise tests among loci can still reveal targets of selection (Mousset et al. 2003). Alternatively, Wright and Charlesworth (2004) formulated a maximum likelihood version of the HKA test that can formally test a model of selection at individual loci against a model without selection. Using whole-genome data, Langley et al. (2012) used Ford and Aquadro's (1996) formulation of the HKA test to calculate the expected numbers of segregating sites and fixed differences in each window using the observed variation and divergence on the rest of the chromosome. Each genomic window was therefore being tested as an outlier compared to a defined background. However, in all of these situations one is assuming that at least some of the loci in the dataset are evolving neutrally. As a counterexample, imagine a dataset in which levels of polymorphism at 99 loci are affected by positive selection at linked sites, but at one locus are unaffected. Any implementation of the HKA test would identify the single neutrally evolving locus as the outlier, and further infer that it was affected by balancing selection. If *all* loci are affected by recent positive selection, or all are affected by balancing selection, there may not even be heterogeneity in levels of polymorphism among loci; in this case the HKA test would not be significant. Nonetheless, a significant HKA test can be taken as evidence against the neutral hypothesis (but see caveats below), and the loci with the highest or lowest levels of polymorphism will represent the best candidates for balancing or positive selection.

Before addressing the assumptions made by the HKA test and their common violations, it is important to reiterate what exactly a significant test means. Because the HKA test is looking for the effects of selection on linked variation, a significant test does not mean that any of the variants you see are under direct selection or are in any way non-neutral. It means that levels of polymorphism (and possibly, but more rarely, divergence) are different than expected under only mutation-drift equilibrium. In the case of balancing selection it is possible that *the* selected polymorphism is in your dataset, but for completed selective sweeps it is the low level of neutral polymorphisms that makes the test significant—the selected allele has already fixed and is not present among the current polymorphisms. The inference that loci are evolving non-neutrally is only a statement about the forces affecting levels of polymorphism, and also does not imply anything about constraint in these regions.

The HKA test makes a number of assumptions that may be violated in real data, some of which make the test conservative and some of which make it anticonservative (i.e., more likely to produce false positives). The two most conservative assumptions are that there is no recombination within loci and that there is free recombination between loci. One reason to use coalescent simulations to generate the null distribution for the HKA test is to relax these assumptions, making the test more powerful. Another reason to use simulations is that the HKA test assumes the population is at equilibrium: nonequilibrium demographic histories can lead to greater variance in polymorphism among loci, leading to false positives. Simulating such histories ensures that this variance is accounted for. Similarly, the HKA test assumes that the polymorphism data come from a single population. Hidden population structure can lead in some circumstances to false positives (Ingvarsson 2004). Possibly the most important assumption of the HKA test is that neutral mutation rates are constant over time at each locus. Neutral mutation rates that are not constant can cause a disparity between polymorphism and divergence. There are multiple possible causes of changes in neutral mutation rates: the underlying mutation rate can be evolving, purifying selection can have gotten stronger in the recent past, or purifying selection can have gotten weaker in the recent past. For instance, selection may be stronger because of recent gain-of-function mutations, or it may be weaker because a gene recently became a pseudogene (e.g., Schrider and Kern 2015). In the former case polymorphism will be low relative to divergence, and in the latter case it will be high. As with the interpretation of all such tests, careful examination of the region of interest is necessary to make accurate biological inferences.

DETECTING SELECTION USING THE ALLELE FREQUENCY SPECTRUM

Thus far we have discussed the effects of linked selection on levels of polymorphism, but not on the exact distribution of variants. One of the most beautiful features of population genetics is that we can predict the allele frequency spectrum of neutral variants in the absence of linked selection (Chapter 6).

We can therefore use deviations from this expected distribution to detect the signal of different types of selection.

The effects of positive selection on the allele frequency spectrum

Once again we will focus on the effects of positive selection and balancing selection on the allele frequency spectrum. We do not need to consider the impact of weak negative selection on the frequency spectrum of linked neutral variants because it has a very small effect (Golding 1997; Neuhauser and Krone 1997; Przeworski, Charlesworth, and Wall 1999; Williamson and Orive 2002); as mentioned above with regard to background selection, strongly deleterious alleles also do not affect the frequency spectrum across much of parameter space (Hudson and Kaplan 1994). As with levels of polymorphism, a genealogical perspective will help us to understand the influence of positive and balancing selection on the frequency spectrum. We begin by focusing on a region completely linked to the selected allele—there has been no recombination between the neutral variants at this locus and the selected site. Regions flanking the selected site, but only partially linked to it, can show subtle yet important differences in their allele frequency spectra.

The genealogy at a locus completely linked to a hard selective sweep is often described as star-like (Figure 8.1C). The rapid rise in frequency of the selected allele from a single copy means that most coalescent events happen at the beginning of the sweep, when its frequency is low, giving rise to the star-like pattern. There is also little time for coalescence among lineages during a sweep, and lineages can coalesce simultaneously when there is positive selection. The shorter the time to fixation—or the less time has elapsed since the end of the sweep—the less time there is for coalescence. When there are mutations along each branch of a completely star-like genealogy they are by definition singletons: there is only one descendant chromosome in the sample with each derived allele. Slower sweeps, or more time since the sweep has ended, will lead to older coalescences and fewer singletons than stronger or more recent sweeps. In either case, the allele frequency spectrum can be skewed toward an excess of low-frequency polymorphisms relative to the neutral spectrum (**FIGURE 8.5**). This left-skewed frequency spectrum is the signature of a selective sweep at a locus (Aguadé, Miyashita, and Langley 1989; Braverman et al. 1995). In the extreme, all lineages have coalesced quite recently and there is no variation in the sample. Such cases are not amenable to methods that use the allele frequency spectrum because there are no segregating sites. However, as variation arises at a locus each new polymorphism appears as a singleton. Therefore, even with extremely strong sweeps the skewed allele frequency spectrum will appear after the sweep has ended, or in flanking regions that have partially escaped selection.

When recombination intervenes between a sweeping allele and linked neutral variants, additional patterns in the frequency spectrum can appear. Of course with enough recombination linked neutral variation will be entirely unaffected by a sweep, and there will be no signal of selection. But neutral loci closely linked to the selected allele—such that only one or a few

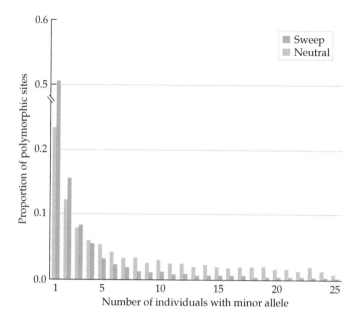

FIGURE 8.5   The effect of positive selection on the allele frequency spectrum at a locus completely linked to the selected allele. The average frequency spectrum at loci evolving neutrally (yellow bars) is compared to one at loci recently experiencing a selective sweep (blue bars), based on simulations. (From Braverman et al. 1995.)

sampled lineages become "unhitched" from it—can also produce an excess of *high*-frequency derived alleles (Fay and Wu 2000). To understand how this can occur, consider the neutral derived alleles that by chance are on the same chromosome as the advantageous allele when it first arises (Figure 8.1A). These alleles will be driven to high frequency by selection. If they recombine onto another background during the sweep, placing the ancestral allele on the selected haplotype (**FIGURE 8.6A**), then they may not reach fixation. Note that since these neutral derived alleles occur on many chromosomes containing the advantageous allele, any single recombination event cannot completely separate them from the sweep. Instead, it is the fact that the ancestral alleles at the same positions become "hitched" to the sweep that prevents their fixation. Genealogically, this process results in a star-like pattern combined with a neutral-like pattern (**FIGURE 8.6B**). The ensuing allele frequency spectrum has an excess of both low- and high-frequency derived alleles (**FIGURE 8.6C**). This pattern is a strong indicator of the "shoulders" of a selective sweep.

The effects of balancing selection on the allele frequency spectrum

As we have seen, the genealogy at a locus under balancing selection can be quite different from a neutral genealogy. The exact form of balancing selection has a large influence on exactly how different these topologies are, as do

FIGURE 8.6 The allele frequency spectrum at loci flanking a sweep. (A) An advantageous mutation (shown as a star) has risen to intermediate frequency when recombination occurs between a chromosome carrying this allele and a chromosome not carrying this allele (illustrated as a dashed line). The genealogy of the flanking region (highlighted in the alignment) is shown, which means that the star does not actually arise on this tree. (B) The advantageous mutation fixes, bringing along the two different backgrounds with which it was associated in this flanking region. (C) The resulting frequency spectrum based on simulated datasets. (A, B based on a graphic made by Dan Schrider; C after Nielsen 2005.)

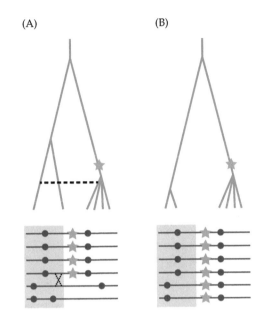

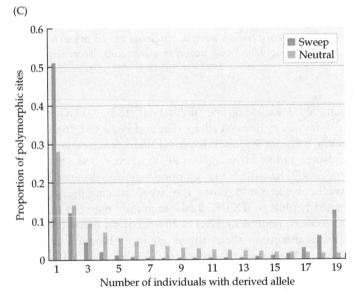

the age of the balanced polymorphism and the frequencies of the balanced alleles (Charlesworth 2006). Here we will consider a simple scenario with a single biallelic site that has been under balancing selection for at least $2N_e$ generations—long enough that variation within each allelic class has been able to build up and reach equilibrium.

The major distinguishing features of a genealogy at a locus under balancing selection are the total height of the tree and the length of the two internal branches (Figure 8.3). The two allelic classes are defined by a balanced polymorphism distinguishing the long internal branches of the tree (noted by stars in Figure 8.3B). Neutral variation has accumulated both within and between classes, so that overall levels of variation (which includes both types of polymorphisms) are much higher than expected under neutrality. The differences in the allele frequency spectrum at such a locus compared to a neutral locus are due solely to the mutations that accumulate on the long internal branches. These polymorphisms are at intermediate frequency in the entire sample, leading to a distinctive "bump" in the allele frequency spectrum.

There are a number of important features of balanced polymorphisms that make them easier or harder to detect. Within each allelic class the genealogies resemble neutral genealogies (albeit from slightly smaller populations), so it is only by sampling chromosomes from both classes that we detect balancing selection. As a consequence, the frequencies of the two selected alleles are key to detecting balancing selection via the allele frequency spectrum: if one class is too low in frequency in the sample then there will be no excess of intermediate-frequency polymorphisms. Biallelic balanced polymorphisms closer to 50% in frequency will therefore be the easiest to detect. In addition, balanced polymorphisms that are very young will be difficult to detect, or at least difficult to ascribe to the action of balancing selection. New alleles subject to balancing selection—for instance, in the form of heterozygote advantage or negative frequency-dependent selection—are expected to have an initial rapid rise in frequency. If a sample is taken at such a locus during the rapid rise there may be a strong signature of selection, but it will resemble a partial or ongoing selective sweep. It will be difficult if not impossible to determine that this pattern is due to a form of balancing selection.

Detecting selection using the allele frequency spectrum: Tajima's D and related tests

Given what we now know about how various forms of selection can perturb the allele frequency spectrum of linked neutral variants, we shift our attention to methods for detecting these perturbations. A straightforward solution would appear to be to use any statistical method that tests for a difference between two distributions. However, linkage among sites means that polymorphisms at a locus are not independent, which violates the assumptions made by almost all such tests. Instead, we will compare different estimators of the population mutation parameter θ in order to detect deviations. This may seem like a roundabout way to study the allele frequency spectrum, but as we will see, it works well in both practice and theory.

In Chapter 3 we learned about a number of different summary statistics of variation, all of which were also expected to be estimators of θ under the standard neutral model (which includes an equilibrium population history). These statistics include the commonly used π and θ_W (Equations 3.2 and 3.5, respectively), as well as the less commonly used but no less informative ξ_1, η_1, and θ_H (Equations 3.8–3.10). Additional estimators of θ are available when there are multiple alleles at a site (Equations 3.13 and 3.15) or when sequencing errors are common (Equations 3.16 and 3.17); a larger number of summary statistics are also possible but are rarely used in practice (Fu 1995). Because all of these statistics estimate the same quantity, they should have the same value under the standard neutral model; as a result, the expected difference between π and θ_W should be 0. We will denote the difference between any two estimators of θ with the letter δ (it is usually denoted d, but there are too many other quantities called d), so that $E(\delta) = 0$.

Tajima (1989a) used δ to construct the first test to detect disparities between the allele frequency spectrum in a sample and the one expected under neutrality. His statistic, D, was defined as:

$$D = \frac{\delta}{\sqrt{Var(\delta)}} = \frac{\pi - \theta_W}{\sqrt{Var(\pi - \theta_W)}}$$ (8.4)

where $Var(\pi - \theta_W)$ can be calculated as (Tajima 1989a, eq. 38):

$$Var(\pi - \theta_W) = \left[\frac{\frac{n+1}{3(n-1)} - \frac{1}{\sum_{i=1}^{n-1}\frac{1}{i}}}{\sum_{i=1}^{n-1}\frac{1}{i}} \right] S$$

$$+ \left[\frac{\frac{2(n^2+n+3)}{9n(n-1)} - \frac{n+2}{n\sum_{i=1}^{n-1}\frac{1}{i}} + \frac{\sum_{i=1}^{n-1}\frac{1}{i^2}}{\left(\sum_{i=1}^{n-1}\frac{1}{i}\right)^2}}{\left(\sum_{i=1}^{n-1}\frac{1}{i}\right)^2 + \sum_{i=1}^{n-1}\frac{1}{i^2}} \right] S(S-1)$$ (8.5)

with n and S denoting the sample size and number of segregating sites at a locus, respectively.

Fu and Li (1993a) created statistics quite similar to Tajima's D for data in which segregating polymorphisms are either defined as ancestral/derived or are not so defined (i.e., using the unfolded or folded frequency spectrum). The statistics using polarized data both use the estimator ξ_1, while the statistics using unpolarized data both use the estimator η_1. These are known as Fu and Li's D, F, D^*, and F^*:

$$D = \frac{\pi - \xi_1}{\sqrt{Var(\pi - \xi_1)}}$$ (8.6)

$$F = \frac{\theta_W - \xi_1}{\sqrt{Var(\theta_W - \xi_1)}}$$ (8.7)

$$D^* = \frac{\pi - \eta_1}{\sqrt{Var(\pi - \eta_1)}}$$ (8.8)

$$F^* = \frac{\theta_W - \eta_1}{\sqrt{Var(\theta_W - \eta_1)}}$$ (8.9)

The variances for all of these statistics can be found in Fu and Li (1993a), though note that there is a typo in the formula for the variance in F^* in this paper; the correct expression can be found in Simonsen, Churchill, and Aquadro (1995).

Under the standard neutral model all of these test statistics are expected to have a mean of 0, though with low or no recombination the values are actually slightly negative (Thornton 2005). The tests were originally constructed to resemble z-scores, with the mean divided by the standard deviation (i.e., with a mean of 0 and a variance of 1). Using this normalization scheme it was hoped that the statistics might fit a normal distribution, with loci showing deviations from neutrality lying in either tail of such a distribution. However, none of these test statistics fit a parametric distribution very well, and right from the beginning computer simulations were used to define the null distribution (Tajima 1989a). This means that the values in the denominators are irrelevant to any significance testing, as we would find the same null distributions if we calculated them from the δ-values alone (note, however, that the power of each test is dependent on this underlying variance). In practice, all of these statistics are still calculated as shown above, if only to make numerical results comparable across studies. We will discuss one additional δ-based test shortly (Fay and Wu's H: $\delta = \pi - \theta_H$), noting here only that it is similar to all of the above statistics but is calculated without the denominator. These tests are just some of many similar ones created, of the many possible combinations of θ estimators that are possible (Fu 1996, 1997; Zeng et al. 2006; Achaz 2009).

Neutrality tests of the allele frequency spectrum are quite easy to carry out. A researcher must only collect a random sample of chromosomes from nature (at least $n = 3$ chromosomes) and be able to estimate allele frequencies. Only variable sites at each locus are needed, and the number of invariant sites—and hence per site values—do not figure into any calculations. This means that these statistics can be calculated from any type of data, not just full sequences, though ascertainment bias will still be an important issue as it has major effects on the frequency spectrum (Figure 6.6). The possible values taken by these test statistics are to some degree limited by the number of segregating sites; because of this, further normalization can be done to make comparisons among loci (Schaeffer 2002). Given the amount of data being collected by most modern genomic studies, however, these normalization steps are largely unnecessary. No information about haplotypic phase is necessary, and Tajima's D as well as Fu and Li's D^* and F^* do not need polarized data, so no outgroup sequence is necessary. The allele frequency spectrum is quite sensitive to sampling design and to the fact that all chromosomes come from a true population (e.g., Przeworski 2002). This characteristic points to the larger issue of alternative explanations for deviations from the expected neutral-equilibrium allele frequency spectrum, a point to which we return later in greater depth. However, because of their ease of use and their power to detect selection, these tests have become some of the most popular in population genetics.

Interpreting Tajima's D and related tests of the frequency spectrum

All of the tests using the allele frequency spectrum described thus far produce very similar numerical values under selective sweeps and under balancing selection. Put simply: these values are all negative when there has been a sweep, and they are all positive when there is balancing selection. A good heuristic is that Tajima's D and Fu and Li's D, F, D^*, and F^* are usually significant when the values obtained are either greater than $+2$ or less than -2. The exact thresholds depend on sample size, number of polymorphisms, and the conditions under which simulations are run, but anything $>|2|$ is improbable under the neutral-equilibrium model and is likely to be statistically significant. Here we examine why the tests generate these positive and negative values.

Selective sweeps lead to an excess of low-frequency neutral variation surrounding a selected site (Figure 8.5). When there is an excess of low-frequency polymorphisms, summary statistics such as θ_W, ξ_1, and η_1 will all be greater than π. To see why this is, remember that π is calculated from allele frequencies, while the other statistics are calculated only from the numbers of segregating sites. Just after a selective sweep all polymorphisms are low in frequency, which means that π will be much lower than expected, while statistics based on counts of segregating sites (like θ_W) will be much closer to their expected values. If π is less than θ_W then $\delta (= \pi - \theta_W)$ will be less than 0, and consequently Tajima's D will be negative (**FIGURE 8.7A**). Similarly, as polymorphism levels rebound after a sweep a large fraction of low-frequency variation will be singletons, meaning that ξ_1 and η_1 will approach their expected values faster than π and even somewhat faster than θ_W, which is based on all of the segregating sites in a sample (see **FIGURE 8.8**). Consequently, Fu and Li's D, F, D^*, and F^* will all also be negative after a selective sweep has occurred.

Balancing selection leads to an excess of intermediate-frequency neutral variation surrounding a selected site (Figure 8.3B). When there is an excess of intermediate-frequency polymorphisms, π will be greater than θ_W, ξ_1, and η_1. Again, this is because π is calculated using allele frequencies and is the sum of site heterozygosities—since heterozygosity is maximized by intermediate-frequency polymorphisms, π can be higher than expected. Summary statistics based on counts of polymorphisms are not greatly affected by their higher frequencies. If π is greater than θ_W then $\delta (= \pi - \theta_W)$ will be greater than 0, and consequently Tajima's D will be positive. Similarly, both π and θ_W will be greater than either ξ_1 or η_1 when there is an excess of higher-frequency variants. Consequently, Fu and Li's D, F, D^*, and F^* will all also be positive under balancing selection.

It is worth asking why the various estimators of θ tell us about the allele frequency spectrum. A simple answer is that *for π to be equal to θ_W, the frequency spectrum must match the neutral expectation*. That is, both of these statistics (and all of the others mentioned here) are only estimators of θ when all variation is at mutation-drift equilibrium. While this is a good proximate clarification for when the statistics might differ, it does not provide any enlightenment about why they differ. An insightful explanation comes from Achaz (2009): all summary statistics that are also estimators of θ are weighted combinations of the number of polymorphisms at each frequency in the allele frequency spectrum. Recall that in a sample of n chromosomes, polymorphisms are found at frequencies from $1/n$ to $n - 1/n$ (for the unfolded spectrum; similar arguments

(A)

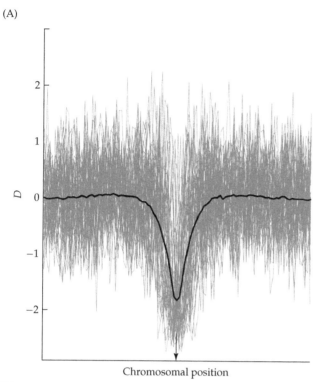

Chromosomal position

FIGURE 8.7   The effect of a selective sweep on summaries of the allele frequency spectrum. For a positively selected variant that has just fixed in the center of the region (marked by an arrow), the values of (A) Tajima's D and (B) Fay and Wu's H are shown. Recombination occurs across the region, so the frequency spectrum is unaffected at positions far away from the selected site. The black line represents the mean value of each statistic across 2,000 simulated datasets, and individual results from 50 simulations are shown in gray.

(B)

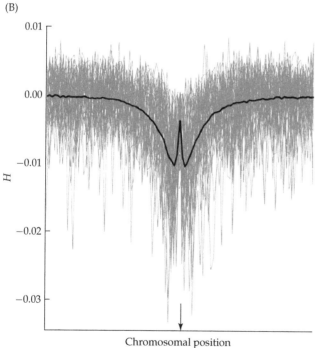

Chromosomal position

(A)

(B)

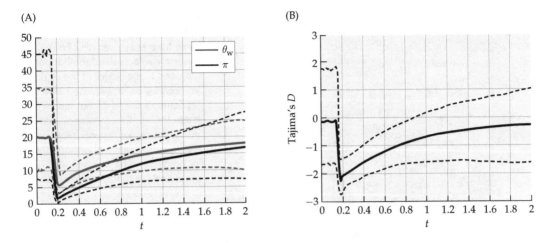

FIGURE 8.8 The effects of a sweep on measures of polymorphism. In both panels the simulated advantageous allele arises at $t = 0$ and takes approximately $0.6N_e$ generations to fix (where t in the graph is measured in units of $2N_e$ generations). (A) The means and 95% confidence intervals of θ_W and π over time (where $\theta = 20$ prior to the sweep). (B) The mean and 95% confidence interval of Tajima's D over the same time period. (From Simonsen, Churchill, and Aquadro 1995.)

apply to the folded spectrum). The allele frequency spectrum is simply a count of the number of polymorphisms at each of these frequencies, and it can be represented as a vector of the counts. Achaz (2009) showed that our summary statistics can all be computed by multiplying these counts by a set of weights (also a vector) specific to each statistic (see fig. 1 in Achaz 2009). The simplest example of this relationship can be understood by considering ξ_1, in which all of the weight in the statistic is placed on the segregating sites present on one chromosome; none of the other frequency bins contribute to ξ_1. While some statistics are heavily weighted toward low-frequency polymorphisms (e.g., θ_W), some are relatively uniformly weighted (e.g., π), and some are weighted toward high-frequency polymorphisms (e.g., θ_H). Combining statistics with different weightings into a test (i.e., a δ-value) makes them sensitive to distortions in the allele frequency spectrum. Considering the combinations of weights used in each test provides a conceptual basis to understand why tests can differ in their sensitivities toward different types of selection.

Finally, it is important to stress that any interpretation of tests of the frequency spectrum strongly rely on assumptions about the individual neutrality of the polymorphisms in the sample. In the above explanations I have described the expected effects of linked selection on neutral variants—none of the polymorphisms were themselves directly affected by selection. Problems would arise, for instance, if a sample contained multiple slightly deleterious polymorphisms, leading to a skewed frequency spectrum (see Figure 7.6). In fact, Tajima (1989a) originally applied his test to exactly this kind of data, concluding that a negative D-value was evidence for weakly deleterious variation.

To ensure that we are detecting linked selection Tajima's D and similar tests are often run using only synonymous or noncoding polymorphisms, although in genomic datasets the fraction of all variants in a window under direct selection can be quite low. My colleagues and I previously showed how to explicitly test for slightly deleterious variation at a locus by comparing D-values calculated separately for nonsynonymous and synonymous polymorphisms (Hahn, Rausher, and Cunningham 2002). This *heterogeneity test* can be used for any situation in which different selective processes may be differentially affecting subsets of the data, and it can specifically test for the effects of linked selection even in the presence of non-neutral variation.

Detecting selection using the allele frequency spectrum: Fay and Wu's H

Another commonly employed test of neutrality also uses deviations from the allele frequency spectrum but is most sensitive to detecting the shoulders of selective sweeps (i.e., the regions flanking the sweep). Fay and Wu (2000) defined their test statistic as:

$$H = \pi - \theta_{\mathrm{H}}$$

(8.10)

which is similar to the other tests of the frequency spectrum we have seen, but without a denominator. The H test is best at detecting an excess of high-frequency derived polymorphisms: recall that θ_{H} is defined using the unfolded spectrum and heavily weights high-frequency polymorphisms (Equation 3.10). This test is not especially sensitive to the excess of intermediate-frequency polymorphisms associated with balancing selection.

At a non-recombining locus surrounding a recently fixed advantageous mutation we do not expect to see an excess of high-frequency derived mutations. Any neutral derived mutations that hitchhiked along with the advantageous one have also been fixed, so we should not generally expect Fay and Wu's H to be significant. However, at a neutral locus flanking the selected locus there can be an excess of high-frequency derived alleles, as long as there has been some recombination between the loci. As was discussed earlier in the chapter, when neutral derived alleles that are initially associated with the advantageous mutation become "unhitched" during the passage to fixation, they can be left at high frequency. Recombination during the sweep therefore produces genealogies that contain an excess of both high- and low-frequency polymorphisms (Figure 8.6). Because so few high-frequency polymorphisms are expected in the neutral allele frequency spectrum, Fay and Wu's H is better able to detect this deviation. When a sweep occurs, therefore, H will be negative (**FIGURE 8.7B**).

Importantly, H will be lowest in the shoulders of a completed sweep, not directly at the site of selection (Figure 8.7B). Care must therefore be taken when identifying the locus that is the direct target of selection. Much of the interpretation of H will depend on exactly on how large the windows being analyzed are relative to the recombination rate. If the recombination rate is low and the windows are small enough, one may be able to detect a pattern of paired windows with very negative H-values on either side of a window with low levels of polymorphism but $H = 0$. If the recombination rate is high or the windows are much larger, the target of selection may reside within a single significant window with negative H-values. The resolution of any analysis is dependent on many parameters outside the control of individual researchers, and so interpreting results always requires some careful thought.

In the next section we consider the statistical power to detect selection using tests of the allele frequency spectrum. Before moving on, though, I would like to preview a topic to which we will return after this: partial or incomplete sweeps. Before a sweep has completed—that is, before the advantageous allele has fixed—the sweeping haplotype will rise in frequency, possibly bringing with it multiple linked neutral variants. *In cases like this, there can be an excess of intermediate- or high-frequency derived neutral alleles directly around the target of selection.* Now the task of the researcher becomes even harder: Are we looking at the shoulders of a completed sweep or the center of a partial sweep? Genealogically, the signature of a partial sweep (Figure 8.6A) can be quite similar to the shoulder of a completed sweep (Figure 8.6B) and can also resemble a soft sweep (Schrider et al. 2015). Although these possibilities are difficult to disentangle, later in this chapter and again in Chapter 10 we will consider statistical tools that can help us to determine the exact forms of selection acting in nature. Methods that explicitly consider the physical arrangement of low-frequency and high-frequency derived alleles will help to detect and pinpoint targets of selection.

The power to detect natural selection using tests of the frequency spectrum

The advantage of tests looking at patterns of linked neutral polymorphism—whether at the total amount of variation or at the frequency spectrum of this variation—is that they make it possible to detect natural selection on even a single selected mutation. These tests do not depend on the repeated fixation of multiple advantageous changes, or on the maintenance of more than two balanced alleles. The fixation of a single advantageous mutation affects linked neutral variation in a predictable way, enabling us to determine the mode of selection. But how strong does selection have to be to leave behind detectable signals? And how long after a sweep (or balancing selection) begins can we detect these signals? These questions have been addressed multiple times in the literature, using many different statistical tests (e.g., Simonsen, Churchill, and Aquadro 1995; Fu 1997; Przeworski 2002), and here I give only a brief overview of the most important conclusions from these studies.

In regions of normal recombination, the power to detect balancing selection decreases as balanced polymorphisms become very old. This decline occurs because recombination will wear away signals at linked sites (e.g., Figure 8.4), making it harder to detect older balancing selection (Hudson and Kaplan 1988; Wiuf et al. 2004; Gao, Przeworski, and Sella 2015). With increasing age and recombination, only the nucleotide site containing the balanced polymorphism will have the selected genealogy. Under such conditions it is not possible to detect patterns of balancing selection from linked neutral polymorphism. Earlier in the chapter we discussed the difficulties in determining whether balancing selection is acting on newly arising balanced polymorphisms. Now we can see that we are limited both at the early and at the later stages of balancing selection, possibly restricting the number of cases detected in nature.

As an advantageous allele rises in frequency it sweeps away linked neutral polymorphism, and after it fixes levels of polymorphism immediately begin to return to their expected values. The time window in which selection can be detected is therefore limited: too early during the sweep and neither the levels nor frequencies of linked variation will be perturbed; too late after the sweep has ended and both levels and frequencies of variants will have returned to normal. As with balancing selection, our ability to detect selective sweeps is thus also bounded.

We can see the effects of selection most clearly by tracking the values of π and θ_W after an advantageous mutation appears (Figure 8.8A). For scenarios involving relatively weak selection, in which the advantageous allele takes approximately $0.6N_e$ generations to fix, there is essentially no signal of selection for approximately the first $0.3N_e$ generations (Figure 8.8A; remember that time in the figure is measured in units of $2N_e$ generations). Neither π nor θ_W are initially reduced because the selected allele is still at low frequency. As it rises in frequency it begins to depress levels of linked polymorphism and to distort the frequency spectrum, leading to deviations from the neutral expectations (Figure 8.8B). Just prior to the fixation of the advantageous allele and its associated haplotype, Tajima's D takes on its minimum value. This may seem surprising given the earlier discussion of the processes that generate the excess of low-frequency polymorphisms, as very few new alleles will have arisen. Instead, what is producing this signal is the rapid extinction of derived alleles found on non-advantageous backgrounds. As the last non-advantageous haplotypes are driven from the population they are found at low frequencies and can carry with them a relatively large number of derived alleles (see, e.g., Figure 8.1B). These low-frequency variants are what cause Tajima's D to be so low, and they are easily distinguishable from post-fixation patterns because all of the low-frequency alleles are on one or a few haplotypes. Good empirical examples of this pattern can be found at the *TRPV6* gene in humans (Stajich and Hahn 2005) and at the *janus-ocnus* locus in *Drosophila simulans* (Parsch, Meiklejohn, and Hartl 2001).

The power to detect selection is determined by the magnitude of reduction in our test statistics and the length of time that they stay reduced. For the example shown in Figure 8.8B, Tajima's D is reduced well below the levels expected under neutrality, meaning that there is power to detect selection, at least at the peak of this signal. As π rebounds to its expected value (or at least returns within the bounds of its expected value) the difference between π and θ_W disappears, and therefore the power to detect deviations in the frequency spectrum disappears. Although we will not examine explicit power calculations for the HKA test, note that its power should approximately follow the value of θ_W alone. When θ_W is severely reduced the HKA test will have power to detect selective sweeps, but as θ_W returns to values that are consistent with a neutral locus the test will no longer be significant (in Figure 8.8A this loss of significance coincides with a point approximately $0.8N_e$ generations after the beginning of the sweep).

The power to detect selection differs among the different tests we have considered. **FIGURE 8.9A** shows the power of several tests of the allele frequency spectrum to detect selection. Here the power of the tests is expressed as the fraction of simulated datasets with test statistic values outside 95% of values simulated under the neutral-equilibrium null hypothesis. We can see that the greatest power in the tests coincides with the time when the test statistics are lowest (compare Figure 8.9A with 8.8B): there is close to a 90% rejection rate when Tajima's D reaches its minimum, though power quickly declines with time. By N_e generations after the sweep has started there is essentially no statistical power to reject neutrality—this point defines the end of the window in which we can detect selection. If selection is stronger there may be more power to reject neutrality, and power may peak sooner after the sweep begins. However, the signature of selection still will not usually extend more than N_e generations after the start of the sweep (Simonsen, Churchill, and Aquadro 1995). Among the tests shown here, Tajima's D has consistently higher power than either Fu and Li's D^* or their F^* (Figure 8.9A). Regardless of which test is used, there is at least as much variation in power explained by sample size. **FIGURE 8.9B** shows that we can increase our power by increasing the number of chromosomes sampled. Though there is not a lot of power gained by increasing sample sizes from $n = 50$ to $n = 100$, we cannot reach significance for more than 50% of the cases with selection when $n = 20$ or less. The increased power comes from the increased resolution in the allele frequency spectrum afforded by larger sample sizes: we are simply better able to distinguish deviations between distributions with an increased number of frequency bins into which polymorphisms can be placed.

Our ability to detect sweeps is also determined by the distance between our sampled loci and the location of the selected site. To a first approximation, the

(A)

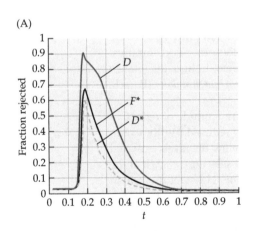

(B)

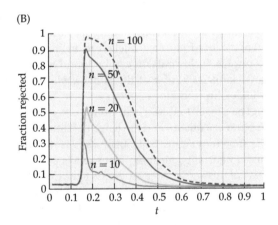

FIGURE 8.9 The power to detect selective sweeps. In both panels the simulated sweeps are the same ones shown in Figure 8.8. (A) The power of different test statistics to reject neutrality over the period of time since the sweep began (where t is measured in $2N_e$ generations). (B) The power of Tajima's D to reject neutrality given different numbers of sampled chromosomes. (From Simonsen, Churchill, and Aquadro 1995.)

effect of distance is much like the effect of time: just as power begins to decline after a sweep ends, so it also declines as we move away from the selected site (e.g., Pennings and Hermisson 2006b). This reduction in power is easily understood by considering the effects of recombination on our ability to detect deviations in the allele frequency spectrum. As recombination unhitches linked backgrounds from the selected allele, more sampled chromosomes will come from non-selected haplotypes. When there are enough non-selected haplotypes in our sample we will no longer see any deviation in the frequency spectrum. Move far enough away on a recombining chromosome and there will be no signal of selection at all (Figure 8.7). For similar reasons, Fay and Wu's H test will show a steep decline in power moving away from the selected site, as the excess of high-frequency alleles generated by selection is present only in an exquisitely fine window flanking a sweep (e.g., Przeworski 2002).

DETECTING SELECTION USING PATTERNS OF LINKAGE DISEQUILIBRIUM

One advantage to all of the methods discussed thus far is that haplotypic phase is not needed for any of them. As long as you have information on the number and frequency of variants observed in a sample—and sometimes information on sequences from closely related species—no additional phase information is needed. While full sequence data are usually required, single diploid individuals with unresolved haplotypes will work just as well as fully phased data for methods based on levels of diversity or the allele frequency spectrum. Pooled sequencing data is also amenable to those methods.

However, natural selection does have an effect on the associations between neutral alleles linked to a selected site. Therefore, ignoring this information means that we could be losing power to detect selection and, more importantly, to distinguish between modes of selection. One of the earliest tests of neutrality (the Ewens-Watterson test; Ewens 1972; Watterson 1977, 1978) used haplotype information, though this was due more to a technical limitation than to any push to increase power: allozyme technologies did not permit the resolution of nucleotide sequences, so protein haplotypes ("alleles," in their terminology) were the unit of variation being studied. The Ewens-Watterson test is still one of the most powerful methods for detecting certain kinds of selection (Zeng, Shi, and Wu 2007), and so it and other tests that use haplotypic phase remain relevant today. Here we describe several of the most popular such tests, starting with a discussion of how various modes of selection affect both the frequency distribution of haplotypes and general patterns of linkage disequilibrium.

The effects of balancing selection on linkage disequilibrium

To understand the effects of selection on haplotypic diversity, recall that there are a number of ways of summarizing this variation (Chapter 3): we can count the number of unique haplotypes in a sample (K) or the expected haplotype homozygosity, F:

$$F = \sum_{j=1}^{K} p_j^2 \qquad (8.11)$$

where p_j is the frequency of the jth haplotype at a locus (out of K unique haplotypes; see also Equation 3.21). From this we can calculate the expected heterozygosity, $H \, (= 1 - F)$, as well. (I apologize for the two Fs and two Hs in this chapter—it cannot be helped.) We can also introduce two new measures: M, defined as the frequency of the most common haplotype, and C, a vector of haplotype frequencies, analogous to the allele frequency spectrum for individual segregating sites but calculated using haplotypes. The expected frequencies of all unique haplotypes in C under the standard neutral model were given by Ewens (1972) in his eponymous *Ewens sampling formula*. With no recombination, K is at most $S + 1$. With free recombination there can be up to 2^S unique haplotypes. In a population conforming to the standard neutral model, the number of unique haplotypes we expect to find should be positively correlated with S (and therefore with n). Likewise, the haplotype homozygosity should be negatively correlated with K (Watterson 1977). Recombination will increase the number of haplotypes, leading to higher K and lower F.

Balancing selection can perturb these expectations, though perhaps not too much. A balanced polymorphism at a single site with two alleles leads to an extreme genealogy with two long internal branches (Figure 8.3B). While there can be many polymorphisms on these branches, each additional polymorphism after the first to appear on either internal branch does not create an additional unique haplotype—it just further differentiates the two allelic classes. Therefore, the value of $K \, | \, S$ (the number of unique haplotypes given the number of polymorphic sites) will be lower than expected under a neutral genealogy. Likewise, we expect that the haplotype frequency spectrum, C, could be perturbed relative to the neutral expectation, with a deficit of low-frequency haplotypes.

Linkage disequilibrium is maximized when polymorphisms arise on the same branch of a genealogy (Chapter 4). Therefore, we expect that the average LD at a locus—measured by statistics such as Z_{nS} (Equation 4.5)—will increase under balancing selection as many variants arise on the two long internal branches. Of course this prediction, like the ones above regarding haplotype frequencies, is based on a situation with no recombination. When there is recombination, balancing selection has the effect of generating even more haplotypic diversity than expected, as the ancient age of balanced alleles means that they are associated with many more recombination events. Indeed, DeGiorgio, Lohmueller, and Nielsen (2014) proposed that the site of a balanced polymorphism may resemble a recombination hotspot, with much less pairwise linkage disequilibrium across this locus than otherwise expected. As a consequence, the effects of balancing selection on patterns of haplotype diversity and linkage disequilibrium are expected to be mixed, or at least strongly dependent on the age of balanced polymorphisms and rates of recombination.

The effects of completed sweeps on linkage disequilibrium

Positive selection can have very strong effects on haplotypic diversity and linkage disequilibrium, especially when compared to balancing selection (negative selection will have negligible effects on haplotypes, beyond limiting the number and frequency of segregating polymorphisms). First, we consider a recently completed sweep. In this case there are two distinct effects of positive selection. In the non-recombined region directly surrounding the advantageous mutation, most variation has been removed or is made up of polymorphisms that arose during the sweep. Therefore, as we saw above when considering tests based on the allele frequency spectrum, there will be a star-like genealogy with many low-frequency variants (Figure 8.1C). In this case there is very little haplotype structure and almost no linkage disequilibrium. However, there is a deviation from the expected patterns of haplotype diversity, as there can be more haplotypes than expected: the value of $K \mid S$ is higher than expected (though K may be lower than the equilibrium expectation). This is because each low-frequency polymorphism is creating a new haplotype. Similarly, haplotype homozygosity will be lower than expected because few sampled chromosomes share the same states at polymorphic sites.

The effects of completed selective sweeps on haplotype diversity will be stronger in the regions flanking the fixed advantageous mutation (Figure 8.6). These regions are driven up in frequency by selection but do not reach fixation because of recombination off the selected background. The rapid increase in frequency of a single haplotype strongly affects the haplotype diversity in such regions and increases linkage disequilibrium within them (Przeworski 2002; Kim and Nielsen 2004; McVean 2007). The rapid rise in frequency means that there is a single haplotype that is more common than expected under neutrality. As a consequence, the number of haplotypes (K) will be lower than expected, the expected homozygosity of haplotypes (F) will be higher, the haplotype frequency spectrum (C) will be skewed, and, most importantly, the frequency of the most common haplotype (M) will be much higher than expected. Linkage disequilibrium will be increased on either side of the sweep, but there will not be higher LD across the selected site (i.e., between polymorphisms on either side of the fixed allele; Kim and Nielsen 2004). LD is higher in the shoulders of the sweep because many polymorphisms can arise on the long branch connecting the neutral background with the one initially associated with the advantageous allele (Figure 8.6B); these polymorphisms will be in strong linkage disequilibrium with each other.

The effects of partial sweeps on linkage disequilibrium

There can be many reasons we may have sampled an advantageous allele while it is still polymorphic. It may be part of an ongoing sweep, with the advantageous allele destined for fixation. In this case we happen to have taken a sample of chromosomes in the short window of time when such alleles are passing through the population. Alternatively, the beneficial allele may have risen rapidly in frequency as a result of the advantage it gave in an environment that recently changed. That is, it may no longer be advantageous, but the effect of its rapid rise may still be detectable in the population. Similarly, multiple forms of balancing selection (such as negative frequency-dependent selection) will drive low-frequency alleles up in frequency until they become too frequent, at which point they will lose their advantage. Because we cannot be sure of the reasons why we are sampling a polymorphic sweep in a population, I prefer the more agnostic term *partial sweep* to the oft-used *ongoing sweep* or *incomplete sweep*.

If we are lucky enough to catch a sweeping allele while it is still polymorphic, we can expect to find the strongest effects on haplotype diversity and linkage disequilibrium. Like the shoulders of completed sweeps, partial sweeps drive a single haplotype up in frequency (Figure 8.1B). This means that again the number of haplotypes will be lower, the expected homozygosity of haplotypes will be higher, the haplotype frequency spectrum will be skewed, and the frequency of the most common haplotype will be higher than expected given the number of polymorphic sites in the sample.

As an alternative way to think about the effects of partial sweeps, note that the frequency of an allele is indicative of its age. As briefly discussed in Chapter 6, in general new (derived) alleles are low in frequency, while older alleles are higher in frequency (Watterson and Guess 1977; Griffiths and Tavaré 1998). This relationship is important because the age of an allele is proportional to the amount of neutral variation associated with it: we expect allelic classes defined by new alleles (i.e., the set of haplotypes carrying this allele) to have less variation within them than allelic classes defined by older alleles. Therefore, if we observe an allelic class with much less variation than expected given the number of times it occurs in our sample, this lower level of variation may be due to selection driving the allelic class up in frequency. For instance, the allelic class defined by the star in Figure 8.1B is at high frequency (many lineages carry the "star" allele), but the total length of the genealogy associated with it is very small because of positive selection.

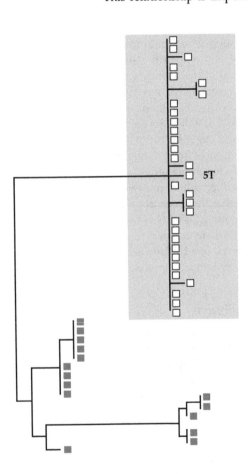

While this may seem like just another way to describe the effects of partial sweeps on the summaries K, F, C, and M, it also suggests alternative ways to detect the patterns of such sweeps. Instead of counting unique haplotypes, we can ask about the amount of variation associated with allelic classes. The variation associated with these classes can also take on multiple forms, further expanding our repertoire. The most obvious form of variation associated with an allelic class is nucleotide polymorphism. When selection drives an allele up in frequency rapidly there will be few nucleotide polymorphisms that arise within this class—that is, there will be fewer mutations occurring in the sub-tree defined by the advantageous mutation. Therefore, we may find one single haplotype at higher frequency than expected or one allelic class with a small number of associated polymorphisms (e.g., **FIGURE 8.10**). This pattern will be the basis of a test introduced in the next section.

FIGURE 8.10 Relationships among haplotypes at the *MMP3* locus in humans. Haplotypes carrying the "5T" functional allele are boxed; the remaining haplotypes carry the "6T" allele. (From Rockman et al. 2004.)

Recombination is another form of variation that can be associated with allelic classes. As with nucleotide mutations, the number of recombination events in a genealogy is proportional to the total length of the genealogy (Equation 6.12). Therefore, as an advantageous allele rises in frequency, there should be fewer recombination events associated with it than expected—that is, there should be fewer recombination events occurring in the sub-tree defined by the advantageous mutation. The lack of such recombination events will be manifested as long segments without recombination. There are multiple ways to measure the length of non-recombined segments, but one simple and widely used method is to consider the length of a region flanking the selected site in which all polymorphic sites are homozygous (**FIGURE 8.11**). This statistic is usually calculated using haplotype frequencies, such that we are still measuring the *expected* homozygosity at each position based on sample frequencies (i.e., *F*). When there is strong selection driving a haplotype up in frequency we can maintain homozygosity over very long distances. Note, however, that in Figure 8.11 tracts of *observed* homozygosity in each individual are plotted.

An important experimental advantage to some methods that measure the amount of recombination is that they do not require full sequencing of individuals—genotype data can be used. While genotype data would obviously not suit our purposes if we were trying to measure the amount of nucleotide variation associated with individual haplotypes, when measuring recombination we have no such ascertainment problem. Our ability to detect

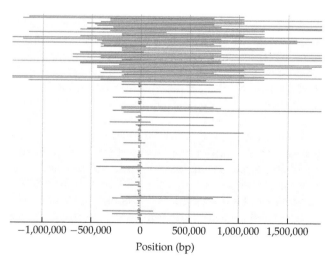

FIGURE 8.11 Haplotype homozygosity associated with different alleles at the *LCT* locus in humans. Position 0 distinguishes individuals homozygous for the lactase persistence allele (red) from individuals homozygous for the non-persistence allele (blue). The length of tracts homozygous at polymorphic sites flanking the functional alleles are shown. (From Tishkoff et al. 2007.)

recombination events is unaffected (or affected only a little) by pre-ascertained polymorphisms. This means that recombination-based tests for partial sweeps can be used with datasets that are not amenable to other analyses.

Detecting selection using linkage disequilibrium: Haplotype summary statistics

In describing the effects of selection on haplotypic diversity and linkage disequilibrium we have largely covered the general approaches one would use to detect selection. However, there are a number of tests that should be explained explicitly, as well as best practices for several others.

Given a number of unique haplotypes, K, a number of segregating sites, S, and a sample size, n, Depaulis and Veuille (1998) defined a *haplotype number test*. As with almost all of the tests introduced here, exact results for the expected value of K are often not possible, and when they are it is only with the assumption that there is no recombination. Therefore, coalescent simulations with the appropriate amount of recombination must be used to generate confidence intervals around K. Similar tests by Strobeck (1987) and Fu (1997) use π instead of S, with the latter calculating the probability of K alleles using the Ewens sampling formula (Ewens 1972). Andolfatto, Wall, and Kreitman (1999) showed how these methods could be used while controlling for the multiple implicit tests that are often done when choosing regions to analyze. Confidence intervals for all methods are found by simulation.

Tests that use haplotype homozygosity, F, or haplotype heterozygosity, H, can be conditional on either K or S. That is, we can test for unusual values of either $F|K$ or $F|S$. The former test—haplotype homozygosity conditional on the number of unique haplotypes—is called the *Ewens-Watterson test* (Ewens 1972; Watterson 1977, 1978). For no apparent reason $F|S$ is not generally used: instead, we use $H|S$, with heterozygosity instead of homozygosity (the *haplotype diversity test*; Depaulis and Veuille 1998). Related to these tests are ones based on M and C. The test $M|K$—the frequency of the most common haplotype conditional on the number of unique haplotypes—was first proposed in Ewens (1973), where he also proposed tests based on the number of haplotypes that only appeared once in a sample. Slatkin (1994a, 1996) proposed a test using $C|K$, while Innan et al. (2005) introduced a test based on $C|S$. Again, in all of these tests we should use simulations in order to generate confidence intervals, though it is important to note that standard fixed-S methods for carrying out coalescent simulations (Chapter 6) may not be appropriate (Markovtsova, Marjoram, and Tavaré 2001; Wall and Hudson 2001).

A test introduced in Hudson et al. (1994) is used instead of $M|S$, though $M|S$ can be considered a special case of the test. This test is often referred to as *Hudson's haplotype test* (*HHT*). Hudson's haplotype test again asks whether there is a high-frequency allelic class in the sample, and whether it is associated with lower than expected levels of nucleotide variation. For example, in the example using the *MMP3* regulatory region shown in Figure 8.10, the locus contains 22 identical haplotypes among 46 sampled chromosomes (and $S = 35$ segregating sites). The HHT asks whether this is an unusual pattern using coalescent simulations: after generating many neutral genealogies with

46 chromosomes, we "throw" 35 mutations on each genealogy and ask how many times we observe 22 or more identical haplotypes. In this case we observed only 95 out of 10,000 simulated datasets that had 22 or more identical haplotyes (so $P = 0.0095$; Rockman et al. 2004). Though this example explicitly uses the number of identical haplotypes (M), we could also use a focal polymorphism of interest to define allelic classes. In *MMP3* our polymorphism of interest defines a class of 31 haplotypes carrying the "5T" allele and 15 haplotypes carrying the "6T" allele. The 31 5T haplotypes only have $S = 7$ polymorphisms among them, while the 15 6T haplotypes contain the other 28 polymorphisms. Hudson's haplotype test can be used to ask the probability of obtaining a partition as extreme or more extreme than observing 31 haplotypes with 7 segregating sites. If there is no *a priori* focal polymorphism, one can also partition the data to find the maximum number of haplotypes with the minimum number of segregating sites between them (Parsch, Meiklejohn, and Hartl 2001; Meiklejohn et al. 2004). This approach is unbiased and can be used to scan genomes for regions with unusual haplotype structure, whether at the center of a partial sweep or in the shoulder of a completed sweep.

Detecting selection using linkage disequilibrium: Measures of pairwise LD

As was discussed earlier, positive selection can greatly increase values of pairwise linkage disequilibrium, but this effect varies both along chromosomes and with time since a sweep began. For completed sweeps, LD will be high in flanking regions that have partially escaped fixation (Figure 8.6B) but will not be elevated directly around the selected site. Partial sweeps will be accompanied by increased LD surrounding the selected site.

To measure these increases in LD, one simple approach is to use a summary statistic like Z_{nS} that averages pairwise LD across all pairs of segregating sites in the sample. One simply needs to calculate Z_{nS} in windows along a chromosome to see locally elevated (or possibly depleted) values surrounding selected variants. As with many of the haplotype-based summaries discussed here, the choice of window size in which to calculate this statistic matters a lot: too large and the lack of LD around the selected site (for a completed sweep) will be missed; too small and larger patterns may be missed.

As an alternative way to detect the signature of LD around a completed sweep, Kim and Nielsen (2004) proposed a statistic that looks explicitly for a spatial pattern of high LD followed by low LD followed by high LD. Their statistic, ω, is defined as:

$$\omega = \frac{\left(\binom{l}{2} + \binom{S-l}{2}\right)^{-1} \left(\sum_{i,j\in L} r_{ij}^2 + \sum_{i,j\in R} r_{ij}^2\right)}{(1/l(S-l)) \sum_{i\in L, j\in R} r_{ij}^2} \tag{8.12}$$

where we have divided our S segregating sites arranged linearly along a chromosome into two groups, one to the left (L) and one to the right (R) of the ith polymorphism (the left-hand group includes position i). For whichever value of i we choose, we calculate LD (using r^2; Equation 4.4) between all the sites on the left, and separately between all the sites on the right (the two sums in the numerator). We allow i to go from 2 to $S - 2$ so that there are at least two variable sites in each group. In the denominator of this equation we calculate LD pairwise between pairs of sites in the left and right groups, but not within groups. The statistic in Equation 8.12 is also sometimes calculated without singleton polymorphisms (Jensen et al. 2007).

The statistic ω is designed so that it is maximized when there is high LD within the two groups (i.e., in both shoulders of the sweep) and low LD between groups. In this way it matches the expectations for patterns of linkage disequilibrium surrounding a completed sweep if we imagine that the selected site was just to the right of the ith polymorphism. In practice we do not know where the selected site is, so we choose the value of i that maximizes this statistic; we refer to the resulting value as ω_{max} (Kim and Nielsen 2004). Using the spatial pattern of LD expected after a sweep can be more powerful than overall summaries of LD (like Z_{nS}), but now we are assuming that the selected site is located physically within the range of polymorphisms considered. If the selected site lies outside the variants used to calculate ω, we do not expect this statistic to have very much power to detect selection—fortunately, this means that it can actually be used to help localize the targets of selection (see Chapter 10). Fast and computationally efficient software exists to calculate ω centered on every position in a genome (Alachiotis, Stamatakis, and Pavlidis 2012).

Detecting selection using linkage disequilibrium: Haplotype homozygosity

In order to find regions where partial sweeps have taken place we can also look for derived alleles associated with large blocks of very little recombination. There are many methods to measure regions with little recombination (e.g., Slatkin and Bertorelle 2001; Hanchard et al 2006; Wang et al. 2006), but the most common way is to assess the expected haplotype homozygosity associated with new derived alleles. There are also many different ways to use measures of haplotype homozygosity to construct a test (e.g., Sabeti et al. 2002; Toomajian et al. 2003; Voight et al. 2006; Ferrer-Admetlla et al. 2014). Here we describe several of the most commonly used methods, starting with the *extended haplotype homozygosity* (*EHH*) test of Sabeti et al. (2002).

All of the approaches we describe start with a focal or "core" polymorphism with two alleles, one ancestral and one derived (in their original paper Sabeti et al. [2002] used a core haplotype, but the test is now generally applied to single core sites). We divide the haplotypes in a region into those carrying the ancestral allele and those carrying the derived allele at this variable site. If there are n total chromosomes sampled, there are n_A ancestral and n_D derived haplotypes ($n_A + n_D = n$). Among all the haplotypes carrying the derived allele at the core in a region, we can then calculate the expected haplotype homozygosity, F (Equation 8.11); we can also do the same for the chromosomes carrying the ancestral allele. The calculation of F for each haplotype class is equivalent to calculating EHH for each, with some additional wrinkles described below. As a test statistic we could compare the relative EHH (REHH) of the derived class to the ancestral class (i.e., F_D/F_A), with possible values going from 0 to infinity. We would expect selection on the derived allele to increase the value of this statistic (**FIGURE 8.12**).

One issue with the way I have described the test thus far is how to choose a window size over which to calculate F. It is obvious that F starts at 1 (when only the focal allele is used) and must decay when one uses larger window sizes, as it becomes more and more likely that recombination events have occurred. How do we know how large a window to use as a test statistic? Sabeti et al. (2002) calculated EHH and REHH at multiple distances away from the core polymorphism of interest, determining P-values via simulation for each distance used. This approach results in a collection of P-values (one for each window size), which may still be difficult to interpret. As an alternative, Voight et al. (2006) proposed a method for integrating EHH over different window sizes to get an *integrated haplotype homozygosity (iHH)*. This single statistic is easily associated with each core polymorphism tested.

In order to calculate iHH we now must keep track of the polymorphisms to either side of our core polymorphism, as well as their genetic distances apart (iHH requires a fine-scale recombination map of the organism of interest). We refer to the position of our focal polymorphism as x_0, the position of the closest polymorphism as x_1, and so on; we repeat all calculations moving to the left and right (i.e., both upstream and downstream) of the core site, so x_1 can

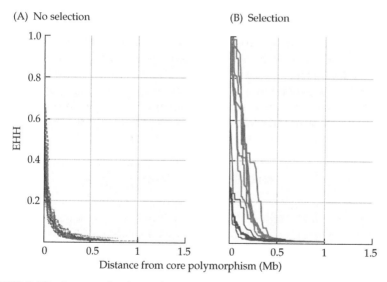

FIGURE 8.12 Decay in haplotype homozygosity (EHH) away from a core polymorphism. In both panels, each line represents results from a different simulated dataset. (A) EHH for backgrounds associated with derived and ancestral alleles (red and blue lines) when there is no selection. (B) EHH for derived and ancestral alleles when the derived allele is advantageous and at frequency 0.5. (From Voight et al. 2006.)

refer to the closest polymorphism in either direction. We start by calculating EHH for the haplotypes carrying the derived allele out to polymorphism x_i as:

$$\mathrm{EHH}_D\left(x_i\right) = \sum_{j=1}^{K_D} p_j^{\,2} \tag{8.13}$$

which is exactly the same calculation as that used for F in Equation 8.11, with K_D defined as the number of unique haplotypes associated with the derived allele at the focal location. The homozygosity associated with the ancestral allele out to polymorphism x_i, EHH_A, is calculated in the same manner, using the unique haplotypes associated with the ancestral allele at the focal site (K_A).

We can now calculate the integrated haplotype homozygosity by integrating over EHH values at different values of x_i using the trapezoidal rule. Again, we calculate iHH separately for haplotypes associated with the derived and ancestral alleles at the core site (borrowing notation from Szpiech and Hernandez 2014):

$$\begin{aligned}
\mathrm{iHH}_D = {} & \sum_{i=1}^{L} \frac{1}{2}\left[\mathrm{EHH}_D\left(x_{i-1}\right) + \mathrm{EHH}_D\left(x_i\right)\right] * g\left(x_{i-1}, x_i\right) \\
& + \sum_{i=1}^{R} \frac{1}{2}\left[\mathrm{EHH}_D\left(x_{i-1}\right) + \mathrm{EHH}_D\left(x_i\right)\right] * g\left(x_{i-1}, x_i\right)
\end{aligned} \tag{8.14}$$

Here we calculate the area under the EHH_D curve going out both to the left (L) and right (R) of the core site. In each direction we take the average value of EHH for adjacent markers and multiply this by the genetic distance between them, $g(x_{i-1}, x_i)$. A question arises as to how far upstream and downstream ("left" and "right") to calculate EHH values—generally these calculations are carried out until $\mathrm{EHH}_D(x_i) < 0.05$.

Like EHH_D values on their own, iHH_D values without comparison to iHH_A are relatively uninformative. After calculating iHH_A (again using Equation 8.14), we can construct an analog of REHH using iHH values. This is referred to as the *unstandardized integrated haplotype score (iHS)*:

$$\text{unstandardized iHS} = \ln\left(\frac{\mathrm{iHH}_D}{\mathrm{iHH}_A}\right) \tag{8.15}$$

As in Szpiech and Hernandez (2014), we modify this definition relative to the one given in Voight et al. (2006) so that it is a ratio of derived to ancestral iHH, in the same way that REHH is a ratio of derived to ancestral EHH. When there is positive selection on the derived allele, unstandardized iHS becomes larger and positive.

Although the unstandardized iHS values in combination with simulations could be used to generate P-values, it has become common practice to instead report "standardized" iHS values. These values are standardized simply by comparing the unstandardized scores at our focal site to unstandardized scores for polymorphisms at similar frequencies across the genome. We need to use only those scores with similar frequencies because unstandardized iHS can vary with derived allele frequency. We calculate this value as:

$$\text{iHS} = \frac{\ln\left(\dfrac{\text{iHH}_D}{\text{iHH}_A}\right) - E_p\left[\ln\left(\dfrac{\text{iHH}_D}{\text{iHH}_A}\right)\right]}{SD_p\left[\ln\left(\dfrac{\text{iHH}_D}{\text{iHH}_A}\right)\right]} \tag{8.16}$$

where the expectation ($E_p[\bullet]$) and standard deviation ($SD_p[\bullet]$) of unstandardized iHS are estimated from the empirical distribution calculated around polymorphisms at the same derived allele frequency, p, as the core site under consideration. In other words, iHS expects that you have calculated unstandardized iHS for every polymorphism in the genome (often excluding those with a minor allele frequency under 5%), and that you have estimated the mean and standard deviation among all such polymorphisms at every frequency, or in small frequency bins. As a result, iHS should be approximately normally distributed, with a mean at 0. This statistic does not have P-values attached to it but instead can be used to identify extreme values across the genome.

Let us take a step back to ask: Why are we doing all of these calculations with haplotype homozygosities? Are these tests any better than the tests outlined earlier in the chapter? In fact, EHH and iHH are quite similar in conception and practice to Hudson's haplotype test, and they come with the same caveats (they also appear to be quite similar in power; Zeng, Shi, and Wu 2007). In all methods we split our data into two sets based on a polymorphism of interest, asking about the amount of variation associated with each set. In Hudson's haplotype test we consider the nucleotide mutation events associated with each focal allele, expecting that selection will lower the resulting nucleotide variation in the selected set of haplotypes because of the rapid coalescence among lineages. Similarly, in the EHH and iHH tests we are indirectly considering the recombination events associated with each focal allele, measured as the length of homozygous haplotypes.

The big advantage to the EHH and iHH tests is that they, unlike HHT, can be used with genotyped samples—there is no need for full sequences. The ascertainment bias inherent in genotype data (Chapter 2) means that such data cannot be used to ask about levels of nucleotide variation. However, there is no such ascertainment with respect to recombination events, and therefore genotype data can be used in tests that are looking for signatures of reduced recombination. The EHH/iHH tests *can* be used with full sequence data, and their power and robustness can be improved slightly by using genomic distances instead of recombination distances (Ferrer-Admetlla et al. 2014). But in these cases it may be easier and more powerful simply to use HHT or other similar tests because they will always provide greater power to detect mutation events relative to recombination events, so tests based on nucleotide variation should be more sensitive. We can also make comparisons of

EHH/iHH values between populations at the same positions (referred to as either *Rsb* [Tang, Thornton, and Stoneking 2007] or *XP-EHH* [Sabeti et al. 2007]), something that is not possible using the *P*-values from HHT. Such comparisons can tell us about population-specific sweeps using patterns of haplotype homozygosity.

Interpreting tests based on haplotype structure and linkage disequilibrium

We have already introduced many of the caveats and interpretations of the tests mentioned here as we have covered them. But there is one point regarding the effect of recombination worth discussing a little further.

Because the amount of recombination at a locus has a large effect on the haplotypes that are observed, it is important to ensure that simulations use the amount of recombination appropriate to the locus being studied. Although most site-based tests (like Tajima's *D*) are more conservative with low recombination, haplotype-based tests have a more complicated relationship with recombination (Wall 1999; Zeng, Shi, and Wu 2007). This is largely because we may be testing for either an excess or a deficit of haplotypes. Low recombination always means fewer haplotypes, so if we are testing for a relative excess of haplotypes (such as occurs after a completed sweep), simulations with low recombination will be anticonservative. Conversely, more recombination always means more haplotypes, so tests looking for a deficit of haplotypes (such as with a partial sweep) will also be anticonservative if we simulate with high recombination. It is therefore important to accurately estimate recombination at each locus being tested, and to use locus-specific recombination rates in simulations. Simply using an average recombination rate when there is variation across loci can result in false positives.

Recombination rate variation is largely controlled for using the REHH or iHS tests (but not completely; see Ferrer-Admetlla et al. 2014). This is because we are always comparing homozygosity between haplotypes associated with the ancestral and derived alleles at a single locus. We expect that regions with high (or low) recombination will have high (or low) recombination within both the ancestral- and derived-carrying chromosomes, such that among-locus variation does not affect our results. Even when this assumption does not hold—such as when there is a site affecting recombination rate that is polymorphic in a population—the effects on haplotype structure and linkage disequilibrium are minimal (Hellenthal, Pritchard, and Stephens 2006).

Finally, it should be mentioned that the power to detect completed sweeps using haplotype structure or LD declines quite rapidly, even by comparison to frequency spectrum–based tests. Whereas we have some power to detect selection up to about $0.4N_e$ generations after the sweep has ended using Tajima's *D* (Figure 8.9A), using the number of unique haplotypes or Z_{nS} will give us power only until about $0.05N_e$ generations since fixation of the advantageous allele (Pennings and Hermisson 2006b). Despite this greatly reduced power, when used in conjunction with other tests these statistics are often informative about the precise targets or mode of selection.

CAVEATS TO TESTS OF LINKED SELECTION

As was mentioned earlier in the chapter, all tests of linked selection come with a very important caveat: our null expectations are based on neutral-equilibrium populations (sometimes referred to as the *standard neutral model*). This means that both selection and nonequilibrium demographic histories can cause deviations from these expectations, making it very difficult to tease the two apart (we come back to this task in Chapter 10). Nonequilibrium demography can shift both the mean value of our test statistics and the variance expected among loci. As a result, we cannot simply re-center the mean of our data but must also simulate data under nonequilibrium histories to properly capture the expected variance in order to avoid false positives. But because selection can mimic many aspects of nonequilibrium population histories (see Chapter 9), this procedure also leads to an increase in false negatives (Hahn 2008).

It is worth considering why we did not focus on the assumptions of neutral-equilibrium populations in the discussion of tests of selection in earlier chapters. One reason is that many of these tests compare two classes of sites that are assumed to be affected equally by demography. For instance, the McDonald-Kreitman test is explicitly a comparison between nonsynonymous and synonymous changes within the same gene. Similarly, it is often the case that F_{ST} is used to robustly infer the action of selection by comparing values in different classes of polymorphisms (e.g., Barreiro et al. 2008). This does not mean that these tests are unaffected by demography, but they are certainly less affected.

Tests using the amount of variation, the allele frequency spectrum, and even linkage disequilibrium use comparisons against theoretical expectations and can therefore be especially sensitive not only to nonequilibrium histories but also to sampling schemes. If we do not sample from a single population, we can often be misled. The problems with sampling extend to tests using levels of variation, as loci with large between-population levels of differentiation can be incorrectly interpreted as evidence for balancing selection. Similarly, sampling from multiple populations can lead to skews in the frequency spectrum if alleles differ in frequency across populations. Even when we separately consider the frequency spectrum for nonsynonymous and synonymous polymorphisms, our sampling scheme can affect our conclusions about selection. Consider a gene at which different amino acids are fixed because of local adaptation in multiple different populations—if we were to sample only one individual per sample we would incorrectly infer an excess of low-frequency nonsynonymous alleles, consistent with weakly deleterious variation (similar incorrect inferences would be made using the MK test as well). We have also already seen (in Chapter 5) that polymorphisms that are in linkage equilibrium within populations can appear to be in linkage *disequilibrium* when sampling across multiple populations. Again, this may lead us to infer the wrong types of selection when we interpret our patterns of LD in terms of our expectations within single populations.

Overall then, all methods for detecting linked selection must be accompanied by a careful consideration of the demographic history of the populations being sampled. In addition, the exact sampling design of experiments that set out to detect linked selection must be thought out in advance, or population structure needs to be accounted for before tests are conducted. In the next chapter we discuss the effects of nonequilibrium demographic histories on patterns of variation, returning in Chapter 10 to a synthesis of the effects of selection and demography on molecular variation.

DEMOGRAPHIC HISTORY

9

Genetic variation data are widely used to infer the demographic history of populations and species. As populations grow, contract, split, and come back together, the variants carried across individuals record these changes. Many methods have been developed to infer demographic events from sequence data and to estimate their magnitude. There are three main demographic parameters that researchers attempt to estimate: effective population size (N_e), migration rate (m), and the time of population divergence (t). Multiple elaborations of each of these parameters is possible—for instance, the value of N_e as it changes through time and when these changes occurred—but these three parameters form the building blocks of almost all more complex models. Often we can only estimate these parameters jointly (e.g., $N_e\mu$, $N_e m$), and their values can depend on one another (e.g., t can be expressed in units of N_e generations). This means that obtaining the numerical value of any single parameter is dependent on external estimates of others. Most often we use external estimates of the mutation rate, as there are experimental methods for doing so; we also use estimates of the generation time to convert time from units of N_e to units of years.

Any discussion of the methods used to study population demography must begin by clarifying what exactly it is that we are able to estimate. The most important thing to consider is that patterns of variation are determined by the joint effects of population history and natural selection. This means that our estimates also include the joint effects of both of these processes. The effects of selection are apparent across the vast majority of almost all genomes (e.g., Corbett-Detig, Hartl, and Sackton 2015), and it is therefore very difficult to obtain estimates solely influenced by demographic history (but see Voight et al. 2005 and Gazave et al. 2014 for attempts to do exactly this). The problem is compounded by the fact that most methods assume there is no effect of selection, as violations of this assumption can result in the inference of demographic events (such as migration) when none has occurred (e.g., Mathew and Jensen 2015; Roux et al. 2016).

Demographic inference is a very important part of modern molecular population genetics. Most of the methods described in this section have been used many times to inform our knowledge of population histories. The caveats listed above should not detract from this use but should give us pause in interpreting every parameter value produced by our analyses. Though the big-picture outlines of inferred population histories may be correct, very few studies are able to verify the precise parameter estimates with known histories (e.g., McCoy, Garud, et al. 2014). Newer methods are being developed that can co-estimate demography and selection (Chapter 10), but more needs to be done to determine how robust our inferences are to violations of the most common models. For simplicity, however, throughout this chapter I will refer to methods that aim to infer population histories as *demographic*, recognizing that in real data many forces may affect the results.

THE DEMOGRAPHIC HISTORY OF SINGLE POPULATIONS

The simplest inferences of demographic history use only a single population. In a single population there is one evolutionary parameter to be estimated—N_e—and the only demographic events that can occur are changes in this effective population size. As mentioned above, however, the value of N_e can change through time, and so we may want to estimate different values of N_e at multiple time points in the past. Below we build to an understanding of demographic inference for single populations by discussing the histories that are possible, how variation is affected by nonequilibrium histories, and the different ways genealogical measures can be used to infer specific histories. We move on to estimating parameters when multiple populations are considered in the next part of the chapter.

Nonequilibrium demographic histories

Our null model in demographic inference is usually a Wright-Fisher population at equilibrium (**FIGURE 9.1A**). Equilibrium means that the population has had a constant size of N_e (or $2N_e$) for such an extended period of time that the entire history of our sample has only experienced this size. There is no linked selection, and neutral mutations have arisen at a constant rate so that the allele frequency spectrum matches our expectations under mutation-drift equilibrium.

Given this equilibrium history, there are only two nonequilibrium events that can occur when considering a single population: N_e can go up (looking forward in time) or N_e can go down. We refer to these events as population expansion and population contraction, respectively (**FIGURE 9.1B,C**). All further embellishments to create more complex histories in a single population are combinations of, or can be reduced to, expansions and contractions. The one exception in discussions of more complex models is that we often talk about population "bottlenecks" as a unique type of history, even though they are just a combination of a contraction followed by an expansion (**FIGURE 9.1D**). But the bottleneck model occurs often enough in population genetics that it is helpful to consider on its own.

A simple model of either expansion or contraction has three parameters (Figure 9.1B,C): the ancestral effective population size (denoted N_A), the time before the present at which the population size instantaneously changed

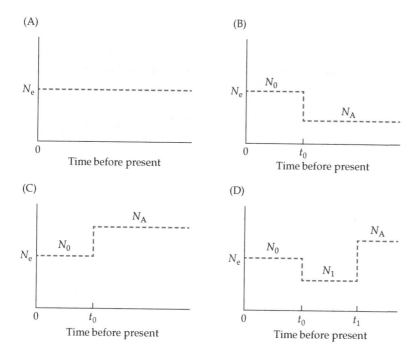

FIGURE 9.1 Demographic histories of single populations. In all panels the current day is all the way to the left on the x-axis. (A) Equilibrium history. (B) Population expansion. (C) Population contraction. (D) Population bottleneck.

(t_0; measured in $2N_0$ generations), and the current effective population size (N_0). Although we usually consider population size changes to be instantaneous, this assumption is clearly not realistic—it is used solely for computational and mathematical convenience. A more realistic size-change model that is commonly used has N_e growing or shrinking exponentially. Exponential models also have three parameters: N_0, t_0, and the exponential growth rate, r (where r can be negative to specify population contraction). Here t_0 represents the time that the population begins to grow or shrink exponentially. We do not need to specify the ancestral effective population size because it is given by N_0/e^{rt_0}. Exponential size changes may still not be the most realistic demographic model, but again they are relatively easy to work with. One rarely sees models in which population size changes linearly with time.

A standard bottleneck model has five parameters (Figure 9.1D): N_A, N_1, N_0, t_1, and t_0. Essentially these are a combination of the initial contraction and subsequent expansion parameters. The added parameters represent the size of the population during the bottleneck (N_1), the time before the present that the population contracted (t_1), and the time before the present that it expanded (t_0). It is also common to see these events represented by the "strength" of the contraction during the bottleneck ($= N_1/N_A$) and the length of the bottleneck ($= t_1 - t_0$), with no change in the number of parameters that need to be specified. There is no requirement that the population be exactly the same size post-bottleneck as pre-bottleneck—the ancestral size can be either above or below the current size—though for simplicity they are often considered to be equal.

The term *bottleneck* is an ambiguous one in the literature. When explicit inference of population histories is being carried out there is no ambiguity, and all of the different bottleneck parameters must be specified. But *bottleneck* is often used more loosely as a synonym for either expansion *or* contraction. That is, authors will use the term to mean more expansion-like histories or more contraction-like histories. Because bottlenecks are a combination of the two, individual outcomes of population bottlenecks *can* resemble either more of an expansion history or more of a contraction history, depending on which of these two events has had the dominant effect on patterns of variation. While this lax language should not be condoned, the conflation of terms does point to the difficulty we have in distinguishing among similar demographic histories.

The effect of changing population size on gene genealogies

Population expansions, contractions, and bottlenecks all affect patterns of nucleotide variation. These effects can best be understood by first considering the genealogy at a locus (or set of loci) that has experienced such events.

Because nonequilibrium histories in a single population are represented only by changes in N_e, we must consider what changing the value of N_e does to the underlying genealogy. Recall from Chapter 6 that the rate of coalescence is determined by the population size. With a constant-sized diploid population the probability that two lineages will coalesce in a single generation is $1/2N_e$, and therefore the average waiting time until the next coalescent event in a genealogy is directly proportional to N_e (Equation 6.2). With changing population size the same expectations hold, but now the rate of coalescence will be proportional to the existing population size within any time window. In periods when population sizes are large the rate of coalescence will be relatively low, while in periods when population sizes are small the rate of coalescence will be relatively high. Given continuously changing population sizes—as in exponentially growing or shrinking populations—we also have expectations for the rate of coalescence at any time point (Slatkin and Hudson 1991). Alternatively, we can think of the coalescent process for nonequilibrium populations as following the typical process for a population of constant size, noting that times in the tree are proportional to N_e generations. Changes in population size are then represented by the stretching or compressing of branch lengths in the intervals corresponding to larger or smaller N_e, respectively. No matter how we view this process, we expect to see fewer coalescent events in periods experiencing large population sizes and more coalescent events in periods experiencing small population sizes.

Examples of how changes in N_e can affect genealogies are shown in **FIGURE 9.2**. Relative to the genealogy from an equilibrium population with the same current effective population size, genealogies from expanding populations have short internal branches and long external branches (Figure 9.2B). This is because the rate of coalescence in the most recent epoch is slower as a result of population growth, while the rate in the previous epoch is faster as

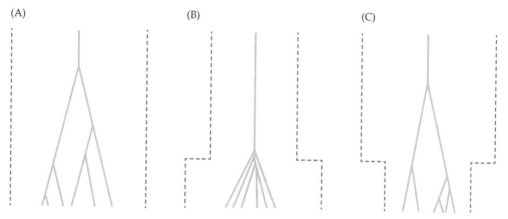

FIGURE 9.2   Representative genealogies from different population histories. In all panels the size of the population is denoted by the distance between the dashed lines. (A) Genealogy from an equilibrium population. (B) Genealogy from a population undergoing an expansion. (C) Genealogy from a population undergoing a contraction.

a result of the smaller population size. In exponentially growing populations the rate of coalescence increases exponentially going *back* in time. One further effect of the increased rate of coalescence going back in time is that genealogies at different loci will all tend to have their most recent common ancestor in a small window of time when the population started to expand (Slatkin and Hudson 1991). As a result, the variance in tree height among loci (and therefore the variance in total tree length) will be lower in expanding populations than in equilibrium populations. With greater population growth the genealogies will become even more star-like, and the variance in tree height among loci will become even smaller.

For populations undergoing a contraction, the opposite effect is seen: longer internal branches and shorter external branches (Figure 9.2C). In this case the rate of coalescence in the most recent epoch is higher because of the small population size, while the rate in the previous epoch is slower as a result of the larger population size. Here, the severity of the contraction (N_0/N_A) and the time since the contraction started (t_0) are also crucial to the expected tree shape. If the contraction is too severe or lasts too long, all lineages will coalesce in the most recent epoch. Consequently, severe contractions may be more difficult to detect than moderate ones (Voight et al. 2005). In order to observe a genealogy like the one shown in Figure 9.2C, two or more lineages must "escape" the contraction, instead coalescing in the prior epoch. There are no specific values of these parameters that will always lead to trees with longer internal branches, as the probability that any two lineages will escape the contraction without coalescing is just a function of the joint parameter N_0/t_0. As a result of this stochasticity in tree shape, the variance in tree height among loci (and therefore the variance in total tree length) will be *higher* in contracting populations than in equilibrium populations. That is, some loci will resemble the genealogy in Figure 9.2C, while others will resemble very short neutral genealogies, completely coalescing in the most recent epoch.

Genealogies from bottlenecked populations may resemble those from either an expanding or a contracting population (or both), depending on the timing of the constituent contraction and expansion. If the initial contraction was relatively recent or relatively severe, then the genealogical history at most loci will be dominated by this event. In contrast, if the contraction was relatively further in the past and not as severe, but the expansion was rapid and large, then genealogies will be dominated by the expansion event. For any arbitrary history combining contractions and expansions, the shape of genealogies from such populations will reflect the changing rate of coalescence at many different time points. Whether or not any single demographic event dominates, the shape of the tree should be an accurate guide to the history of changes in N_e.

The effect of changing population size on the allele frequency spectrum

Now that we have gained an intuition for the effect of changes in N_e on the shape of genealogies, we would like to apply this insight to understanding these effects on patterns of sequence variation. Dating back to the beginning of molecular population genetics, researchers have attempted to examine the effects of nonequilibrium histories on the average heterozygosity and number of alleles in allozyme data (e.g., Nei, Maruyama, and Chakraborty 1975; Maruyama and Fuerst 1985). There are many different ways to summarize patterns of sequence variation, and many of the different statistics discussed in this book will be affected by nonequilibrium histories. DNA sequence data clearly allow us to go beyond the very coarse inferences available with allozymes, and here I begin with a discussion of the behavior of the allele frequency spectrum. In the next section I will cover some other commonly used statistics in demographic inference, as well as several approaches to measuring sequence variation that have not been covered before now.

The effects of nonequilibrium histories on the allele frequency spectrum should be apparent from the shape of the genealogies shown in Figure 9.2. With longer external branches come more singleton polymorphisms, and therefore a skew in the allele frequency spectrum toward low-frequency alleles. Population expansions generate star-like trees with relatively long external branches, resulting in exactly this sort of skew (Figure 9.2B). Therefore, we expect summaries of the allele frequency spectrum such as Tajima's D or Fu and Li's D (Chapter 8) to be negative under population expansion (Tajima 1989b). Conversely, with longer internal branches come more intermediate-frequency polymorphisms and a commensurate skew in the allele frequency spectrum. Population contractions can generate exactly these sorts of genealogies (Figure 9.2C). Again, we expect summaries of the allele frequency spectrum to reflect this skew, resulting in positive values of Tajima's D and similar statistics (Tajima 1989b). As an alternative to this genealogical view, we can also try to understand the effects of nonequilibrium demography by taking a forward-in-time view. In an expanding population the addition of new chromosomes means that more mutations can be generated every generation. While these mutations will eventually result in higher overall levels of variation, the new polymorphisms will all start out at low frequency, leading to a skewed frequency spectrum. In a contracting population many chromosomes are lost, taking rare polymorphisms with them. These rare alleles will be lost disproportionately, leading to an excess of intermediate-frequency polymorphisms and a skew in the frequency spectrum.

The effects of a bottleneck on the allele frequency spectrum can vary. As stated above, the genealogical consequences of a bottleneck can be either more expansion-like or more contraction-like, and this behavior is reflected in the allele frequency spectrum. If the expansion event dominates, then the frequency spectrum will be skewed toward low-frequency alleles. If the contraction event dominates, the spectrum will be skewed toward intermediate-frequency alleles. These dynamics are demonstrated in **FIGURE 9.3**, where a brief but very strong bottleneck starts at varying times in the past, returning to the same size post-contraction. When the contraction portion of the bottleneck occurs longer ago, Tajima's D is negative, reflecting the fact that it is the subsequent expansion that has the major effect on the allele frequency spectrum. As the contraction becomes more recent it begins to dominate patterns of variation, such that very recent bottlenecks have a positive Tajima's D (Figure 9.3). There is also a point in the past at which there is no overall skew in the frequency spectrum, and Tajima's $D = 0$. Here the effects of the contraction and expansion have exactly canceled each other out, cautioning against the use of only a single summary statistic for demographic inference. However, note that in Figure 9.3 only the mean value of Tajima's D is reported. A bottleneck will also generally increase the variance in D among loci (Voight et al. 2005; Stajich and Hahn 2005), suggesting that there is more information contained even in this simple summary statistic.

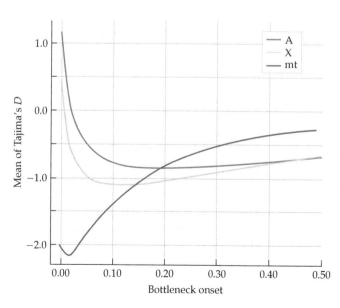

FIGURE 9.3 Frequency spectrum after a bottleneck. The mean value of Tajima's D across 20,000 simulated datasets that have a bottleneck with a 95% reduction in N_e for $0.05N_e$ generations. The figure shows the effects of the bottleneck for loci on the autosomes (A), X chromosome (X), and mitochondrial genome (mt). Time on the x-axis is measured in units of $2N_e$ generations. (From Gattepaille, Jakobsson, and Blum 2013.)

As an added complication, Figure 9.3 shows the effects of the same bottleneck history on loci located on the autosomes, the X chromosome, and the mitochondrial genome. The main difference among these loci is that the effective population size of each is $2N_e$, $1.5N_e$, and $0.5N_e$, respectively. But this seemingly small difference in population size can result in very different genealogical outcomes: for the exact same population history, loci found in the same individuals can give either very positive or very negative Tajima's D values (Fay and Wu 1999). This phenomenon occurs because the smaller N_e of mtDNA means that the rate of coalescence during the contraction phase of the bottleneck is even higher, while the higher mutation rate of this genome means that the recovery of variation can be faster. As a result, all genealogies find their most recent common ancestor toward the beginning of the contraction, with no lineages escaping into the older epoch with larger population sizes. Variation is quickly reintroduced because of the higher mutation rates, and therefore the history of mtDNA is dominated by the subsequent expansion phase of the bottleneck. In general, nonequilibrium population histories can affect genealogies at loci in distinct genomic compartments (i.e., autosomes, X, mtDNA) quite differently, and so we should expect to see differences in patterns of variation (e.g., Fay and Wu 1999; Hey and Harris 1999; Wall, Andolfatto, and Przeworski 2002; Pool and Nielsen 2008).

The effect of changing population size on other measures of variation

The allele frequency spectrum is just one summary of variation data that enables us to make inferences about changing population size, and it is in fact limited in its resolution (Myers, Fefferman, and Patterson 2008; Terhorst and Song 2015). Fortunately, there are many other measures of variation that capture additional information about the shape of genealogies from nonequilibrium histories, including the genealogies themselves.

Patterns of linkage disequilibrium are affected by changes in N_e. Recall that a genealogical interpretation of LD considers the branches of a genealogy on which mutations have arisen (Chapter 6). Higher levels of LD—as measured by statistics such as r^2—occur when mutations have occurred on the same branch of the tree, and this branch is an internal branch. If mutations arise on different branches of the genealogy, or arise on the same external branch, LD will be lower. Given these considerations, it should be clear that the average level of LD will be positively correlated with the average length of internal branches of a genealogy. Histories of population expansion are characterized by extremely short internal branches relative to external branches (Figure 9.2B). Therefore, expansion events are expected to generate lower levels of LD (Slatkin 1994b). We can also view this pattern in light of the average allele frequency, which in an expansion is lower than expected. The large number of new mutations in expanding populations will not be consistently associated with one another, and therefore LD will not be as high as in the equilibrium case. In contrast, histories of population contraction (or bottlenecks dominated by the contraction phase) are characterized by long internal branches relative to external branches (Figure 9.2C). Therefore, contraction events can generate higher levels of LD (e.g., Tishkoff et al. 1996; Kruglyak 1999; Andolfatto and Przeworski 2000; Reich et al. 2001; Wall, Andolfatto, and Przeworski 2002).

Summaries of haplotype variation (such as K, F, or M; Chapter 8) will also be affected by nonequilibrium histories, and in fact much of the theory used when studying allozymes (i.e., the infinite alleles model) can be applied to these data. Because haplotype diversity reflects the joint patterns contained in the allele frequency spectrum and LD, it may carry additional information not seen when analyzing either one alone. However, there are relatively few studies that use only these haplotype summary statistics for demographic inference (e.g., Lohmueller, Bustamante, and Clark 2009). One exception is the so-called *mismatch distribution* (Rogers and Harpending 1992). As its name implies, the mismatch distribution is an accounting of pairwise differences between all haplotypes at a locus—these are the k_{ij} values in Equation 3.3. Assuming there is no recombination, these pairwise differences will be approximately geometrically distributed for an equilibrium population: most pairs of haplotypes will be expected to have very few nucleotide differences between them, with fewer and fewer pairs having more and more differences (Slatkin and Hudson 1991). Under population expansion, the underlying tree will be starlike, and therefore all haplotypes will be approximately equally related (e.g., Figure 9.2B). Consequently, there will be a unimodal mismatch distribution at such loci (Slatkin and Hudson 1991; Rogers and Harpending 1992). This genealogical feature of population expansions allows researchers to estimate the onset of the expansion phase (given an estimated mutation rate), as long as the data come from a non-recombining locus. Because of the strong assumptions concerning the lack of recombination, the use of the mismatch distribution has largely been confined to analyses of mtDNA and Y chromosomes. In addition, the mismatch distribution seems to be used in the literature solely to study population expansions—not contractions. The reason for this is likely the nonidentifiability of contractions relative to equilibrium populations. Although in the limit the mismatch distribution should be geometric for populations at equilibrium, for any single tree the observed distribution is often multimodal, reflecting the longer internal branches separating sub-genealogies containing more closely related haplotypes (e.g., Figure 9.2A; Slatkin and Hudson 1991; Rogers and Harpending 1992). The resulting mismatch distribution will be highly similar to the one expected under a contraction.

The discussion thus far should make it clear that a major goal of demographic inference in a single population is to infer the coalescence times among sampled chromosomes. If we can estimate the coalescence times, then we can estimate the rate of coalescence during any period of time; this in turn represents an estimate of N_e. The main tool researchers have used to obtain these coalescence times is drawn from the sample genealogies themselves. By *genealogy*, here we mean a topology and an associated set of branch lengths representing time, such that all of the coalescence times are known (**FIGURE 9.4A**). In order to estimate such a tree we require that there have been no recombination events and enough nucleotide substitutions to estimate all branch lengths sufficiently. When these conditions are met, the average time to go

(A)

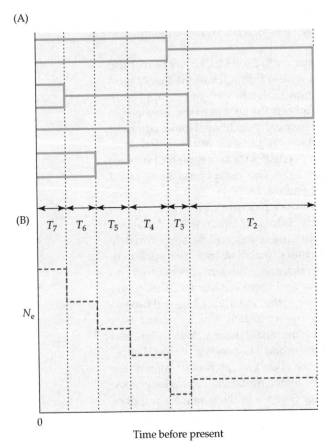

(B)

T_7 | T_6 | T_5 | T_4 | T_3 | T_2

N_e

0

Time before present

FIGURE 9.4 Estimating N_e from coalescence times. (A) A representative genealogy with branch lengths representing time. The vertical dashed lines demarcate periods with i lineages in them, and the times between dashed lines are denoted by the T_is. (B) For each interval we can use the coalescent time (T_i) to estimate N_e, using Equation 6.2. (After Ho and Shapiro 2011.)

from $i \rightarrow i-1$ lineages (for all i) can be used as a maximum likelihood estimate of θ (Felsenstein 1992; Fu and Li 1993b). If the mutation rate, μ, is known, these coalescence times (the T_is in Figure 9.4) can therefore be used to estimate N_e. The more interesting implication of this general approach is that *each* time between coalescent events is informative about N_e during that time window (**FIGURE 9.4B**). In an equilibrium population the rate of coalescence in each interval should be the same, with the time until the next coalescent event simply proportional to $1/[i(i-1)]$ (Equation 6.2). However, as N_e changes so does the rate of coalescence and, consequently, the time between events. Approaches using coalescent time intervals estimated from non-recombining loci can therefore be used to infer broad nonequilibrium histories (Nee et al. 1995) or specific estimates of N_e through time (Pybus, Rambaut, and Harvey 2000; and see below).

The presence of recombination poses major constraints on methods that use full genealogies. When there is recombination the inferred tree is a mixture of multiple (correlated) genealogies and may not accurately reflect any single marginal tree. The average tree will tend to be more star-like because the average coalescence time between all pairs of lineages will be the same (Schierup and Hein 2000). As with the mismatch distribution, this means that such methods have been limited to analyses of completely non-recombining loci such as mtDNA and Y chromosomes. In order to deal with recombination we could estimate genealogies from blocks of autosomal sequences where no recombination is detected, and many methods take this approach (e.g., Gattepaille, Günther, and Jakobsson 2016). However, even when there are no detectable recombination events (i.e., when the four-gamete test is not violated), recombination can still have an effect on the reconstructed tree (Hudson and Kaplan 1985). In addition, such methods generally assume that there is free recombination between all non-recombining blocks, which is also unlikely to be the case with even a moderate number of loci.

Multiple approaches estimate coalescent times in the presence of recombination by modeling the correlation between tree topologies at neighboring loci. Some of the most common methods take advantage of the sequentially Markovian coalescent (SMC; McVean and Cardin 2005)—or the updated version of this model, SMC´ (Marjoram and Wall 2006)—to infer population sizes through time. Recall that this model describes how genealogies change as you move along a chromosome: when there is a recombination event the tree may change slightly from the one preceding it (Chapter 6). Using the SMC for inference therefore allows you to gain information from both coalescence times and recombination, as well as to account for correlations among trees. The simplest SMC-based methods use coalescent times between only a pair of chromosomes (usually in a single diploid individual) but include information from tens of thousands of loci or more. This approach provides rich information about the rate of coalescence at many periods in the past. A description of these methods can be found below.

Finally, the demographic history of a population can leave a signal in the length of sequences identical between individuals. The two main approaches leveraging such information consider either the length of sequences shared without an intervening recombination event (denoted *identical-by-descent*, or *IBD*) or the length of sequences shared without an intervening mutation (denoted *identical-by-state*, or *IBS*; also sometimes referred to as *runs of homozygosity*). Before considering how these measures are used, it is important to point out that the terms *identical-by-descent* and *identical-by-state* can have multiple meanings. At a single nucleotide position, we often wish to specify whether sequences share a common allele because it was inherited from their common ancestor (IBD) or because convergent mutations to the same allelic state occurred on multiple lineages (IBS). This usage of these terms is most often found in a phylogenetic context, in which multiple mutations at a site are a concern. *Identical-by-descent* is also used in fixed pedigrees to denote loci that are shared between individuals because they are inherited from the same reference individual. Here we are applying the terms to samples of (relatively) unrelated individuals in a single population.

The effects of population history on the length of sequences that are either IBD or IBS are the same, and a major distinction between the two is whether the relevant model is applied to genotype data (IBD) or full sequence data (IBS). For simplicity I will first explain the underlying idea for IBS at a nonrecombining locus because only one process (mutation) needs to be considered (**FIGURE 9.5**). Assuming that mutations are Poisson-distributed along the genome, the length of sequences that are IBS between two chromosomes is expected to be exponentially distributed with mean $1/(2\mu T)$, where T is coalescence time. Because the two sequences are identical except for the $2\mu T$ mutations that have accumulated between them, the length of these identical tracts (the IBS tracts) contains information about coalescence times. These lengths are inversely correlated with T: very recent coalescences will leave less time for mutation and therefore larger average IBS tracts, and vice versa. Similarly, if we consider only recombination, the length of IBD tracts is informative about coalescence times, with more recombination resulting in smaller

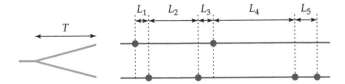

FIGURE 9.5 The length of tracts that are identical-by-state. For a given time to coalescence between two lineages (T, shown at left) $2\mu T$ mutations will accumulate between them (derived mutations shown as circles in the alignment at right). The lengths of tracts that are identical-by-state between the two chromosomes are denoted by the L_is, which are on average shorter given larger values of T. The full distribution of these lengths is also informative about changes in N_e over time.

average tracts. From the full distribution of tract lengths we can also infer changes in N_e over time. Detailed population histories can be gleaned from the distribution of IBD tracts (e.g., Palamara et al. 2012), but they are limited to very short periods of the recent past (Chapman and Thompson 2002). IBS tracts offer the ability to examine changes in N_e across much longer periods of time, but both mutation and recombination have to be taken into account. Fortunately, recent work has been able to derive the expected IBS tract length distribution under arbitrary population histories (Harris and Nielsen 2013).

Types of data and computational methods used in demographic inference

Understanding how nonequilibrium population histories affect genetic variation helps us to understand how to infer these histories from data. For every measure of sequence variation there are at least a few (and sometimes many more than a few) methods for turning these data into estimates of demographic history. Many early methods focused on rejecting equilibrium scenarios using summary statistics (e.g., Fu 1997; Ramos-Onsins and Rozas 2002), but the resulting answers were often limited simply to whether there had been a recent expansion or contraction or some other deviation from the standard neutral model. More recent work has focused explicitly on parameter estimation, with the values of N_e at various points in time as parameters. Of course the simplest inference comes from setting one of the estimators of θ (Chapter 3) equal to $4N_e\mu$ and solving for N_e using an external estimate of the mutation rate. If selection has had no effect on levels of variation, then the value of N_e from such calculations roughly represents the harmonic mean of population sizes across the recent past (see Crow and Kimura 1970, pg. 110).

More complex inferences of changes in N_e over time are possible using many different approaches that use sample sizes ranging from $n = 2$ to thousands of chromosomes, and sequence data organized in multiple different ways. Here I will focus on three data types that are commonly used, whether in analyses involving a single population or those involving multiple populations. Even when the underlying sequences are collected in the same manner, the data can be presented and analyzed in different ways using different assumptions. These three data types generally represent the different ways that linkage is treated in the downstream analyses, ranging from completely ignoring relationships among sites to fully taking into account these correlations.

The first data type considers each polymorphic site separately, under the assumption that they are independent from one another. This approach can provide millions of data points from a single dataset containing up to thousands of sequenced individuals, though the only information it captures is allele frequency. Data collected from genotyping platforms are included within this type, recognizing that the ascertainment scheme must be accounted for in order to make accurate inferences. The second data type considers a large number of short sequence blocks, each containing a moderate number of polymorphisms. All demographic reconstruction methods using such data assume free recombination (i.e., independence) between blocks but differ as to whether there is no recombination or some recombination within blocks. Because results can be very sensitive to the specific recombination rates used (Ramírez-Soriano et al. 2008), it is commonly assumed that there is no recombination within blocks. This assumption does not improve inferences but is made for convenience. The datasets to be analyzed can have thousands of sequence blocks and dozens to hundreds of individuals, and they are readily collected using reduced representation approaches such as RAD-seq or transcriptome sequencing (Chapter 2). The third data type considers whole genomes or long contiguous stretches of a genome at once. The correlations induced by linkage between sites or between short windows along a chromosome must be accounted for when analyzing such data. Methods using these data are often the most computationally demanding, limiting the sample sizes that can be used to only a very small number of individuals.

In addition to using diverse types of data, methods for carrying out demographic inference use many different computational and statistical approaches for parameter estimation. Closed-form mathematical expressions to determine the probability of sequence data given a set of parameters are available only for simple histories (Ewens 1972) or for very small samples from more complex histories (Lohse, Harrison, and Barton 2011). Therefore, a wide variety of approaches from computational statistics have been used in order to estimate the detailed demographic histories of populations. Likelihood, Bayesian, and heuristic methods are all employed, often with slight variations. A comprehensive review of these approaches is beyond the scope of this book, but here I attempt to familiarize the reader with at least the basic vocabulary necessary to understand specific methods described in the next section.

The most commonly used computational approaches use likelihood or Bayesian methods (see ch. 8 in Wakeley 2009 for a good overview). It is important to note at the outset that Bayesian methods employ likelihood: the key distinction between the two is that Bayesian methods apply prior probabilities to likelihoods, providing posterior probabilities as output. To a first approximation, Bayesian methods that employ uniform (i.e., uninformative) priors will give results equivalent to those produced by likelihood methods, in that the maximum *a posteriori* estimates will correspond to the maximum likelihood estimates.

Common approaches for calculating likelihoods in population genetics can broadly be divided into two types: those that use likelihood functions (*full likelihood methods*) and those that use simulations to approximate the likelihood function (*approximate likelihood methods*). Analytical functions for the likelihood of sequence data are available for a range of models, often under the assumption of either no recombination or free recombination, and likelihoods can be calculated via numerical methods. A well-known example calculates the likelihood of a tree topology given sequence data (Felsenstein 1981a), and indeed such methods are used in evaluating coalescent genealogies from non-recombining loci (e.g., Kuhner, Yamato, and Felsenstein 1995). Approximate likelihood methods were first described earlier in this book (Chapter 4): by simulating data under a range of parameter values (often using coalescent simulations) we can approximate the likelihood of a model's parameters, allowing us to find the maximum likelihood estimates (MLEs) of our parameters of interest. Approximate methods are therefore quite useful, as they can be used to evaluate complex histories or summary statistics for which we do not have a likelihood function (Weiss and von Haeseler 1998).

The distinction between full likelihood methods and approximate methods is not always clear, with many methods falling somewhere in-between. For instance, even when likelihood functions are given it may be difficult to find MLEs of parameters because of the enormous state space that must be searched. This is especially true for models with multiple parameters, including scenarios in which a large space of tree topologies must be explored. In these cases, we can use methods that efficiently explore parameter space but that do not guarantee maximum likelihood estimates, including *importance sampling* and MCMC methods (reviewed in Wakeley 2009, sec. 8.4). MCMC is widely used in population genetics (e.g., Kuhner, Yamato, and Felsenstein 1995; Hey and Nielsen 2007) and is often required by "full" Bayesian methods because it is almost impossible to find maximum *a posteriori* estimates without such approaches. Another method that is sometimes described as approximate involves *composite likelihood* calculations. Often we have full likelihood functions for one part of the data—for instance, single polymorphic sites or pairs of sites—but not for all the data together. The idea of composite likelihood is that we can combine the likelihoods from each individual data partition by assuming they are independent. This approach may provide the correct MLEs of our parameters, though the confidence associated with each parameter value must be calculated via resampling or the Godambe information matrix (Coffman et al. 2016). Again, composite likelihoods are commonly used in population genetics (e.g., Nielsen 2000; Hudson 2001; Wooding and Rogers 2002; Polanski and Kimmel 2003; Marth et al. 2004). We can even calculate the individual likelihoods of each data partition using simulations (i.e., approximate likelihood) and then combine these likelihoods to get a composite likelihood estimate (e.g., Adams and Hudson 2004; Excoffier et al. 2013).

Finally, approximate Bayesian computation (ABC) methods have more recently found wide use in demographic inference (they are originally due to Tavaré et al. 1997). ABC combines the flexibility of approximate likelihood with the ability to specify prior distributions on parameter values. ABC simulates a large number of datasets, calculating a collection of summary statistics for each one. By sampling parameter values that produce simulated datasets with a close fit—assessed via regression—to the observed data, ABC produces an approximate sample from the posterior distribution (Beaumont, Zhang, and Balding 2002). Because ABC often uses sampling methods, it also offers the ability to estimate values for a large number of parameters.

Inferring nonequilibrium population histories

A large number of software packages are available for demographic inference, implementing an equally large number of methods (see Schraiber and Akey 2015). Rather than a tedious description of all these methods, here I describe the major classes of methods and highlight some of the more commonly used packages that implement them. These methods generally follow the types of data described in the previous section.

The allele frequency spectrum is often used on its own to infer demographic histories. There are multiple ways to find the parameters of a demographic model that best fit this spectrum, but one important similarity between all methods is that they are limited to histories with $n - 1$ epochs, as this would be the number of frequency bins in the data if we used polarized mutations. As we will see, methods that do not depend on the allele frequency spectrum can subdivide histories into many more epochs, each with its own value of N_e (many methods can also infer histories that do not have discrete epochs). Estimates of N_e are often obtained by estimating the rate of coalescence in each specified time period. But the allele frequency spectrum does not have finer resolution than the number of frequency bins in the sample, so it cannot provide estimates with more resolution. Regardless, in practice such methods are used to estimate N_e in many fewer than $n - 1$ time periods. Estimating histories from the frequency spectrum is also almost always done using composite likelihood. This is because there are multiple ways to calculate the most likely demographic history from individual allele frequencies, but not from the allele frequency spectrum as a whole. Assuming all polymorphisms are independent from one another allows for the efficient identification of the maximum likelihood parameter values. The efficiency (and low cost) of using only the allele frequency spectrum also enables the analysis of datasets containing thousands of individuals, which in turn opens up the opportunity to study very rare alleles. Exactly these sorts of studies have revealed the demographic history of even the last few thousand years of human existence (e.g., Coventry et al. 2010; Keinan and Clark 2012; Nelson et al. 2012; Tennessen et al. 2012).

One common class of methods uses diffusion theory to predict the allele frequency spectrum at individual sites. Fully solved for the case in which only mutation and drift are acting by Kimura (1955), diffusion theory gives the probability density of polymorphisms in a population. Williamson and colleagues (2005) were the first to show how numerical solution of diffusion approximation equations could be used for arbitrary demographic models. This general approach is implemented in the popular software package $\partial a \partial i$ ("diffusion approximations for demographic inference"; Gutenkunst et al. 2009). $\partial a \partial i$ can be used on either single populations or multiple populations and can infer constant or exponentially expanding or shrinking population sizes in multiple time periods. While there are other ways to solve the diffusion equations (e.g., Lukić and Hey 2012), *∂a∂i* has become a very widely used package for demographic inference.

Another common class of methods connects the observed allele frequency spectrum to a tree topology. Such methods present an intuitive way to understand the connection between a particular demographic history, the genealogies produced by this history, and the resulting frequency spectrum. Theory originally presented in Fu (1995) and Griffiths and Tavaré (1998) allows us to connect the probability of finding a segregating site at a particular frequency in a sample with the coalescence times in a genealogy. This theory and related computational methods have been extended in multiple ways (e.g., Wooding and Rogers 2002; Polanski and Kimmel 2003; Marth et al. 2004) and are implemented in the *stairway plot* (Liu and Fu 2015) and *fastNeutrino* (Bhaskar, Wang, and Song 2015) software packages.

As discussed earlier, treating polymorphisms from different sites as if they are completely independent does not take linkage into account and therefore does not carry all of the information contained in sequence data. Although they are still not "full" sequence data, multiple different types of methods use short sequence blocks for analysis. These methods generally assume no recombination within blocks and free recombination between blocks; again, these assumptions mean that we can calculate tractable likelihoods on each individual block and then multiply these likelihoods together to get an overall composite likelihood.

One class of methods based on sequence blocks estimates population histories from gene genealogies at each locus. The coalescence times in such genealogies are informative about population sizes at many periods in the past. One of the first such methods to provide a rich description of these histories is referred to as the *skyline plot* (Pybus, Rambaut, and Harvey 2000). The method is called this simply because plotting estimates of N_e against time can resemble the skyline of a city when there are many changes in population size over time (as in Figure 9.4B). Skyline plots demand a lot from the data: that the genealogies be accurately inferred and that they be ultrametric (i.e., contain information about time), which also means that the substitution model must be accurate. Individual trees and locus-specific substitution rates are rarely known perfectly, so the skyline plot has been extended in multiple ways to incorporate this uncertainty (reviewed in Ho and Shapiro 2011). The main extensions have been to apply Bayesian methods in order to integrate over a large number of compatible gene genealogies (Drummond et al. 2005) and to combine information from multiple loci (Heled and Drummond 2008). Bayesian inferences from skyline plots are still limited in the number of loci that can be used—and make important assumptions about the lack of recombination within each locus—but under ideal conditions can provide accurate reconstructions of population histories (**FIGURE 9.6**). These methods are implemented in the software package *BEAST* (Drummond and Rambaut 2007).

Many approximate methods are used to infer demographic histories from short sequence blocks because they can be applied to summary statistics calculated from each block. Coalescent simulations are used to generate the same summary statistics from a wide range of demographic histories, and the combination of simulated parameter values that produce the best fit to the observed data is taken as the estimated history. As an example, Voight et al. (2005) estimated the demographic histories for multiple human populations using sequence data collected from 40 loci. In order to capture information about levels of polymorphism, the allele frequency spectrum, and linkage disequilibrium, the authors calculated the average of Tajima's D, Fu and Li's D^*, S, π, and ρ_{CL}, as well as the variance in Tajima's D across loci. Considering a family of bottleneck models for non-African populations and a growth model for their sampled African population, they used approximate likelihood to simulate thousands

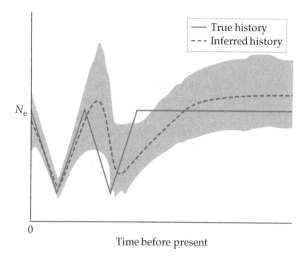

FIGURE 9.6 Example of inference using the skyline plot. The extended Bayesian skyline plot applied to a simulated history (solid line). The input data consisted of 32 loci, each with $n = 16$ sampled sequences. The dashed line is the median history from the posterior, while the gray area represents the 95% central posterior density. (After Heled and Drummond 2008.)

of replicate datasets (each consisting of 40 loci) for each set of parameter values across a grid of parameter values. Loci were also simulated with a range of recombination rates. The fit of each set of simulations was assessed, and a small region of parameter space that fit the data well was identified.

Approximate likelihood inferences are easy to carry out, but often only a small range of possible histories can be explored. In order to expand this range to include more parameters (e.g., more possible population size changes), ABC methods are commonly used. Multiple programs have been developed to allow easy implementation of ABC methods for all types of data and population histories (e.g., Hickerson, Stahl, and Takebayashi 2007; Cornuet et al. 2008; Lopes, Balding, and Beaumont 2009; Thornton 2009; Pavlidis, Laurent, and Stephan. 2010; Wegmann et al. 2010; Boitard et al. 2016). There is still some skill involved in choosing which summary statistics to include in model selection—and some knowledge required to specify the even wider range of histories to be considered—but ABC methods represent a very flexible and accessible set of tools. It should also be noted that recent theoretical advances make it easy to calculate the expected values of summary statistics under a range of histories (Gao and Keinan 2016), which may further lower the bar for carrying out demographic inference using short sequence blocks.

The final set of methods to consider use very long stretches of chromosomes as the basis for demographic inference. Once again, the goal is to use coalescence times across the recent past to estimate values of N_e in multiple discretized time windows. But using this type of data necessarily means modeling both recombination and mutation, as well as extracting signal from patterns produced by both processes. Here I briefly discuss one of the first methods to do this, the *pairwise SMC* (or *PSMC*) *model* (Li and Durbin 2011). PSMC only examines coalescence times between pairs of chromosomes (i.e., $n = 2$), usually in a single diploid individual. This constraint means that haplotypic phase is no longer required and that there is only one possible tree topology, making the inference problem much easier. This also means that the single coalescence time within each tree is the time to the most recent common ancestor (T_{MRCA}) at each locus. Within this single individual are thousands of individual loci, each with a coalescence time and a different length (**FIGURE 9.7A**). These observations provide information about the population size at different times in the past. Unlike methods based on the allele frequency spectrum, for which there is a limit to the number of historical epochs in which N_e can be estimated, there is no natural number of epochs for PSMC. Instead, PSMC discretizes times in the past so that any coalescences within the same short period are assumed to come from a population with the same constant size during this epoch, and multiple approaches can be used to determine the boundaries of these time windows (Li and Durbin 2011). PSMC uses a hidden Markov model (HMM) run along the genome to infer the time periods in which each locus coalesces (the "hidden" states in the HMM), providing a detailed history of population size through time (**FIGURE 9.7B**).

One drawback of the PSMC method is that studying only two sequences at a time does not provide a lot of information about population sizes in the very recent past. This is because few coalescent events are expected to have occurred during this time period, which means that there can be a lot of uncertainty associated with estimates of recent N_e (Li and Durbin 2011). Therefore, multiple methods have been developed to include more sequences in the analysis while still modeling the dependencies among linked loci, as larger samples mean more coalescent events in the recent past. One such method is the *multiple sequential Markovian coalescent*, or *MSMC, model* (Schiffels and Durbin 2014). MSMC can handle multiple (phased) sequences at a time but focuses solely on the first coalescent event in the sample, no matter which two sequences this happens between. As with PSMC, an HMM is run along each chromosome to infer the time to first coalescence at each locus, along with the pair of sequences that coalesced. In this way MSMC is most informative about events in the recent past and is in some ways complementary to PSMC. The software *MSMC* implements this method, along with PSMC using the updated SMC´ model (Schiffels and Durbin 2014); this latter implementation is the recommended one for PSMC inference.

Two methods that can be applied to modest numbers of individuals—without making only pairwise comparisons—are implemented in the software *diCal* (Sheehan, Harris, and Song 2013) and *ARGweaver* (Rasmussen et al. 2014). Both of these methods differ in important underlying ways from PSMC and MSMC, and both are statistically and computationally elegant solutions to a very difficult inference problem. *diCal* uses the conditional sampling distribution (CSD) to carry out its calculations. The conditional sampling distribution "describes the probability, under a particular population genetic model, that an additionally sampled DNA sequence is of a certain type, given

(A)

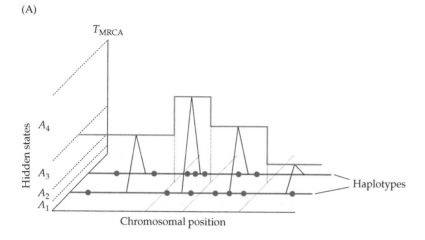

(B)

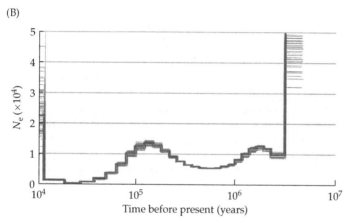

FIGURE 9.7 The PSMC model and method. (A) The PSMC model infers the time to coalescence (in discretized "hidden" states, A_1, A_2, etc.) for pairs of sequences in each non-recombining segment of the genome. Because the sample size is $n = 2$, the time to coalescence also represents the T_{MRCA} of each segment. Recombination events (shown as light dotted lines along the chromosome) allow for transitions between different T_{MRCA}s. Filled dots represent mutations that have accumulated on each haplotype and are used to help infer the T_{MRCA}. (B) The PSMC method applied to the autosomes of a Korean individual. The solid red line is the estimate from all the data, with each gray line representing one of 100 bootstrapped estimates. Note the log-scale on the x-axis, converted to years before present. (A adapted from Dutheil 2017; B adapted from Li and Durbin 2011.)

that a collection of sequences has already been observed" (Paul, Steinrücken, and Song 2011, pg. 1115). In this way, the CSD can be used to describe the joint probability of all haplotypes in the sample under an arbitrary demographic history quickly and easily. This property makes it ideal for estimating the optimal history. As yet another alternative, Palacios, Wakeley, and Ramachandran (2015) have shown that Bayesian methods can be used with the SMC′ model and known gene genealogies without discretizing past time periods, improving demographic inference (similar to what has been done for skyline plots; Minin, Bloomquist, and Suchard 2008). Currently such methods using the SMC′ depend on accurately inferred gene genealogies as input, but Bayesian methods can naturally be extended to deal with genealogical uncertainty in the future.

Finally, one limitation of single-population studies must be mentioned before moving on to multiple populations. Methods that aim to infer N_e values in the past from polymorphisms sampled in the present are obviously constrained in how far back they can "see." If there are no coalescent events in the distant past (and no polymorphisms from these periods), then our data contain no information about population sizes. Indeed, the expected T_{MRCA} is only $2N_e$ generations for a sample of size 2, approaching $4N_e$ for larger samples (Equation 6.3). This average is still not reflective of the limits of our approach, as it does not tell us the upper limit to observing coalescences. Fortunately, the probability distribution of the T_{MRCA} for a sample of size n is known (Tavaré 1984; Takahata and Nei 1985). Using this distribution, we expect that 95% of genealogies (i.e., loci) will have found their T_{MRCA} by $6N_e$ generations ago when $n = 2$. Even for large sample sizes (e.g., $n = 100$), 95% of loci will have a T_{MRCA} less than about $8N_e$ generations ago. This means that our data contain almost no information about population sizes further back in time. It must also be pointed out that our estimates of N_e are almost certainly more error-prone as we move further into the past, as demographic history is being inferred from fewer lineages within each locus and from fewer loci (those that have not yet reached their MRCA). This is one reason that the timescale in PSMC is measured on the log scale, such that larger intervals are in the more distant past. Unfortunately, this sort of uncertainty in estimates of demography is not accounted for in every method and every representation of results.

THE DEMOGRAPHIC HISTORY OF MULTIPLE POPULATIONS

Thus far we have only been concerned with the history of a single population, ignoring the complexity that arises when considering more than one population at a time, or even the existence of other populations. Single populations have only one demographic parameter that can vary—N_e. The histories of multiple populations can also include the time of population divergence between pairs of populations (t) and the amount of migration that may be occurring between them (m). Many of the underlying models and ideas used when studying multiple populations were discussed in Chapter 5, and anyone wishing to focus on inference under population structure should start there. Many of the same ideas, types of data, and computational approaches used in single populations are also relevant to multiple populations, so I will not repeat these descriptions again. Instead, my aim is to extend these ideas and to introduce additional ways of thinking about data that are not collected from a single, panmictic population.

It is also important to point out that most of the methods described here for the study of multiple populations can be used to study multiple species, with little change in implementation or interpretation. We may expect that results will be different when studying different species (because there will be larger t and smaller m), but the inferences are made in much the same way. When methods are not appropriate for use with multiple species, I will be sure to note this. For simplicity, however, I will continue to refer to models of "populations" for the rest of the chapter.

The effect of population structure on gene genealogies

In the simplest case of a history involving two populations, there are six different parameters that must be specified (**FIGURE 9.8A**): the current effective population sizes (N_1 and N_2), the ancestral effective population size (N_A; often assumed to be an average of the current sizes), the split time between the two populations (t), and migration rates in both directions (m_1 and m_2; sometimes assumed to be the same). We can further layer on changes in the effective size in each population over time, changes in the ancestral size, and many elaborations to the migration parameters (such as allowing migration only at certain times in the past). Every further population considered adds an additional N_e and t parameter, an additional N_A parameter, and multiple additional m parameters

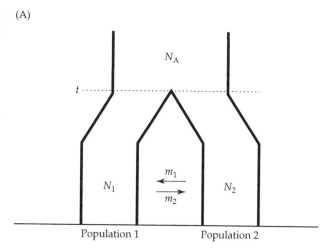

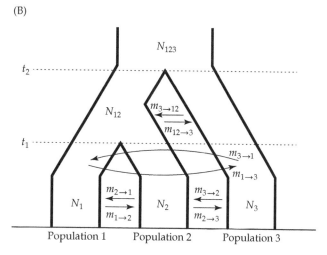

FIGURE 9.8 Models of population structure. (A) In the most basic model there are six parameters to be estimated: three effective population size parameters (N_1, N_2, N_A), two migration parameters (m_1, m_2), and time since the populations split (t). The split time between populations is often measured in units of $2N_e$ generations, where N_e can be either the average of the two current effective population sizes or the ancestral population size. (B) In models with more populations, each additional split adds one split time, two effective population sizes, and six migration parameters (because every population can exchange migrants with all other contemporaneous populations).

(**FIGURE 9.8B**). For instance, if there are K sampled populations, there are also $K - 1$ ancestral populations that must be modeled, as well as $K - 1$ splits between populations. Migration parameters can also exist between all pairs of extant *and* ancestral populations—these ancestral populations can exchange migrants with any other population with which they coexist (Figure 9.8B). For $K = 10$, this implies that there can be up to 19 N_e parameters, 9 t parameters, and 162 m parameters (Hey 2010). Therefore, in these situations it can be difficult both to describe the expected patterns of diversity and to infer histories from sequence data.

Characteristic genealogies are generated when there is population structure. A major effect of structure is that coalescence is not allowed between lineages in different populations, only within populations. If there is no migration, this means there can be no coalescence between these lineages up until time t, when (looking backward in time) the two populations become one again (**FIGURE 9.9A**). With migration, lineages can switch populations with probability m, the fraction of the population made up of migrants, again looking backward in time. After switching populations, these lineages can then coalesce with lineages in the other population (**FIGURE 9.9B,C**). In general, then, subdivision will result in long internal branches developing between populations as t gets larger (holding N_e constant), with any migration that occurs allowing for more recent common ancestry. Because the effects of migration can vary across loci—whether stochastically or because of selection against migrant alleles at specific genes—we also expect there to be much more variation in tree height among loci when there is migration (Figure 9.9; Wakeley 1996a).

Variation in N_e within lineages controls the rate of coalescence in each, which determines how fast population structure becomes manifest in genealogies. Slower rates of coalescence within populations mean that many lineages persist back until time t, which means there is little effect of subdivision on the underlying genealogies. Faster rates of coalescence can result in all lineages reaching their MRCA within each population before time t, so that the only remaining branch in the genealogy is the one connecting the two populations

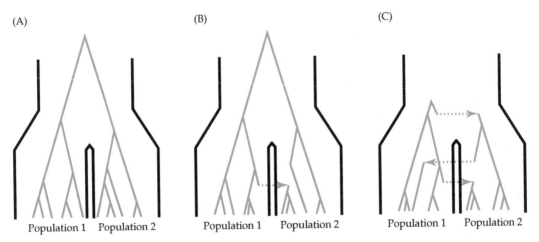

(A)   (B)   (C)

Population 1   Population 2   Population 1   Population 2   Population 1   Population 2

FIGURE 9.9 Representative genealogies under population structure. The time since divergence in all three panels is the same, with migration rates varying between them. (A) No migration. (B) Moderate migration (denoted with the dashed arrow switching populations). (C) High migration.

(Figure 9.9A). The case in which not all lineages have coalesced within populations before they join together in an ancestral population is called *incomplete lineage sorting* (Avise et al. 1983). One result of incomplete lineage sorting is that lineages within a population can be most closely related to lineages in other populations, rather than only to lineages in the same population (e.g., Figure 5.7D). We will return to this important concept later in the chapter. For now, the important take-home message is that t and N_e can have opposite effects on the probability of coalescence in the ancestral population of lineages from different subpopulations. Larger t and smaller N_e both make it less likely that there will be between-population coalescence (apart from the two final lineages that must find a common ancestor between the two populations), while smaller t and larger N_e make it more likely. Measuring t in units of $2N_e$ generations (or any multiple of N_e generations) accommodates this dependency; alternatively, the split between populations can be expressed as the compound parameter $t/2N_e$, where t is time in generations.

The effect of population structure on the allele frequency spectrum

A genealogical view helps us to understand how populations will differ as a result of structure and how this will affect patterns of variation. Obviously the genealogies themselves will reflect different population histories, and these genealogies are often used in demographic inference when there are multiple populations being considered (see next section). But it is also helpful to connect these genealogies to summaries such as the average normalized difference in allele frequencies between populations (i.e., F_{ST}) and the allele frequency spectrum, as they represent simple ways to visualize and understand the data.

As we saw earlier in the book, patterns of allele sharing are strongly dependent on population structure. Under models without gene flow, the average value of F_{ST} depends on N_e and t (Equation 5.15). With migration added, differences in allele frequencies can depend on N_e, t, and m. In general, more closely related populations—having low t, large N_e, or large m—will have more similar allele frequencies across loci, and less closely related populations will have greater differences (Figure 5.7). This allele sharing (or conversely, difference in allele frequency) is determined genealogically by the probability that lineages in different populations will coalesce with each other sooner than they will with other lineages in the same population. We can even express F_{ST} as a function of the average time to coalescence within populations relative to between populations (Slatkin 1991, eq. 8), such that smaller differences between these two times imply less average differentiation. Coalescence between populations can be due to either incomplete lineage sorting or migration, as both lead to more recent shared ancestry and therefore more shared polymorphisms.

The allele frequency spectrum is also strongly affected by population structure, though we must now distinguish among different ways to measure this spectrum. Two such methods are straightforward: the choice is simply whether to report the allele frequency spectrum from each population separately (e.g., populations 1 and 2 in Figure 9.8) or to report the "combined" spectrum from all individuals sampled from both populations together. It may seem obvious that combining populations before analyzing data is not the ideal approach; however, this may be a common occurrence when we do not know there is population structure in the first place. It is therefore important to understand the effect this hidden structure can have on our summaries of the allele frequency spectrum. Recall that in Chapter 5 we also addressed how hidden population structure can affect linkage disequilibrium (the "multilocus Wahlund effect"; Figure 5.11), and we will not discuss this effect further here.

In cases in which there is no migration between two populations, the individual allele frequency spectrum in each population will match the one expected in a single population without structure. However, when taking the combined allele frequency spectrum from all samples at once there can be large deviations from expected values. The time since population splitting, t, is the main determinant of this deviation (**FIGURE 9.10**). This time (measured in $2N_e$ generations) is key in determining the presence and length of the last branch in the genealogy. When the branch is long, polymorphisms accumulate in each population along this branch and are shared among all individuals within each population. As a consequence, when data are combined there will be an excess of intermediate-frequency polymorphisms, and summary statistics such as Tajima's D will be positive (Figure 9.10). Importantly, this genealogical structure and pattern of variation looks very much like the one generated under either balancing selection (Figure 8.3) or a population bottleneck (Figure 9.2C): long internal branches and an excess of intermediate-frequency polymorphisms (Nielsen and Beaumont 2009; Peter, Wegmann, and Excoffier 2010).

The patterns observed when samples from multiple populations are mistakenly combined are also strongly dependent on the balance of sampling and the number of subpopulations sampled. The example shown in Figure 9.10 has only two populations and equal numbers of sampled chromosomes from each population ($n = 25$ in each). If we had instead sampled 49 chromosomes from the first population and 1 chromosome from the second, we might still have captured the long internal branch separating populations, but all of the polymorphisms unique to the second population would have been present as singletons. Alternatively, we might have sampled 50 populations, taking

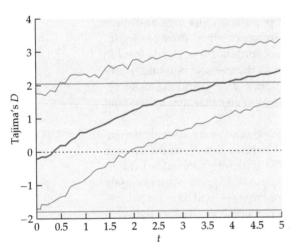

FIGURE 9.10 The effect of population structure on the allele frequency spectrum. Two populations are simulated with different split times between them, t, measured in $2N_e$ generations. Tajima's D is calculated based on the combined sample, with equal numbers of chromosomes from each population ($n = 25$ from each). The thick line shows median D based on 1,000 simulations at each time point, with 2.5% and 97.5% simulated values shown as thin lines. The dashed horizontal line shows the expected value of D in each population, with upper and lower horizontal lines showing the critical values for rejecting the null hypothesis. (After Simonsen, Churchill, and Aquadro 1995.)

one chromosome from each. In this case the pattern of population structure would also likely have been apparent as an excess of low-frequency polymorphisms (Ptak and Przeworski 2002), mimicking either a population expansion (Figure 9.2B), a selective sweep (Figure 8.1C), or slightly deleterious variation (Figure 7.6).

When there is migration between populations, the allele frequency spectrum can behave in similarly complex ways. If samples are only taken from a single population, migration into this population can affect the spectrum, generally increasing the number of common alleles at the expense of rare alleles (De and Durrett 2007). This implies that unsampled populations (also called "ghost" populations; Beerli 2004) may often have effects on our inferences, even when we believe we are studying solely the history of a single population. Conversely, if samples are taken at random from a large collection of populations evolving under the island model or stepping-stone model with large m (Chapter 1), then there is not expected to be a deviation from the equilibrium allele frequency spectrum (De and Durrett 2007). Finally, sampling schemes somewhere in-between these two extremes (e.g., combined from a smaller number of populations) can lead to an excess of low-frequency variants, as in the no-migration case (Gattepaille, Jakobsson, and Blum 2013). Overall, then, we should be careful to identify any population structure in our samples before summarizing the data (see Chapter 5 for methods to do this). Care must also be taken in interpreting our data, as patterns may be generated by either non-neutral evolution or various nonequilibrium population histories.

A quite different approach uses a third way to summarize the allele frequency spectrum—the so-called *joint allele frequency spectrum* (also frequently referred to as the *joint site frequency spectrum*, or *JSFS*). If the sample sizes from two populations are n_1 and n_2, then S_{ij} represents the number of derived alleles on i chromosomes from population 1 and j chromosomes from population 2. The joint allele frequency spectrum summarizes all of the polymorphisms in the two samples in this way:

$$\begin{bmatrix} S_{00} & S_{01} & \cdots & S_{0n_2} \\ S_{10} & S_{11} & \cdots & S_{1n_2} \\ \vdots & \vdots & & \vdots \\ S_{n_10} & S_{n_11} & \cdots & S_{n_1n_2} \end{bmatrix}$$

Values in the first column represent polymorphisms unique to population 1, while values along the top row represent polymorphisms unique to population 2 (S_{00} and $S_{n_1n_2}$ are by definition always 0, as invariant sites are not included). All other entries represent polymorphisms shared by the two populations. The joint spectrum can also be represented as a higher-dimensional matrix when more than two populations are considered, and minor allele frequencies can be used if polymorphisms cannot be polarized.

The joint spectrum was first used by Wakeley and Hey (1997) to estimate the demographic history of a pair of populations without migration. Even when assuming there is no migration, the joint spectrum contains all of the information present in the allele frequency spectrum from a single population (and therefore information about changes in N_e), as well as information about split times between populations (e.g., Li and Stephan 2006; Keinan et al. 2007; Garrigan 2009). But the joint spectrum is especially sensitive to migration, as there will be more shared polymorphisms than expected without migration—especially shared low-frequency polymorphisms (even though there are fewer low-frequency polymorphisms overall). Gutenkunst et al. (2009) introduced a method for efficiently estimating parameters from the joint frequency spectrum using a diffusion approximation. Their method includes migration and can therefore estimate the full set of parameters describing the demographic history of multiple populations.

Inferring demographic histories for multiple populations

As with inferring the history of single populations, there are a large number of software packages currently used to infer the joint history of multiple populations. The programs differ in the types of data they accept, as well as the maximum number of individuals and populations that can feasibly be analyzed at the same time. As with single-population inferences, these methods accept data from unlinked (or presumed unlinked) single SNPs, blocks of full sequence (often assuming no recombination within blocks and free recombination between blocks), or chromosome-scale sequences. The similarity in the methods used also extends to their approximate run times and computational demands, both of which generally increase as assumptions about linkage are relaxed.

Composite likelihood methods have the ability to analyze both thousands of individuals and thousands of polymorphisms at a time using the joint allele frequency spectrum. The program $\partial a \partial i$ (Gutenkunst et al. 2009), introduced when we discussed the history of single populations, is again one of the most popular programs of this type. While $\partial a \partial i$ is extremely flexible and can handle very large samples, it can only model relationships among up to three populations at a time. A similar method using the joint spectrum to analyze short sequence blocks is implemented in the software *Jaatha* (Naduvilezhath, Rose, and Metzler 2011), but this program can only handle two populations. These are considerable limitations given the number of populations regularly collected in current datasets. An alternative method, implemented in the software *fastsimcoal2* (Excoffier et al. 2013), uses coalescent simulations to calculate the approximate likelihood of the joint frequency spectrum. The use of simulations means that quite complex models involving many different populations can be considered, and various types of ascertainment schemes for the data can be accommodated. Putting aside possible limitations in the number of populations, the inferences made by $\partial a \partial i$ are quite robust to different types of errors: the original implementation corrects for errors made in the misidentification of the ancestral state, while newer versions can account for sequencing errors (Gravel et al. 2011). Accounting for sequencing errors—especially biased false positives (those that consistently make the same incorrect base call; e.g., Meacham et al. 2011)—is extremely important. Because much of the signal about migration in the joint allele frequency spectrum is contained within shared low-frequency alleles, this type of error would lead to incorrect inferences about migration.

There are many methods for making inferences about the history of multiple populations from short blocks of sequence, almost all of which use Bayesian computational approaches. Although such approaches are generally computationally demanding, they also offer more flexibility in the types of mutation models that can be applied to sequences or to other types of data (e.g., microsatellites). As discussed in Chapter 5, the *IM* program and its successors (Hey and Nielsen 2004, 2007; Hey 2010) have been widely used to infer N_e, t, and m from two or more populations; the latest version of this program, *IMa2* (Hey 2010), can handle up to 10 extant populations. Some methods ignore ancestral relationships among populations, assuming an island model in which only N_e and m parameters need to be estimated. While it is not clear how many empirical examples in which populations have reached drift-migration equilibrium there will be, the software *Migrate* can estimate these parameters if the assumptions of the model are met (Beerli and Felsenstein 1999, 2001). The software *G-PhoCS* (Gronau et al. 2011) extends *MCMCcoal* (Rannala and Yang 2003) to estimate migration in addition to N_e and t. *G-PhoCS* does not require phased sequences, unlike most other methods of this type. Rather, it takes unphased data as input, integrating over possible haplotypes as part of the MCMC search. Finally, ABC methods are also commonly used when analyzing short blocks of sequence, and essentially all of the ABC methods mentioned when considering single populations are flexible enough to handle the analysis of multiple populations.

There are relatively few software packages that can make inferences from multiple populations using very long stretches of sequence data. The *MSMC* program discussed earlier measures a *relative cross coalescence rate* between populations (Schiffels and Durbin 2014). This rate compares the timing of coalescence between individuals in different populations to coalescence between individuals in the same population. Moving backward in time, the relative cross coalescence rate goes from a value of 0 for a pair of well-separated populations—because almost all coalescence occurs within populations—to a value of 1 when the two populations have reached their common ancestor. As such, this measure is not explicitly estimating either t or m, and in fact it may better capture the gradual differentiation of populations that do not have an instantaneous split time. A newer method, implemented in the program *SMC++* (Terhorst, Kamm, and Song 2017), combines the sequentially Markovian coalescent with an extension of the conditional sampling distribution used by *diCal*. *SMC++* uses the CSD to describe the probability of the allele frequency spectrum in the sample and can be used with multiple populations. Its main advantage over *MSMC* is the ability to handle hundreds of unphased genomes. However, it assumes that populations split instantaneously with no subsequent migration, and so is only estimating N_e and t.

Multi-population histories as networks

Thus far we have assumed either that the hierarchical relationships among populations are known, or that there are no such relationships (i.e., the island model). While there is only one possible relationship between two populations (Figure 9.8A), for three or more populations we must specify a *phylogeny* (or *population tree*). There are many methods for constructing phylogenies, with the earliest using allele frequencies in each population (Cavalli-Sforza and Edwards 1967; Thompson 1973; Felsenstein 1981b). As long as drift is the only force acting to change allele frequencies, populations that are more closely related are expected to have more similar allele frequencies. Therefore, the length of branches in such trees will be proportional to t/N_e along the lineage, and the distance between populations will represent the amount of drift separating them. Only population allele frequencies are needed to construct these trees—with no requirement for individual genotypes—and large numbers of polymorphisms can be easily combined to provide high-resolution results. The *Contml* program (Felsenstein 1981b) in the *PHYLIP* package (Felsenstein 1989) builds such trees directly from allele frequency data.

Bifurcating trees are the most appropriate representation of relationships among populations related solely by splitting events. However, when populations exchange migrants (and they often do), the relationships among them are no longer solely bifurcating. *Reticulate phylogenies*—or *networks*—have connections between lineages after splitting events. These connections can be viewed as the merger of ancestral populations or as the contribution of genetic material from one lineage to another, either as a single event or continuously over time. The result of such reticulations is that individual gene genealogies evolving inside the network may follow one of multiple paths through time. In fact, the presence of non-zero migration parameters in the standard representation of population trees (e.g., Figure 9.8) implies a reticulate phylogeny. In this sense the estimation of migration parameters is equivalent to asking how much reticulation is in a tree, no matter how the tree is drawn. Much of the excitement over population networks, however, has come from newer methods that can construct reticulate trees solely from allele frequencies. In order to explain these methods, we must first make a digression into an area of lexical ambiguity.

The term *admixture* has made a recent resurgence in population genetics. It is most often taken to refer to the merger of individuals from two parental populations to create a new *admixed* (or *hybrid*) population (Figure 9.11A). Well-known examples of admixed populations include African Americans and Latinos/Hispanics in the Americas (e.g., Bryc et al. 2010). The genetic study of admixture goes back to at least Bernstein (1931; cited in Chakraborty 1986), with many newer methods for estimating the demographic parameters of admixed populations drawing from large numbers of markers (e.g., Bertorelle and Excoffier 1998; Chikhi, Bruford, and Beaumont 2001; Wang 2003; Sousa et al. 2009). More recently, however, the term *admixture* has become synonymous with gene flow of any type, including the creation of new populations (Figure 9.11A) and the contributions of migrants to preexisting populations (Figure 9.11B). Genealogically (and biologically) there are important differences between these scenarios, the main one being whether genetic drift has occurred along a lineage after the split with its close relatives but before the migration event (dashed branch in Figure 9.11B). This difference will result in slightly larger coalescence times for genealogies connecting P_3 and P_1 compared to those uniting P_3 and P_2 (Figure 9.11B). While I prefer to restrict the term *admixture* to its more traditional meaning—using *migration*, *gene flow*, or *introgression* (when between different species) for contributions to existing populations—it is important to note the fluidity with which these terms are used in the current literature.

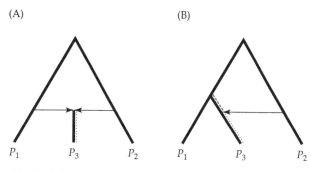

FIGURE 9.11 Two representations of reticulate phylogenies (networks). Both trees show relationships among three populations, P_1, P_2, and P_3, where P_3 is the population accepting migrants. (A) A tree demonstrating the traditional meaning of admixture. P_1 and P_2 are the parental populations that mix to create P_3. The branch with dots is the one measured by $f_3(P_3; P_1, P_2)$. (B) A tree with migration from P_2 into the preexisting P_3 population. The main difference from the tree in (A) is that drift can occur along the branch with dashes prior to the arrival of migrants. The branch with dots is again the one measured by $f_3(P_3; P_1, P_2)$.

Given this background, we can now discuss several popular methods for inferring *admixture graphs* (or *admixture trees*). Such graphs are networks that resemble either of the trees in **FIGURE 9.11**, representing both splitting and reticulation among populations. Allele frequency data are all that are needed to infer admixture graphs, with the branch lengths again representing the amount of drift along a lineage. Each branch of the tree also has a weight denoting its proportional contribution to its descendant's ancestry, such that the sum of weights of all branches leading to a particular node equal 100%. In this way admixture graphs explicitly represent the ancestry of "admixed" populations (i.e., those that have received any migrants), including the contributions of any ancestral or contemporaneous populations.

The two most popular methods for building admixture graphs are *TreeMix* (Pickrell and Pritchard 2012) and *MixMapper* (Patterson et al. 2012; Lipson et al. 2013). The methods differ somewhat in their details, but both use only population allele frequencies—as well as their variances and covariances—to build the graphs. Both begin by building a bifurcation-only tree to use as a backbone, followed by the addition of reticulate branches (cf. Lathrop 1982). *TreeMix* builds the initial tree with all the populations being considered, while *MixMapper* only uses populations with no evidence of admixture. The methods proceed by the sequential addition to the network of those admixture/migration events (or admixed populations in the case of *MixMapper*) that significantly improve the fit of the data. Neither *TreeMix* nor *MixMapper* can distinguish between true admixture and gene flow into preexisting populations. Instead, any genetic drift that has occurred along an independent lineage prior to the influx of migrants is subsumed into estimates of the amount of drift that has occurred subsequent to migration (e.g., fig. 1D in Patterson et al. 2012).

The method implemented in *MixMapper* uses a series of *f*-statistics introduced by Reich, Patterson, and colleagues (Reich et al. 2009; Green et al. 2010; Patterson et al. 2012). These statistics are related to the *F*-statistics introduced by Sewall Wright (Chapter 5) but are not exactly the same (see Peter 2016 for a discussion of the similarities and differences). There are three different statistics—denoted f_2, f_3, and f_4—that use allele frequencies from either two, three, or four populations, respectively. The f_2 statistic measures the distance between two populations (call them P_1 and P_2) and is proportional to t/N_e. With sample allele frequencies p_1 and p_2 and sample sizes n_1 and n_2, f_2 can be calculated as:

$$f_2(P_1, P_2) = (p_1 - p_2)^2 - \frac{h_1}{n_1} - \frac{h_2}{n_2} \tag{9.1}$$

where

$$h_1 = \frac{n_A n_a}{n_1(n_1 - 1)} \tag{9.2}$$

and n_A and n_a are the counts of the two alternate alleles at a site in population 1, such that $n_A + n_a = n_1$ and $n_A/n_1 = p_1$ (Patterson et al. 2012). A similar calculation is made for h_2, recognizing that n_A and n_a should reflect the count of alleles in population 2. The average value across all polymorphic sites is then used as an estimate of f_2 between populations. As described here, f_2 is similar to many other measures of genetic distance (see Chapter 5 in this book, as well as Nei 1987, ch. 9) and does not seem to be a necessary addition to the population genetics toolbox. However, combinations of f_2 among more than two populations can be used to form the f_3 and f_4 statistics (Reich et al. 2009; Patterson et al. 2012), and these measures have become invaluable tools in the field for detecting gene flow.

I mentioned above that *MixMapper* builds its initial tree using only populations for which there is no evidence of admixture. It identifies such populations using f_3 as a test statistic. Consider three populations, P_1, P_2, and P_3, all of which we wish to test for admixture. We calculate f_3 for P_3 as:

$$f_3(P_3; P_1, P_2) = (p_3 - p_1)(p_3 - p_2) - \frac{h_3}{n_3} \tag{9.3}$$

where again h_3 is calculated using n_A, n_a, and n_3 from P_3 (using Equation 9.2), averaging across polymorphisms (Patterson et al. 2012). Similar calculations can be made for populations P_1 and P_2 by rearranging the values in Equation 9.3—that is, by calculating $f_3(P_1; P_2, P_3)$ and $f_3(P_2; P_1, P_3)$ using the appropriate allele frequencies and sample sizes.

In cases in which there is admixture or gene flow, f_3 can be negative for the population that has accepted migrants but will not be negative for either of the other populations. Why is f_3 negative, and what does it measure exactly? $f_3(P_3; P_1, P_2)$ can be thought of as the length of the external branch leading to population P_3 (dotted branches in Figure 9.11); rearrangements of f_3 for P_1 and P_2 are similarly interpreted as the external branches leading to those populations. If the tree relating the three populations has no reticulations, then this branch must be positive (or at least non-negative); this is a requirement of bifurcation-only trees (Reich et al. 2009). However, if population P_3 does have mixed ancestry from two other populations, then this branch length can be negative. In this sense, f_3 is a test of *treeness* among P_1, P_2, and P_3—asking whether they form a bifurcating phylogenetic tree. An alternative way to think about f_3 is to note that p_3 is not expected to be consistently intermediate in value between p_1 and p_2 unless there is gene flow. When p_3 is intermediate, however, the value of f_3 will be negative (Equation 9.3). This statistic is not negative under all migration scenarios, having the greatest power to reject tree-like evolution for recent reticulations between parental populations that split longer ago in the past and that have contributed approximately equally to the admixed population (Peter 2016). Significance testing for f_3 (and f_4) is usually done by resampling the dataset to generate a z-distribution, often sampling blocks of the genome at a time to obviate the effects of linkage disequilibrium.

The f_4 statistic can be employed both in testing for admixture/migration and in estimating admixture/migration proportions; confusingly, these uses involve either four or five populations, respectively. Here I will only describe f_4 as a test; see Patterson et al. (2012) for details on its use as an estimate of ancestry proportions in admixed populations (the f_4-*ratio*). Consider populations P_1, P_2, P_3, and P_4, arranged in either of two ways (**FIGURE 9.12**). Note that these arrangements are the same unrooted tree, and have simply been rooted in different places—either of these trees or their unrooted counterpart can be used in the f_4 test. Following similar notation as used above, f_4 can be calculated as (Reich et al. 2009; Patterson et al. 2012):

$$f_4(P_1, P_2; P_3, P_4) = (p_1 - p_2)(p_3 - p_4) \qquad (9.4)$$

As with f_3, f_4 is a test of treeness for these four populations: when there is no mixing among populations, there is only one internal branch separating the pair of populations (P_1, P_2) from the pair (P_3, P_4) (thick line in Figure 9.12; Peter 2016). In this case the expectation is that $f_4 = 0$ because the distances between the pairs of populations are uncorrelated. When there is admixture or migration (dashed arrows in Figure 9.12), then there will be multiple internal branches in the tree, and f_4 can be non-zero. Although the mixing events in Figure 9.12 have been drawn as migration events with directionality, the test is also sensitive to true admixture, and without regard to the direction of migration. That is, migration from P_2 into P_3, or any intergroup exchange between either of P_1 and P_2 with P_3 and P_4, will also lead to significant f_4 values. For a more genealogical understanding of f_4 statistics, see the explanation of the D statistic (also known as the *ABBA-BABA test*) below.

It also should be mentioned that there are multiple other methods for inferring population and species relationships as networks or graphs. More phylogenetically oriented methods (e.g., Than, Ruths, and Nakhleh 2008; Kubatko 2009; Yu et al. 2014; Solís-Lemus and Ané 2016) often use only a single sequence from each population or species, inferring the reticulated phylogeny directly from these sequences or from gene trees. Other methods present the

(A)

(B)

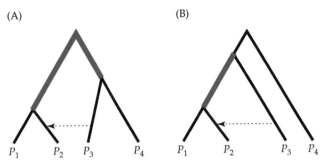

FIGURE 9.12 Networks with four populations. Both trees show relationships among four populations, P_1, P_2, P_3, and P_4. These rooted trees both show the same unrooted topology, containing a single internal branch (thick line) connecting (P_1, P_2) with (P_3, P_4) when there is no migration. (Branch lengths are not to scale.) (A) A symmetric topology with possible migration from P_3 into P_2. (B) An asymmetric topology with possible migration from P_3 into P_2. The test statistic $f_4(P_1, P_2; P_3, P_4)$ can be used to test for the presence of the migration event in either tree representation.

relationship among populations as a graph without hierarchical structure (Dyer and Nason 2004). Such graphs are especially useful for the visualization and analysis of connections (via migration) of populations on a landscape. We explore similar methods for this type of data in the next section.

The analysis of multiple populations in space

Many datasets in population genetics are collected from individuals arrayed across a landscape. Researchers may wish to sample a wide breadth of variation from across a species' range to test for allelic variation associated with environmental factors or to examine possible corridors/barriers to gene flow. This area of research goes by multiple names, including *spatial population genetics, geographical genetics* (Epperson 2003), and *landscape genetics* (Manel et al. 2003). Because of the focus on landscape-scale migration, the models and methods used in studies of this type are generally not applied to multiple species, only multiple populations.

Spatial location is explicitly or implicitly included in such studies, so the infinite island model (Chapter 1) is not appropriate. This is because, when dispersal is limited relative to sampling distance, the infinite island model does not capture the greater genetic similarity expected between individuals or populations that are closer in geographic space. Instead, stepping-stone models (either one- or two-dimensional) or models of individuals continuously distributed across space are used. The distinction between these models is subtle, and the patterns of variation produced by the two are often indistinguishable. Stepping-stone models imagine discrete subpopulations on a lattice that are separated in space by unoccupied regions. Random mating occurs within subpopulations, with migrants exchanged between them; often migration only occurs between neighboring populations (Figure 1.4C,D). Models of continuous populations resemble stepping-stone models, though they permit only single individuals to occupy the vertices of the lattice, and the density of occupied vertices can vary across the landscape. The most important migration-related parameter in this model is σ^2, the variance in distance between parents and offspring (Rousset 2007), akin to a dispersal distance. In both cases we may loosely expect *isolation by distance* (Wright 1940, 1943; Malécot 1948), or the increase in genetic distance with geographic distance (note that this phrase is often shortened to *IBD*, the second use of this abbreviation in this chapter). Although Wright's original model of isolation by distance explicitly involved a continuously distributed population, the term is now more generally used to describe patterns of genetic structure that are due to limited dispersal.

The distinctions between discrete and continuous population models are further muddled by the application of methods for identifying population structure in the first place. The methods introduced in Chapter 5 for placing individuals into populations do not usually take into account sampling location, and they assume that there are discrete populations. When individuals are continuously distributed across a geographic region, the clusters identified may correspond to sampling discontinuities as much as they do to genetic discontinuities (Serre and Pääbo 2004; Rosenberg et al. 2005). Fortunately, models can be constructed that use sampling location as a prior, reducing the problem of creating unnecessary populations (Guillot et al. 2005; François, Ancelet, and Guillot 2006).

There are multiple ways of determining whether isolation by distance is present in a dataset. If individuals are arranged into populations (or they are assumed to be) the standard approach is to regress genetic similarity between pairs of populations against the geographic distance between them. Multiple decisions can be made about how exactly to do this. Often raw values of F_{ST} between populations are not used; instead they are transformed as either $\left(\frac{1}{F_{ST}} - 1\right)/4$ (Slatkin 1993) or $\left(\frac{F_{ST}}{1-F_{ST}}\right)$ (Rousset 1997). Plotting genetic distance versus geographic distance on a log-log plot is expected to produce a linear relationship under IBD (Slatkin 1993). Even when individuals are not in discrete populations, we can still calculate measures of genetic distance between them (Rousset 2000). Regressing such measures against geographic distance will again reveal whether IBD exists. The standard method for determining whether the relationship between genetic distance and geographic distance is significant is a Mantel test (Mantel 1967; Sokal 1979), which uses permutation of population locations (e.g., Jensen, Bohonak, and Kelley 2005). As an alternative to F_{ST}-based statistics, spatial autocorrelation in allele frequencies can be measured via multiple techniques (e.g., Sokal and Oden 1978; Epperson and Li 1996; Smouse and Peakall 1999). Autocorrelation methods ask about the relationship among measurements of a variable (in this case, allele frequency) at different "step" distances apart. While these steps have a clear geographic meaning, there is not necessarily a natural scale over which to choose the size of steps *a priori* to conduct analyses.

Patterns of isolation by distance can be informative about population demographic history. If populations are at migration-drift equilibrium, IBD is expected. In this case, the slope of the regression line is determined by the product of the effective population size or population density and the dispersal distance (Rousset 1997, 2000), and it can therefore be used as an estimate of this product. Absence of a pattern of IBD can imply that populations are not at equilibrium, or that sampling was carried out at distances that do not allow such patterns to appear relative to dispersal distance. Some nonequilibrium population histories—such as range expansion—can also produce IBD over short distances only, though this depends on the details of how organisms move across the landscape (Slatkin 1993). Similar to patterns of IBD, spatial autocorrelation in allele frequencies can be informative about certain demographic parameters (e.g., Barbujani 1987; Hardy and Vekemans 1999; but see Slatkin and Arter 1991).

The methods discussed so far for analyzing landscape-scale genetic data have only provided information about demographic parameters, without attempting to connect these parameters to geographic coordinates. That is, they have not measured quantities such as the direction of migration or areas on a landscape with higher or lower migration. The vanguard attempt to infer such migration patterns across a landscape used principal components analysis (PCA) to summarize allele frequency differences among humans in Europe (Menozzi, Piazza, and Cavalli-Sforza 1978). PCA is a method for dimension reduction in large datasets, with the principal components (eigenvectors) representing orthogonal axes of variation. Menozzi, Piazza, and Cavalli-Sforza (1978) calculated PC 1 (the principal component explaining the greatest variation) for individuals sampled across Europe, interpolating their scores into a geographic map (**FIGURE 9.13**). The authors used this map to make inferences about migration routes and migration timing, concluding that the migration

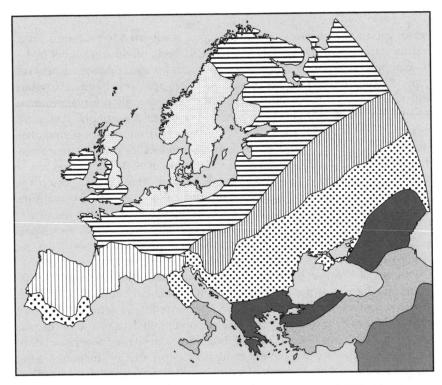

FIGURE 9.13 Interpolated values of principal component 1 on allele frequencies in humans across Europe. Shades indicate different values of PC 1, though there is no direct interpretation of these values. (Adapted from Cavalli-Sforza, Menozzi, and Piazza 1993.)

across Europe proceeded from southeast to northwest. However, patterns arising from the use of single principal components can be artifacts of the constraints imposed by PCA (Novembre and Stephens 2008), and the loading on such maps may actually be orthogonal to the direction of migration (François et al. 2010).

Instead, if two principal components, typically PC 1 and PC 2, are plotted against each other as a PCA scatter plot (Patterson, Price, and Reich 2006; Novembre et al. 2008), a match between genetic and geographic structure can be revealed. In this case individual genotypes are used as the units of analysis, rather than populations. In situations in which isolation by distance exists, the extent of genetic differentiation will be dependent on geographic distance, and PC 1 and PC 2 from genetic data may become surrogates for latitude and longitude (or vice versa). Under these conditions, PCA plots provide a striking match with the geographic locations of sampled individuals (see the cover of this book; Novembre et al. 2008). PCA is still inappropriate for inferring range expansions, but complementary work exists on how to do this appropriately (e.g., Austerlitz et al. 1997; Ramachandran et al. 2005; Peter and Slatkin 2013). Because such plots are easily generated from thousands of polymorphisms in hundreds of individuals (using, for instance, *smartpca* from the *EIGENSOFT* package; Patterson, Price, and Reich 2006), PCA is now a commonly used diagnostic tool in landscape genetics.

PCA produces beautiful images—and principal components are useful in controlling for population structure in association mapping (Price et al. 2006)—but because it is not based on a population genetic model, PCA plots are not directly informative about the underlying migration processes. One problem with PCA is its sensitivity to the details of population sampling across the landscape (Novembre and Stephens 2008; DeGiorgio and Rosenberg 2013), an issue that is unavoidable because of the underlying genealogical relationships among individuals (McVean 2009).

In order to both get around the sampling problem inherent to PCA and directly infer migration patterns across the landscape, several model-based methods have recently been introduced. The ones described here can work with thousands of markers and thousands of individuals, though the individuals are assumed to come from a smaller number of discrete subpopulations. The software *EEMS* (Petkova, Novembre, and Stephens 2016) fits *estimated effective migration surfaces* (hence the name), which can then be used to visualize regions of high or low migration on a landscape (**FIGURE 9.14**). The results from *EEMS* are especially good at capturing local barriers or corridors of gene flow, and the model can handle discontinuities in sampling (Petkova, Novembre, and Stephens 2016). A second method, *SpaceMix* (Bradburd, Ralph, and Coop 2016), places populations in a two-dimensional "geogenetic" space, similar to the way a PCA plots individuals. In *SpaceMix*, however, the distance between populations can be interpreted as proportional to rates of gene flow—closer populations have more migration. An additional important feature of *SpaceMix* is its ability to detect long-range genetic exchange between populations separated on the landscape, as well as the direction of this migration. While this type of

(A) (B)

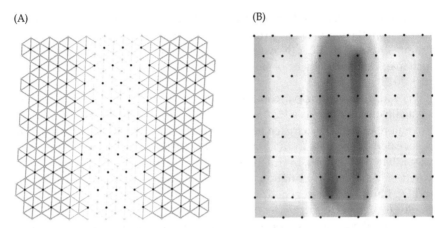

FIGURE 9.14 Estimated effective migration rates on a landscape. (A) Simulations were carried out by sampling individuals from each focal population (black dots), with unsampled populations indicated by white dots at each vertex. Migration connections are drawn using gray lines between population pairs, with the width of the line proportional to the migration rate. There is a region with 10-fold lower migration through the middle of the landscape. (B) Average *EEMS* output from the analysis of 100 datasets simulated as in (A). The black dots represent the location of sampled populations, with the inferred migration rate ranging from highest for dark green to lowest for dark pink. (After House and Hahn 2017.)

admixture can leave a signature in PCA plots (Patterson, Price, and Reich 2006), *SpaceMix* provides clear inferences about the magnitude and direction of gene flow.

Finally, many recent studies have focused on finding *genetic-environment associations*: correlations between specific alleles and environmental variables that reveal the targets of natural selection. Studies of this kind have existed as long as there have been genetic markers (e.g., Dobzhansky 1948), but the deluge of genomic data has made such studies more common and more fruitful (e.g., Hancock et al. 2008, 2011; Fournier-Level et al. 2011; Pyhäjärvi et al. 2013; Yoder et al. 2014; Hoban et al. 2016). The major methodological problem when testing for such associations is in fact IBD: geographically closer populations are both more similar genetically and more likely to share a common environment. Therefore, the shared history among populations can lead to false positives in naïve tests of genetic-environment associations. One of the first attempts to get around genetic non-independence was the use of the *partial* Mantel test (Smouse, Long, and Sokal 1986), which looks for correlations between variables while controlling for genetic distance. Unfortunately, the partial Mantel test can fail in many biologically realistic scenarios (Guillot and Rousset 2013). Fortunately, multiple methods have recently been introduced that can control for population structure while still identifying polymorphisms significantly associated with the environment. Two of the most powerful methods are *Bayenv* (Coop et al. 2010; Günther and Coop 2013) and *LFMM* (Frichot et al. 2013).

POPULATION GENETICS WITHOUT POLYMORPHISM

The final set of approaches to demographic inference discussed in this chapter do not involve any polymorphism data. Instead, they require only a single sequence collected from two or more different populations or species, as in phylogenetic studies. These data do not contain any information about recent demographic history, but they *do* contain information about the history of ancestral populations that pre-date lineage splitting events. They therefore allow us to reason about past population genetic processes without the need for polymorphism data. The approaches described here use what has recently become known as the *multispecies coalescent*, to separate it from predictions made by coalescent theory within species. However, as we will see, many of these methods have been around for more than 30 years and were used to illustrate patterns between species in some of the first papers describing the coalescent model. Below I separate methods using only two species from methods using three or more species, as these are based on quite different predictions.

Ancestral parameters inferred from two species

Given one sequence from each of two species, the two lineages will coalesce in their shared ancestral population (Figure 7.1). Importantly, this coalescence does not occur right at the time of the species' split, t, but instead at some point before it (Gillespie and Langley 1979). For autosomal loci, this time is expected to be $2N_A$ generations before t, where N_A represents the effective size of the ancestral population.

For any single locus we cannot hope to separate the time before the split from the time after. However, given a collection of independent loci, we can begin to tease apart these two quantities. This is because all loci share the fixed amount of time post-speciation (see Equation 7.1), while the coalescence time pre-speciation is exponentially distributed. The variance in these pre-speciation coalescence times is directly proportional to N_A, which led Takahata (1986) to use this variance among loci in order to estimate N_A. Of course we cannot directly observe the coalescence times either, only the amount of nucleotide divergence that has accumulated between two sequences. To account for this, Takahata, Satta, and Klein (1995) introduced a maximum likelihood method that includes the additional variance in divergence contributed by stochasticity in the mutational process.

Estimates of ancestral effective population size are very sensitive to a number of assumptions, including no variation in mutation rate among loci, no selection, no migration, and no recombination within loci (cf. Edwards and Beerli 2000). If the mutation rate differs among loci, or if migration has affected some loci and not others, then the variance in estimated coalescence times will be larger than expected, leading to overestimates of N_A. Yang (1997) introduced a method to control for mutation rate variation among loci, while Innan and Watanabe (2006) introduced a method that can co-estimate post-speciation migration rates and N_A. Selection in the ancestral population can both increase coalescence times (via balancing selection) and decrease coalescence times (via selective sweeps or background selection), inflating the variance in these times among loci. It is not clear how methods to estimate

N_A can account for selective processes, though in the next section we discuss cases in which selection can be positively useful in ancestral inference.

The approaches described thus far assume no recombination within loci and free recombination between loci. Recent methods using HMMs relax both assumptions, in addition to estimating the population recombination rate in the ancestral population (Mailund et al. 2011, 2012). HMMs allow recombination by modeling the transition in states between neighboring segments that individually do not have a history of recombination within them—states in this model represent different discretized coalescence times in the ancestral population. Transitions between states are caused by recombination, such that the frequency of transitions is proportional to the population recombination parameter, ρ ($=4N_A c$). This is actually quite an amazing inference: by simply keeping track of the rate at which sequences change their coalescence times along a chromosome, we can estimate the amount of recombination occurring in the distant past. Further elaborations on these HMM models have included migration between species after the lineage splitting event (Mailund et al. 2012).

Ancestral parameters inferred from gene tree discordance

When the data consist of single sequences from each of four or more species, in addition to coalescence times we can make demographic inferences from the amount and type of *gene tree discordance*. Gene tree discordance occurs when the topologies at different loci disagree with each other, meaning that some will also disagree with the phylogeny describing relationships among species. With a single sample from each species there can be no discordance when only two species are considered—there is only one possible relationship. In order to have discordant relationships with these data we require either four taxa or a rooted three-taxon tree (which almost always requires a fourth species with which to root it). There are multiple causes of discordance, both technical and biological (Maddison 1997; Degnan and Rosenberg 2009). The two main biological causes are incomplete lineage sorting (ILS) and introgression/migration/admixture. I begin with a discussion of ILS, as most methods for ancestral population inference only consider this process.

In order to understand ILS and its relationship to gene tree discordance, consider the phylogeny shown in **FIGURE 9.15A**. If we sample a single sequence from each of the three species or populations, then there are three different rooted topologies describing their relationships: one of these three topologies is concordant with the phylogeny (**FIGURE 9.15B,C**), while the other two are discordant (**FIGURE 9.15D,E**). There are actually two ways that a concordant tree can be generated. If the lineages from P_1 and P_2 coalesce in their most recent common ancestral population (which has size N_A; Figure 9.15A), the resulting tree must be concordant (Figure 9.15B). When this occurs, we say there is *lineage sorting*; when it does not occur, we say there is *incomplete lineage sorting*. The probability that these two lineages do not coalesce in the internode population is approximately $e^{-\tau/2N_A}$, where $\tau = t_2 - t_1$ (Hudson 1983a). If the lineages leading to P_1 and P_2 do not coalesce in this population, they can still be concordant when they coalesce in the ancestral population of all three populations (Figure 9.15C). In this population we expect to find the three possible

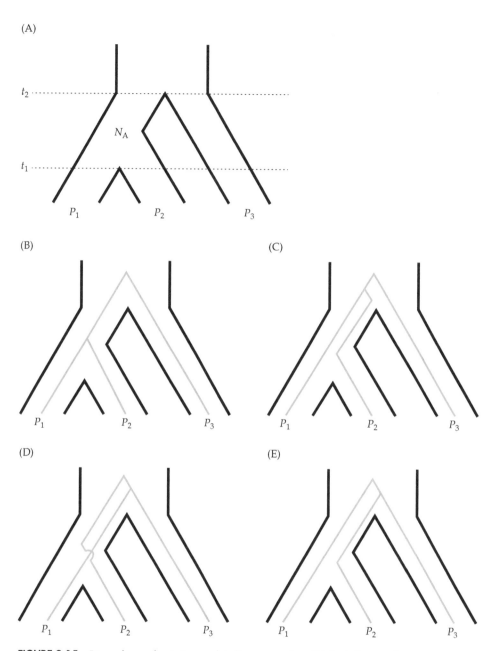

FIGURE 9.15 Discordance due to incomplete lineage sorting. (A) A phylogeny showing the relationships between three populations. The split time between P_1 and P_2 is t_1, and between the common ancestor of $P_1 + P_2$ and P_3 is t_2. The effective size of the common ancestor of P_1 and P_2 is N_A. (B) A gene tree topology (blue lines) concordant with the phylogeny, coalescing in the common ancestor of P_1 and P_2. (C) A concordant genealogy can still occur when there is incomplete lineage sorting, coalescing in the ancestor of all three populations. (D) The discordant genealogy that occurs when P_1 and P_3 coalesce in the ancestor of all three populations. (E) The discordant genealogy that occurs when P_2 and P_3 coalesce in the ancestor of all three populations.

topologies equally frequently, so that there is a one-third possibility of any particular genealogy, including the two discordant topologies (Figure 9.15D,E).

These results imply that the probability of concordance due to lineage sorting is:

$$P(concordance \mid lineage\ sorting) = 1 - e^{-\tau/2N_A} \qquad (9.5)$$

and the probability of concordance due to incomplete lineage sorting is:

$$P(concordance \mid ILS) = \frac{1}{3} e^{-\tau/2N_A} \qquad (9.6)$$

Together, the results provide expectations for the total probability of concordance:

$$P(concordance) = 1 - \frac{2}{3} e^{-\tau/2N_A} \qquad (9.7)$$

as well as the total probability of discordance:

$$P(discordance) = \frac{2}{3} e^{-\tau/2N_A} \qquad (9.8)$$

where each of the two discordant topologies occurs with frequency $1/3e^{-\tau/2N_A}$ (Hudson 1983a; Tajima 1983; Pamilo and Nei 1988).

It is important to understand what these expectations mean and what they imply, as they form the basis for most of the inferences described below. First, the probability of ILS is dependent on two values: the internode ancestral effective population size (N_A) and the internode time ($\tau = t_2 - t_1$). The probability of coalescence within this population is negatively related to effective population size, since the rate of coalescence is lower in larger populations. This probability is positively related to time, since with more time comes more opportunity for coalescence. Second, it is clear that ILS (also called *deep coalescence*; Maddison 1997) is not the same as discordance, though these terms are often used as synonyms in the literature. Genealogies can be concordant even when there is ILS (e.g., Figure 9.15C).

Importantly, all that is needed to generate discordant trees that are due to ILS is two speciation events in rapid succession, no matter how long ago these events occurred. The value of $t_2 - t_1$ is fixed after the second speciation event and does not depend on the magnitude of t_1: the speciation events could have taken place 100 years ago or 100 million years ago. Obviously if the events occurred 100 million years ago there are many other factors that could cause discordance in addition to ILS (e.g., Scornavacca and Galtier 2017). But the discordance due to ILS described here does not ever go away, as the gene trees are "frozen" in their discordant state. In this way, though the process of lineage sorting is the same, ILS when sampling one sequence from each of multiple species is qualitatively different from ILS involving multiple sequences from only two species. In the latter scenario ILS *does* slowly resolve itself over time (e.g., Figure 5.7D–F), eventually reaching the point at which all coalescences (except the last one) occur before the ancestral population (Avise et al. 1983). With enough time or small enough effective population sizes, each of the two populations or species will become reciprocally monophyletic (Hudson and Coyne 2002; Rosenberg 2003). Unfortunately, we do not have different words or phrases for these two cases of ILS, which seems to cause unnecessary confusion in the field.

Demographic inferences from gene tree discordance use the expectations given in Equations 9.5 through 9.8. The basic idea appears to have been first employed by Nei (1986): given a set of inferred gene trees, we can solve for the joint parameter τ/N_A in Equation 9.8 using the fraction of discordant genealogies. Given external estimates of the internode time, τ, we can then solve for the ancestral effective population size, N_A (or vice versa). The same basic approach has been used multiple times, with several embellishments (Wu 1991; Hudson 1992; Takahata, Satta, and Klein 1995; Ruvolo 1997; Chen and Li 2001). The branch lengths of each tree, combined with the expectations laid out above for the two-species case, can also be used to tease apart τ and N_A (Yang 2002). One important improvement to all of these methods has been to take into account some of the errors in gene tree reconstruction, which can lead to increased discordance and therefore increased estimates of N_A. The software *MCMCcoal* mentioned earlier in the chapter can do exactly this (Rannala and Yang 2003; Burgess and Yang 2008), as well as deal with variation in the mutation rate among loci (also see *G-PhoCS*; Gronau et al. 2011). Note that any sources of discordance (including incorrectly inferred gene trees and introgression) that are still unaccounted for by these methods will result in severely overestimated ancestral effective population sizes; estimates from the very distant past should therefore be regarded with caution.

In order to account for recombination, HMMs similar to the ones described for inferences from two species can be used (see Wall 2003 for an alternative approach using approximate likelihood). These HMMs now have states corresponding to each possible topology: two concordant topologies (one with and one without ILS) and two discordant topologies (Hobolth et al. 2007; Dutheil et al. 2009). As before, the HMM is run along a genome-scale alignment of three species plus an outgroup, with transitions between states proportional to the population recombination parameter. Any recombination events taking place in the common ancestor of P_1 and P_2 can lead to state transitions, such that coalescence occurs in a different population without a change in topology (e.g., between the two concordant genealogies) or in the same population with a change in topology. Combining the expectations laid out here and in the previous section using two species, these methods can estimate both effective population sizes and the recombination rate in multiple ancestral populations. Recent extensions to HMM methods have included introgression in the model (e.g., Brandvain et al. 2014; Liu et al. 2014), though these methods are best for identifying genomic regions that have introgressed, rather than estimating population parameters.

To emphasize the point that we are using population genetic reasoning to learn about past populations, consider one additional prediction from models of gene tree discordance. Although the time between the two speciation events, τ, is expected to be constant across the genome, N_A may vary from locus to locus. Most importantly, in the presence of strong selection— either positive or negative—N_A is consistently lower for linked neutral polymorphisms (Maynard Smith and Haigh 1974; Charlesworth, Morgan, and

Charlesworth 1993). Based on the expectations laid out above, we would therefore expect to find less discordance (more concordance) in regions with stronger selection. This hypothesis is supported by results showing lower discordance within regions experiencing strong selection (coding sequences) compared to regions experiencing on average weaker selection (noncoding sequences; Hobolth et al. 2011; Scally et al. 2012). In addition, while strong natural selection may occur at any position in the genome, the effects of this selection are magnified in regions with low recombination, as a much larger region is affected by any selected variant (Chapter 8). Therefore, among a set of randomly chosen loci, the action of strong selection should cause the recombination rate to be negatively correlated with N_A and consequently positively correlated with discordance. Multiple studies have shown that regions of low recombination (in current populations) are associated with less gene tree discordance (Hobolth et al. 2011; Prüfer et al. 2012; Pease and Hahn 2013; Munch et al. 2016), suggesting that they also experienced lower effective sizes in ancestral populations.

Introgression inferred from gene tree discordance

Robust inferences of introgression can be made based on gene tree discordance. Like the approaches described in the previous section, these methods for inferring introgression only require a single sequence from each of four or more populations (at least three populations and an outgroup). Unlike the previous methods, we are now inferring more recent demographic events, occurring on the tip branches of the phylogenies being considered.

For the phylogeny shown in Figure 9.15, the expectation under ILS alone is that the two discordant topologies will be found at equal frequency, and that this frequency will equal $1/3e^{-\tau/2N_A}$. However, when there is migration between lineages there can be a significant imbalance in the number of discordant trees. Specifically, when there is exchange between either P_3 and P_1 or P_3 and P_2, we may observe more of one of the two discordant topologies (either the one in Figure 9.15D or 9.15E, respectively). We cannot detect introgression between P_1 and P_2 in this way because it will not result in any disparity in discordant trees. When exchange does occur between non-sister populations the direction of introgression is not important—for instance, both migration from $P_1 \rightarrow P_3$ and $P_3 \rightarrow P_1$ will lead to an excess of trees with these sequences sister to each other—but there must be more exchange between these two populations or species than between P_2 and P_3; equal migration between the two pairs will result in equal proportions of the two discordant trees. A test for introgression based on this reasoning, using counts of discordant topologies, appears to have first been proposed by Huson et al. (2005).

Rather than use reconstructed gene trees, a straightforward alternative is to use the counts of site patterns that are consistent with the two alternative discordant topologies (Green et al. 2010). A site pattern is simply the pattern of alternative alleles at one site in an alignment. Defining the ancestral state (the state in the outgroup) as A and the derived state as B, the informative site pattern concordant with the phylogeny shown in **FIGURE 9.16A** would be designated BBAA. The discordant topology with P_2 and P_3 sister would lead

(A)

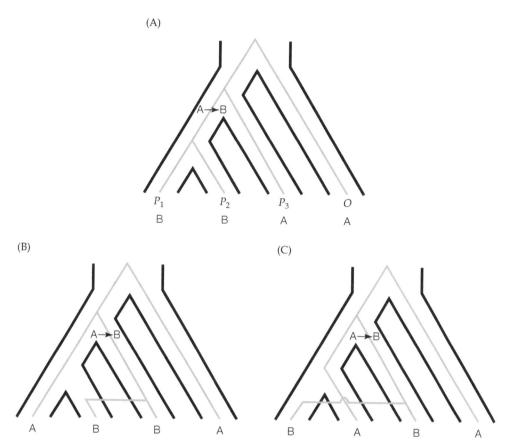

FIGURE 9.16   Basis for the *D* (or ABBA-BABA) test. (A) The species or population phylogeny with a concordant genealogy evolving within it. Given the substitution from A→B shown, the site pattern is BBAA. Note that this is the same phylogeny as in Figure 9.15A, but now the outgroup is shown. (B) A discordant genealogy that is due to introgression from P_3 into P_2. In this case the resulting informative site pattern is ABBA. The same site pattern can also occur under either ILS (see Figure 9.15E) or introgression in the alternate direction. (C) A discordant genealogy that is due to introgression from P_3 into P_1. In this case the resulting informative site pattern is BABA. The same site pattern can also occur under either ILS (see Figure 9.15D) or introgression in the alternate direction.

to the informative site pattern ABBA (**FIGURE 9.16B**), while the one with P_1 and P_3 together would lead to the informative site pattern BABA (**FIGURE 9.16C**).

A test based on counts of these site patterns, n_{ABBA} and n_{BABA}, summed across a large alignment, uses the *D* statistic:

$$D = \frac{n_{ABBA} - n_{BABA}}{n_{ABBA} + n_{BABA}} \tag{9.9}$$

where the expectation of *D* is 0 when incomplete lineage sorting is the only process acting (Green et al. 2010; Durand et al. 2011; Patterson et al. 2012).

Because there are already so many statistics that use the letter D, this test is alternatively known as either the *ABBA-BABA test* or *Patterson's* D, as Nick Patterson is the first author of the section of the supplementary materials of the paper in which it is first described (Green et al. 2010).

Following the same reasoning laid out above for counts of genealogies, migration between P_3 and either P_1 or P_2 will disproportionately increase the frequency of BABA or ABBA site patterns, respectively, since the pair of taxa exchanging alleles should have relatively more shared derived (B) states (Figure 9.16B,C). Positive values of D indicate gene flow in either direction between P_2 and P_3, while negative values indicate gene flow between P_1 and P_3. Values of 0 do not provide any evidence of migration, though again, equal levels of gene flow in the two pairs can lead to $D = 0$.

The D statistic has many similarities to f_4 described earlier and is in fact testing the same null hypothesis (Patterson et al. 2012; Peter 2016). One distinction is that D only works with the asymmetric topology shown in Figure 9.16A or 9.12B and does not work with the symmetric topology shown in Figure 9.12A. The test based on D assumes that the outgroup cannot take part in any introgression and always represents the ancestral state. Though I have explained it here for a single sequence from each population or species, the D statistic can also be calculated using sample allele frequencies (Green et al. 2010; Durand et al. 2011; Patterson et al. 2012). As with f_4, the significance of D is calculated by block jackknife or bootstrap. This calculation results in a genome-wide test statistic, though chromosome-wide values are also informative (e.g., *Heliconius* Genome Consortium 2012). One caveat is that values in smaller windows of the genome can simply reveal sorting of one or the other discordant topology locally; values different from 0 in these cases do not necessarily indicate introgression. Consider, for example, individual non-recombining loci, which can only take on values of $D = 1$ (all ABBA), -1 (all BABA), or 0 (concordant genealogies with all BBAA sites). The expectation of $D = 0$ only holds across many loci averaged across the genome.

The D statistic makes a number of assumptions, though it is generally quite robust to many violations—for instance, it is unlikely that selection could appreciably distort genome-wide site patterns. One important assumption is that there is no population structure in the joint common ancestor of P_1, P_2, and P_3. If there is structure in this population, such that the lineages eventually leading to P_3 were more likely to coalesce with either P_1 or P_2, a significant D value could result (Slatkin and Pollack 2008; Durand et al. 2011; Eriksson and Manica 2012). A second assumption is that there have not been multiple substitutions at a site, as these could also result in an excess of discordant site patterns. For this reason, the D test is generally only used for recent splits between lineages. Tests for exchange among more anciently diverging lineages should use the genealogy-based discordance test of Huson et al. (2005).

With more than four populations or species, additional analyses based on similar ideas can be performed. One useful approach when there are many different populations is to carry out the standard D-test with different subsets of four taxa. In this way, one can reason phylogenetically about where introgression may have occurred on the internal branches of trees (e.g., Green et al.

2010; Martin et al. 2013; Fontaine et al. 2015), though we can still never detect gene flow between sister branches. With five populations the f_4-ratio can be calculated for certain phylogenies to estimate the proportion of the genome affected by introgression (Reich et al. 2009; Patterson et al. 2012). Finally, with four populations in a symmetrical topology (plus an outgroup), the related "D_{FOIL}" statistics can additionally identify the direction of introgression between non-sister lineages (Pease and Hahn 2015).

CAVEATS TO INFERENCES OF DEMOGRAPHIC HISTORY

As was mentioned at the beginning of the chapter, inferences of demographic history come with several caveats. Most importantly, it is crucial to remember that our estimates of population history have also been affected by natural selection and other evolutionary forces—it is almost impossible to obtain "purely" demographic results.

The estimates of N_e returned by all of the methods described here represent the size of a Wright-Fisher population that is undergoing an equivalent amount of drift, but they in no way represent only demographic effects. Changes in N_e over time and among populations are due to many processes—demographic and selective—and should not be taken as numerical estimates of census population size of any kind (Chapter 1). As we will discuss further in Chapter 10, disentangling selection and demography is very difficult, and modern computational methods based on beautiful theory do not avoid this difficulty. As an example, **FIGURE 9.17** shows the effective population size inferred by PSMC when there has been a single selective sweep on a 15 megabase-long chromosome (Schrider, Shanku, and Kern 2016). Although the actual population size (census number of individuals) is constant throughout, selection produces nonequilibrium patterns of N_e. In a very real sense this view of the effective population size is correct: the effect of linked selection has been to reduce N_e for a period of time and to increase the rate of coalescence. But no change in demographic population size has occurred.

Similarly, the meaning of the migration parameter m reflects the realized amount of introgression, and not any biological

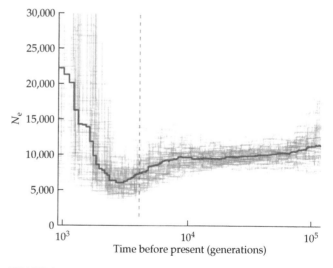

FIGURE 9.17 The effect of selection on demographic inference. PSMC is used to infer N_e when one selective sweep occurs in a population of constant size. Each of 100 replicate simulations are shown (thin gray lines), along with the median across all replicates (thicker blue line). The fixation time of the sweep is shown as a dashed vertical line. (After Schrider, Shanku, and Kern 2016.)

hybridization rate or number of migrant individuals. The "worst sin" in studies of migration across a landscape is to equate non-zero estimates of $N_e m$ between pairs of populations with the actual movement of migrants between these particular populations, rather than the movement of alleles via intervening populations (Rousset 2007, pg. 956). Migration itself can be opposed by selection, but the dependence of many of our estimates on the joint parameter $N_e m$ also means that any processes affecting N_e can affect inferences of the population migration rate.

The fact that many of our inferences rely on joint parameters also means that decisions about one parameter can have large effects on conclusions about others. For instance, because we can often only estimate $N_e \mu$, small changes in the value of the mutation rate used can greatly affect N_e. This in turn can affect estimated times of divergence, since these are often expressed in units of N_e generations (e.g., Sheehan, Harris, and Song 2013). To convert these units into absolute times, estimates of both the generation length and N_e are needed. Our estimates of absolute time can therefore be highly dependent on many parameters, including those affected by selection. All of this simply means that we must be careful to avoid placing too much weight on any particular numerical estimate from demographic inferences.

POPULATION GENOMICS

In the previous chapters we learned about making inferences from DNA sequence data, largely focusing on the effects evolutionary forces have on patterns of variation at single loci. But the field of population genetics has very quickly been transformed into the field of *population genomics*, with whole-genome data being generated for many model and non-model organisms. In *Drosophila*, we have moved from one locus (Kreitman 1983), to two loci (Kreitman and Aguadé 1986), to 20 loci (Begun and Aquadro 1992), to 105 loci (Glinka et al. 2003), to all loci (Begun et al. 2007; Sackton et al. 2009; Langley et al. 2012; Mackay et al. 2012). While this transformation has occurred at different rates in different taxa, we can be assured that the days of Sanger-sequencing a handful of loci in order to make meaningful evolutionary inferences are behind us.

So what exactly is population genomics, and how does it differ (if at all) from population genetics? To some extent, population genomics *is* the ability to query and analyze all the data, one locus at a time. However, if population genomics were simply the capacity to carry out 10,000 McDonald-Kreitman tests, I do not believe there would be so much excitement about it. While there are challenges in doing single-locus population genetics writ large, this approach does not expand the types of inferences we can make. Instead, the most exciting aspect of population genomics is the promise of carrying out analyses that cannot be done on single loci or that are meaningless without genomic context. Often it is this context that provides the real insight when carrying out genome-scale analyses. We have seen a glimpse of these methods when using patterns of linkage disequilibrium to detect targets of selection (Chapter 8).

Population genomics also explicitly addresses the interaction between evolutionary forces, most significantly natural selection and nonequilibrium demographic histories. It has long been recognized that many selective and demographic processes can generate similar patterns of nucleotide variation, and one major justification for collecting data from more loci has been the promise of disentangling these effects. In Chapter 9 we discussed multiple methods that use genome-scale data for inferring demographic histories. Our goal here then is to consider methods for identifying selective effects using genome-scale data, explicitly taking into account the possible complications contributed by nonequilibrium population histories. Although such problems are far from solved, even when using whole genomes, this is an active area of research in which many creative solutions have been proposed.

We begin this chapter with a general discussion of approaches for detecting selected loci in population genomic data, followed by the description of several specific methods for doing so. We end the book with a brief look at several areas of population genomics that promise to be especially active in the near future. These subfields span a range of questions and evolutionary forces—as well as different computational and experimental methods—that simply did not fit neatly into any previous chapter. Further anticipated revolutions in sequencing and computing power ensure that these predictions will necessarily remain only a small part of what molecular population genetics will become in the next decades.

GENOME-WIDE SCANS FOR SELECTION

Motivating questions

The availability of genome-wide data means that we can begin to ask questions about the effects of natural selection on genetic variation across the genome. There have already been many more than 100 studies that have conducted genome-wide scans for selection in a wide variety of organisms (Haasl and Payseur 2016), addressing a large number of questions in biology and population genetics.

Some of the more general questions revolve around the types of genes and types of mutations involved in adaptation. While it is no longer surprising to detect genes involved in immunity and pathogen response evolving under positive selection (e.g., Sackton et al. 2007; Obbard et al. 2009; Fumagalli et al. 2011; McTaggart et al. 2012; Enard et al. 2016; Early et al. 2017)—or genes in the major histocompatibility complex evolving under balancing selection (Hedrick 1998; Meyer and Thomson 2001)—we still do not know whether there are subsets of genes associated with particular functions predictably involved in other forms of adaptation across organisms. We would also like to know the specific types of mutations that are used during adaptation, whether they are nucleotide variants in coding regions or in noncoding regions (Hoekstra and Coyne 2007; Stern and Orgogozo 2008), copy-number variants in either type of region (e.g., Perry et al. 2007), or even the insertion of transposable elements (e.g., Schlenke and Begun 2004; González et al. 2008). We also expect that uncovering adaptively evolving genes will begin to tell us about the very species we are studying—the *reverse ecology* approach (Li, Costello, et al. 2008). By finding the genetic targets of selection, the hope is that the known functions of these loci will implicate the subtle phenotypic differences targeted by selection. One major caveat to this approach is the apparently limitless ability of scientists to tell stories about the possible phenotypes identified by genomic scans (Pavlidis et al. 2012).

Regardless of the specific biological questions asked, the answers require the clear identification of the targets of selection. It is quite hard to pinpoint specific mutations that are under selection—particularly if they have already fixed—and sometimes equally hard to know which of several genes or regulatory elements in a region are targeted, even when we do not need to know the precise mutations. For these reasons it is important to understand the spatial signature of linked selection, as this will enable us to determine whether we have detected the center or the shoulders of a sweep (cf. Hudson et al. 1994; Schrider et al. 2015). Individual tests applied to single loci often do not capture this spatial signal, so we turn to population genomics for more context and more resolution. Even so, the eventual "election" of candidate genes may require a range of further experimental approaches (Stinchcombe and Hoekstra 2008; Pardo-Diaz, Salazar, and Jiggins 2015).

Among the questions that are the focus of population geneticists, those concerned with the frequency and type of natural selection acting across the genome are two of the most important. We would like to be able to estimate, for instance, the fraction of amino acid substitutions that have fixed via positive selection or the fraction or amino acid polymorphisms that are maintained by balancing selection. These questions have been the focus of sometimes acrimonious debate in the field for a long time (e.g., Kimura 1983; Gillespie 1991). Further, of the alleles that do fix via positive selection (whether coding or not), what proportion were already polymorphic when selection started? In other words, is most adaptation driven by new mutation or standing polymorphism? Similarly, it is very important to understand the fraction of the genome affected by linked selection (see next section) and whether the effects are due to hitchhiking or background selection. Such questions require us to distinguish not only between selective and non-selective forces, but among selective models. New computational approaches and genome-wide datasets will both be required to answer these and many other questions in population genetics.

Selection versus demography

In addition to the obvious biological differences between tests examining selection directly on genetic variants (Chapter 7) and those focusing on the neutral variation linked to selected variants (Chapter 8), the two types of approaches differ fundamentally in the manner in which the statistical tests are carried out. Tests of direct selection are almost always conducted by comparing patterns at one type of variant (e.g., nonsynonymous) against the patterns at another type (e.g., synonymous). This comparison between sites—whether within a single gene or across the genome—ensures that any nonequilibrium population history will affect all variants equally. Because of this, tests of direct selection are generally robust to population demography.

By contrast, tests of linked selection are often constructed as comparisons to a theoretical expectation (e.g., the shape of the allele frequency spectrum) and are therefore extremely sensitive to both the sampling scheme and the demographic history of populations. Even tests that appear to be making contrasts among different regions—and therefore might be more robust to nonequilibrium demography (like the HKA test)—are quite sensitive to the variance in levels of diversity among loci, which can be affected by population history. Levels of linkage disequilibrium are also affected by nonequilibrium histories

(Chapter 9), so tests using LD must also consider the demographic history of the population from which the data are drawn. Overall then, it is recognized that many of our tests use a constant-sized, randomly mating population to derive their expected (null) values. As a consequence, the tests are not simply asking whether the data reject a neutral model, but rather whether they reject a compound neutral-equilibrium model. Simply using this equilibrium null model will result in rejection under a wide variety of circumstances (Wares 2010). Therefore, in order to identify true targets of selection, we need some way to separate the effects of nonequilibrium demography from the effects of natural selection.

A favorite aphorism in population genetics is that selection acts locally in the genome while demography acts globally (Lewontin and Krakauer 1973). The implication of this statement is that the patterns of variation found in the majority of the genome represent the demographic history of a population, while any outlier loci represent the targets of selection (**FIGURE 10.1A**). Unfortunately, many forms of selection affect variation in the same way and are occurring throughout the genome. For instance, both background selection (Ewing and Jensen 2016) and hitchhiking (Schrider, Shanku, and Kern 2016) can have similar effects on levels of variation and therefore on demographic inference. If selection is common enough across the genome—and it appears to be in most eukaryotic organisms (Corbett-Detig, Hartl, and Sackton 2015)—then the average pattern of variation represents the average effect of linked selection, not demography (**FIGURE 10.1B**).

This may be obvious from the previous discussion, but it should be stated explicitly: we cannot determine how much variation across the genome is affected by linked selection by naïvely applying tests of selection that are sensitive to demographic history. If we do so we will either overestimate this value (by ignoring nonequilibrium demography) or underestimate this value (by assuming that widespread deviations from the null are due only to demographic history; see below). Instead, the approach researchers have taken to determine the overall effects of linked selection is to test associated predictions of the selection model. One such prediction is that levels of variation will be positively correlated with levels of recombination (Begun and Aquadro 1992). Under neutrality, no relationship between levels of polymorphism and recombination is expected, as the number and frequency of neutral mutations is unaffected by recombination (Hudson 1983b). In the presence of selection, however, levels of polymorphism are reduced by an amount proportional to the strength of selection and the recombination rate (Kaplan, Hudson, and Langley 1989; Charlesworth, Morgan, and Charlesworth 1993; Barton 1998). This means that there will be less polymorphism in regions of lower recombination and more polymorphism in regions of higher recombination (which are more likely to escape the effects of linked selection). In fact, a positive relationship between diversity and recombination now appears to be one of the most universal patterns in population genetics (Hahn 2008; Cutter and Payseur 2013; Corbett-Detig, Hartl, and Sackton 2015), while alternative neutral explanations for this relationship have been excluded (Begun and Aquadro 1992; McGaugh et al. 2012; Pease and Hahn 2013).

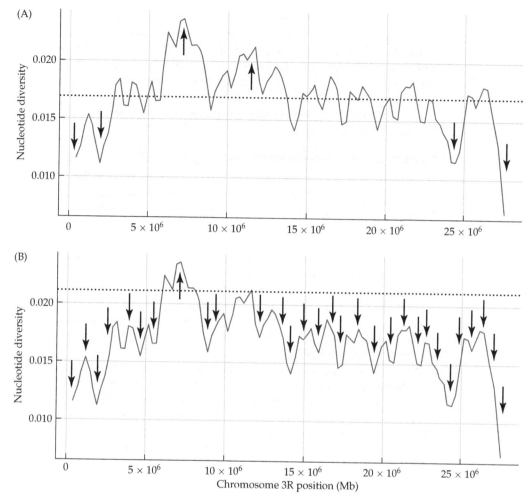

FIGURE 10.1 Variation in levels of polymorphism: (A) the neutralist interpretation, and (B) a selectionist interpretation. The red line in both shows measured values of polymorphism (π) across *D. simulans* chromosome 3R (Begun et al. 2007). Arrows indicate hypothetical effects of linked selection in raising or lowering levels of polymorphism. The dashed line represents the expected mutation-drift equilibrium level of polymorphism under neutrality, given as the average value on chromosome 3R in (A) and as a hypothetical value unaffected by linked selection in (B). (From Hahn 2008.)

The positive correlation between polymorphism and recombination across many species is striking for a number of reasons. First, these results imply that almost no loci are free from the effects of linked selection. Far from being limited to only the regions of lowest recombination, published patterns suggest that all loci but those with the highest rates of recombination are affected by such selection—and even these loci may simply show the least effects of linked selection. Second, recall that in the absence of other forces, the reduction in variation caused by linked selection will rebound to equilibrium levels relatively rapidly (Chapter 8). The fact that polymorphism is correlated with recombination implies that in almost every species examined, at almost every locus, there has recently been a selected allele nearby (whether advantageous or deleterious) such that levels of polymorphism are not at mutation-drift equilibrium. While there are limits to the rate of adaptive substitution (Weissman and Barton 2012), these data appear to be fundamentally incompatible with the view that demographic history is the main determinant of levels and patterns of variation.

General approaches for identifying targets of selection

Discovering that linked selection likely affects a vast majority of the genome is cold comfort for researchers hoping to identify individual loci under selection. In fact, in many ways these two goals act counter to one another: "Ironically, the ability to detect individual instances of selection can decrease as the fraction of the genome affected by linked selection grows" (Haasl and Payseur 2016, pg. 9). While in some cases it may be of interest simply to ask whether a dataset shows signs of any linked selection (e.g., Wright et al. 2005; Caicedo et al. 2007), the goal of many researchers is to find the targets of selection. So how do we identify specific loci affected by positive or balancing selection against a background of rampant linked selection and nonequilibrium population histories? There are two main methods: model-based and outlier ("empirical") approaches.

Model-based approaches generally follow the view that a population's true demographic history can be known, and that we can identify selected loci as those that do not fit this history in some manner. The first step in such an approach is therefore to estimate demographic history, either by using all the data or by using a subset of the data believed to be less affected by selection (e.g., Gazave et al. 2014); alternatively, statistical models without explicit demographic parameters can also be fit to the data (e.g., Whitlock and Lotterhos 2015). Because learning demographic histories is difficult for many reasons other than the presence of selection (Chapter 9), some authors have also used a range of possible histories that have been previously proposed for their species (e.g., Akey et al. 2004). In order to determine the significance of a gene in the tail of the distribution for some statistic (e.g., Tajima's D), most model-based approaches simulate data under the inferred demographic model. A large number of simulated loci generated under this model can be turned into a two-tailed P-value by finding the index of the observed statistic among the simulated values (i) and applying the following formula (Voight et al. 2005):

$$P = 1 - 2 * |0.5 - i| \tag{10.1}$$

For example, if the observed value is the second-highest out of 10,000 simulated values, then $i = 1/10{,}000$ and $P = 0.0002$. Likewise, if it is the second-lowest value ($i = 9{,}999/10{,}000$), the P-value will be the same. However, because population genomics often involves testing a large number of loci, these are only nominal P-values. Care must be taken to correct or account for large numbers of tests (**BOX 10.1**).

■ BOX 10.1
Multiple-testing issues in population genomics

One feature common to population genomic studies is the calculation of many test statistics, and consequently the carrying out of many statistical tests. While there is an ongoing debate over the merits of frequentist versus Bayesian methods in statistics, there is no doubt that frequentist approaches continue to be important in population genetics. As a result, it is still necessary to use P-values and associated concepts in order to justify claims regarding loci of interest.

P-values represent the probability of obtaining data as extreme, or more extreme, than observed if the null hypothesis is true. It is common to use a P-value cutoff of 0.05 as the threshold for assigning significance to a result. If we carry out only a single test, then this threshold corresponds to a probability of $\alpha = 0.05$ that we have rejected the null hypothesis when it is actually true; these are *false positives* or *type I errors*. A problem arises when we carry out more than one test: in this case the P-value cutoff is no longer the same as the value of α. That is, if we want to continue to have a 5%

chance of observing a single false positive, our P-value cutoff must change.

To better understand the problem of multiple testing, it is useful to consider the distribution of P-values when the null hypothesis is true (i.e., when there are no true positives). Under the null hypothesis, P-values are uniformly distributed (Box Figure 10.1A). If we have carried out 1,000 tests, we expect 50 of them to have $P < 0.05$. (Note that this expectation is itself binomially distributed with $n = 1,000$ and $p = 0.05$.) But we also expect 50 of these tests to have $0.35 < P < 0.40$ (or any similar interval). This implies that when we carry out 1,000 tests, even if our null hypothesis is true, we are almost guaranteed to have "significant" results with $P < 0.05$. It would be much more surprising if we did not observe tests with P-values this low.

How then do we ensure that we maintain a probability α of a single false positive when we are carrying out multiple tests? If we think of drawing random numbers from the P-value distribution under the null, our task is to find a

(A)   (B)

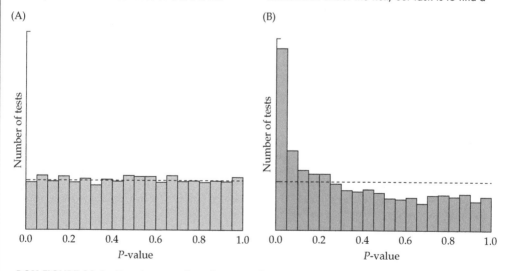

BOX FIGURE 10.1   Distributions of P-values. (A) The distribution of P-values expected when the null hypothesis is true. The dashed line shows the expectation if all tests follow the null. (B) The distribution of P-values when there are true positives in the dataset. (B uses data from Hedenfalk et al. 2001; after Storey and Tibshirani 2003; copyright 2003 National Academy of Sciences, USA.)

(Continued)

■ BOX 10.1 *(continued)*

new *P*-value cutoff (call it α^*) so that we again have only a 5% chance of choosing a value as low or lower given *n* such draws. There are two commonly used methods for doing this. The simplest is called a *Bonferroni correction*, and it was popularized in evolutionary biology by Rice (1989). The Bonferroni-corrected *P*-value cutoff when conducting *m* tests is:

$$\alpha^* = \frac{\alpha}{m}$$

This new cutoff means that individual tests must achieve a *P*-value of less than α^* to be considered significant. The Bonferroni correction is in fact overly conservative and results in a false positive rate of less than α. A more accurate method uses the *Šidàk* (or *Dunn-Šidàk*) *correction* (Šidàk 1967):

$$\alpha^* = 1 - (1 - \alpha)^{1/m}$$

Both the Bonferroni and Šidàk corrections are regularly used in population genomics.

The above approach to dealing with multiple tests is good at preventing a single false positive but can be overly restrictive. For instance, if we have carried out 1,000 tests, then the standard Bonferroni correction would require any single test to have a *P*-value of less than 0.00005 in order to be significant at $\alpha = 0.05$. For some experiments it may be difficult to reach this value, even when there are true positives in the dataset. Rather than have no or only a small number of significant results after correcting for multiple tests, researchers may instead be willing to accept a few false positives if the set of significant tests includes a large number of true positives. An approach that accomplishes this task is to control the *false discovery rate* (*FDR*) instead of the probability of a single false positive.

The false discovery rate is defined as the ratio of the expected number of false positives to the observed number of tests significant at a specific *P*-value (Benjamini and Hochberg 1995). While we cannot determine which tests are false positives, we have already seen that under the null we can easily establish the number of expected false positives at any particular *P*-value: this is just the total number of tests (*m*) times the *P*-value. With *m* = 1,000 we expect 50 false positives at $P < 0.05$, 10 false positives at $P < 0.01$, and so on. Under the null hypothesis we should observe as many significant tests as are expected, at all *P*-values, and therefore the FDR should be 1.0 at all *P*-values.

When there are true positives in a dataset, we should see more significant tests than expected (Box Figure 10.1B). For example, if we observed 200 tests with $P < 0.01$ and *m* = 1,000, then the false discovery rate would be 0.05 because we expected 10 significant tests at this *P*-value but observed 200 (=10/200). We have observed 20 times more significant results than expected by chance, and of all 200 significant tests we expect that only 10 of them will be false positives. We still do not know which are the false positives, but using the FDR has provided us with a list that includes approximately 190 true positives.

The FDR can be used in multiple different ways, and there are slightly different ways of defining it (Benjamini and Hochberg 1995; Storey 2002; Storey and Tibshirani 2003). Starting with a predetermined *P*-value threshold for a dataset, we can report the FDR at this threshold. Alternatively, we can choose an FDR threshold *a priori*, finding the false discovery rate associated with each of our tests in a ranked list of *P*-values and accepting only those tests with an FDR lower than our threshold. The minimum false discovery rate associated with each test is called its *q*-value (Storey 2002), which is a measure of significance in terms of FDR. Many studies in population genomics will use a *q*-value cutoff of 0.05, but even cutoffs of 0.2 or 0.3 may be used in order to amass a large set of true positive tests.

There are several issues with model-based approaches. The main one is that they assume selection is rare, and that the average signal in the data reflects demographic history. Consider data collected from a population at demographic equilibrium but with high rates of adaptive natural selection: because hitchhiking results in an excess of low-frequency mutations, model-based approaches would lead us to infer that a population expansion or bottleneck had occurred, as many loci will show this excess. To find the targets of selection we would then "re-center" the distribution of test statistics by simulating neutral data under the inferred demographic history, whether or not there is any independent (non-molecular) data supporting such a history. Complex models of demographic history that include multiple successive bottlenecks or extra migration parameters may in fact explain all of the variance in our observed test statistics. As might be imagined, this procedure will cause us to miss many or most of the genes undergoing adaptive natural selection, or even to reject those with strong evidence simply because they are in the middle of the distribution. It can therefore be considered an extremely low-powered method for detecting selection.

The alternative outlier approach does not depend on inferring a specific demographic model, though it implicitly makes the same main assumption: that selection is rare. Instead of fitting a particular demographic or statistical model, the outlier approach simply says that the loci in the 1% tails (or some other arbitrary percentile) of the empirical distribution of observed values are targets of selection. There are multiple reasons an outlier approach—often called an *empirical outlier approach*—is used. The main one is that sometimes it is difficult to fit a realistic demographic model, either because of ascertainment bias in the sample (e.g., Akey et al. 2002; Kelley et al. 2006) or simply because the suspected history is more complex than the data allow one to estimate (e.g., Kolaczkowski et al. 2011). An outlier approach can also be much more transparent than model-based approaches, in which P-values are exquisitely dependent on model parameters and choices made during simulation. In the outlier approach there cannot be a shift in the rank of individual loci because of (for instance) the recombination rate used in the demographic simulations.

There are obvious issues with outlier approaches. If there is no natural selection affecting the sampled loci, then outlier approaches will generate false positives. Fortunately, there is now little doubt that selection is affecting large amounts of all genomes, so the assumption that the tails represent selection may not be too erroneous. Conversely, the outlier approach limits us to detecting only a very small percentage of all genes affected by selection, with the number determined by our chosen cutoff value. Even if 50% of genes are targets of positive selection, using the 1% tails ensures that we will find at most 2% of them (though these genes may represent the most extreme cases). There is also ambiguity in assigning modes of selection to our outliers, especially if the mean of the observed distribution does not represent the average effect of demography (cf. Cavalli-Sforza 1966; Lewontin and Krakauer 1973). If positive selection is rampant in the data—and there is no balancing selection—then the lower tail of Tajima's D values may represent these selected loci, while the upper tail may represent the loci *least* affected by selection (and not the targets of balancing selection). As a result, care must be taken in assigning the tails of the observed distribution to specific modes of selection. However, when there is selection in a dataset, outlier approaches can be a powerful way to detect important loci (Kelley et al. 2006; Teshima, Coop, and Przeworski 2006).

METHODS FOR CARRYING OUT GENOME-WIDE SCANS FOR SELECTION

As has been described throughout the book, methods for detecting selection can use a variety of sampling schemes. One of the most fundamental divisions between approaches involves whether samples come from a single population (often including an outgroup species) or multiple populations. Population genomic studies using multiple populations appear to have been more common in the field thus far (Haasl and Payseur 2016). The reason why such studies are more popular is unclear, but it could be both because they can use non-ideal datasets—i.e., those affected by ascertainment biases (e.g., Akey et al. 2002) or based on older genomic technologies (e.g., Turner, Hahn, and Nuzhdin 2005; Turner et al. 2008)—and because multiple-population approaches do not require the genome sequence of a related species. No matter the reason for differences in popularity, I discuss specific methods used in conjunction with the two sampling schemes separately.

Detecting selection using a sample from a single population

Many population genomics methods using single populations do not differ at all from their single-locus counterparts. These methods were some of the first to be applied to population genomic datasets. Because the first such datasets often did not have genomes from outgroups, researchers used methods that did not require divergence data or polarized mutations. These methods include Tajima's D (e.g., Glinka et al. 2003; Akey et al. 2004; Nordborg et al. 2005; Stajich and Hahn 2005) and D-like statistics that can be applied to microsatellites (e.g., Payseur, Cutter, and Nachman 2002), as well as methods based on haplotype homozygosity that are also robust to ascertainment bias (e.g., Sabeti et al. 2002; Voight et al. 2006). With the sequencing of high-quality reference genomes from closely related species came the ability to use methods that incorporate divergence, including the McDonald-Kreitman test (e.g., Bustamante et al. 2005; Shapiro et al. 2007; Li, Costello, et al. 2008), Fay and Wu's H (e.g., Carlson et al. 2005; Ometto et al. 2005), and the HKA test (e.g., Begun et al. 2007). Other than the caveats discussed earlier that are common to all population genomic datasets (e.g., the problem of multiple testing), the implementation and interpretation of these statistics remain the same as in smaller datasets.

In order to take genomic context into account, Kim and Stephan (2002) created a test based on the expected "valley" of polymorphism around recently fixed advantageous alleles (**FIGURE 10.2**). These valleys highlight several important features of selective sweeps.

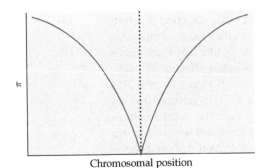

FIGURE 10.2 Effect of hitchhiking on levels of diversity. The expected value of π calculated using equation 13 from Kim and Stephan (2000), with $2Ns = 400$ and $\rho = 0.0004$ for the entire region. The sweep ended $0.005N$ generations before sampling. The vertical dashed line marks the position of the selected site. (After Kim and Stephan 2002.)

They generally contain a region of much reduced polymorphism surrounding the fixed allele, with less reduction as we move away from the selected site. As mentioned earlier in the book, the extent of reduced polymorphism is proportional to s/c, with stronger selection and lower recombination resulting in larger areas lacking variation. The allele frequency spectrum also shows spatial patterning, with an excess of low-frequency derived alleles surrounding the selected site and an excess of high-frequency derived alleles adjacent to the selected site (e.g., Figure 8.7). Although there can be much more stochasticity in individual realizations of this process than are shown in Figure 10.2 (see **FIGURE 10.3**), these general patterns can be used to create a test of neutrality.

The test designed by Kim and Stephan—which is usually referred to as the *CLRT*, for *composite likelihood ratio test*—compares the likelihood of two models: one of hitchhiking and one of neutrality. The hitchhiking model

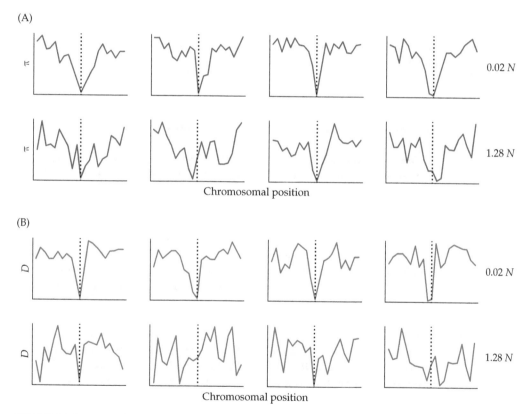

FIGURE 10.3 Example patterns of diversity generated by hitchhiking. (A) Patterns of diversity measured by π when the sweep ended $0.02N$ generations ago (top row) and $1.28N$ generations ago (bottom row). (B) Patterns of diversity measured by Tajima's D when the sweep ended $0.02N$ generations ago (top row) and $1.28N$ generations ago (bottom row). Each panel in (A) and (B) represents an independent simulated dataset, with the vertical dashed line marking the position of the selected site. The parameters used in the simulation are $\gamma = 1000$ and $\rho = 2100$ for the entire region. (From Schrider et al. 2015.)

describes the expected allele frequency spectrum of linked neutral polymorphisms at different distances from the fixed selected allele. The expected number of derived alleles in the frequency interval $[p, p + dp]$ in a region linked to a hitchhiking event is:

$$\phi_A(p) = \begin{cases} \dfrac{\theta}{p} - \dfrac{\theta}{C} & \text{for } 0 < p < C \\[2ex] \dfrac{\theta}{C} & \text{for } 1 - C < p < 1 \end{cases} \tag{10.2}$$

where the parameter C depends on both the strength of selection, $\gamma = 2N_e s$, and the recombination distance between a neutral polymorphism and the selected site, c (Kim and Stephan 2002).

The neutral model describes the allele frequency distribution in an equilibrium population without linked selection. In the neutral model there is no dependence on distance, so the allele frequency distribution is given by (Kimura 1971):

$$\phi_0(p)\,dp = \frac{\theta}{p}\,dp \tag{10.3}$$

This is the continuous analog of the discrete allele frequency spectrum given by Equation 6.8.

The probability of the data under both models is obtained by binomial sampling using the appropriate allele frequency distribution—either the hitchhiking or neutral model (i.e., either ϕ_A or ϕ_0). In order to obtain likelihoods across a region the CLRT multiplies the likelihoods for all sites together (resulting in a composite likelihood; Chapter 9), and the two models can be compared via a likelihood ratio test. The hitchhiking model can also be used to estimate the value of γ. The original implementation of the test calculates this likelihood ratio for the data centered on a candidate target of selection, with P-values generated via neutral simulations. This procedure can be repeated moving along the genome, treating each window or stretch of sequence as the candidate target of selection.

As described, there are several limitations to Kim and Stephan's CLRT. The method assumes both that mutation and recombination rates are known (or can be estimated, usually via θ and ρ in background regions) and that the null model is a neutral-equilibrium model. The method is therefore sensitive to variation in mutation and recombination, as well as nonequilibrium demographic histories. In fact, Jensen, Kim, and colleagues (Jensen, Kim, et al. 2005) have shown that the CLRT has a very high false-positive rate under a wide range of nonequilibrium histories. There are multiple ways to account for the assumption of equilibrium populations when carrying out the CLRT (see Pavlidis, Hutter, and Stephan 2008 for an in-depth review). Jensen, Kim, et al. (2005) proposed a goodness-of-fit test, comparing the fit of the data under the hitchhiking model to the fit under a model of arbitrary complexity without the same spatial patterning as the hitchhiking model. A better fit to the hitchhiking model implies that it is favored over nonequilibrium histories and should still be sensitive to true selective events.

A related method is implemented in the program *SweepFinder* (Nielsen et al. 2005). *SweepFinder* carries out the composite likelihood ratio test of Kim and Stephan (2002) by comparing a hitchhiking model to a "background" model in which the allele frequency spectrum is estimated from the whole genome, rather than assuming a neutral-equilibrium model (this is referred to as "test 2" in Nielsen et al. 2005). Essentially, the test replaces ϕ_0 with a distribution estimated from the data. The same background allele frequency spectrum is used in the CLRT for every region tested, so that regions with unusually large and significant values can be identified. As with other model-based approaches described above, *SweepFinder* controls for demographic history by effectively re-centering the distribution of allele frequencies such that targets of selection strongly differ from the genomic average. Because *SweepFinder* uses this genomic background model it can be quite robust to nonequilibrium demographic histories. However, simulations are still required to calculate the significance of individual values of the CLRT; a high rate of false positives may still occur if these simulations are carried out using an incorrect demographic model (Pavlidis, Jensen, and Stephan 2010; Crisci et al. 2013; Huber et al. 2016).

The original version of *SweepFinder* controlled for variation in the mutation rate by only considering polymorphic sites. Both the folded and unfolded frequency spectrum could be used, meaning that no outgroup genome was necessary, and various ascertainment schemes could be accommodated. However, not including invariant sites results in reduced power to detect selection (Nielsen et al. 2005), as one of the major effects of a sweep is to remove all variation. The problem with including this information is that regions may also lack variation because they have a low mutation rate. In order to address this issue, *SweepFinder2* (DeGiorgio et al. 2016; Huber et al. 2016) has the ability to include sites that are invariant but that represent fixed differences from a closely related outgroup. Related software (*SweeD*) that can apply the *SweepFinder* CLRT to thousands of individuals—and with arbitrary background demographic models—can also use invariant sites (Pavlidis et al. 2013). Including such sites greatly increases the power of the test without generating false positives as a result of regions of low mutation.

The hitchhiking model described here is one of a hard sweep (Chapter 8), with sequences collected very soon after fixation of the advantageous allele. If partial or soft sweeps have occurred—or a longer time has passed since fixation, even for a hard sweep—there may be both reduced power to detect selection and possibly misidentification of the selected site using the CLRT and related methods. It should be clear why partial and soft sweeps could cause us to miss selective events: the skew in the allele frequency spectrum may be much less severe with such events than with completed hard sweeps (Przeworski, Coop, and Wall 2005; Pennings and Hermisson 2006b). We may therefore have little power to reject neutrality, especially when there is a nonequilibrium history. Although significant overlap has been observed between the targets of selection identified by *SweepFinder* and the iHS test (Williamson et al. 2007), it is not clear whether this occurs because the same partial sweeps identified by iHS are also identified by *SweepFinder*, because the same loci are hit repeatedly by multiple selective events, or because the signatures of partial sweeps are really just the shoulders of the completed sweeps (Schrider et al. 2015).

The theoretical expectations of the patterns generated by a hitchhiking event—with the lowest levels of diversity always directly at the site of selection and a nice symmetry to the recovery of diversity (Figure 10.2)—do not always hold. The shape of these valleys of diversity can vary quite a bit as a result of the stochasticity of the coalescent process and recombination during the sweep (Kim and Stephan 2002). If the region surrounding the selected substitution is not always the one with the lowest diversity, we may be misled as to the selected site. Two factors play a major role in determining whether the target of selection is correctly identified. First, the less time that has passed since a sweep has ended, the more accurate our inferences will be (Figure 10.3). If variation was assayed 0.00125N generations after the completion of a sweep, π reaches its minimal value in the window directly around the advantageous allele 95.8% of the time, which would likely result in accurate identification of this site (Schrider et al. 2015). If instead the data are collected 1.28N generations after the sweep has ended, π reaches its minimal value in this window only 61.9% of the time (Figure 10.3A). The accuracy of such measures goes down as more time passes and can be worse at all time points for measures of the allele frequency spectrum such as Tajima's D (Figure 10.3B). The second major factor determining the accuracy of our inference is the demographic history of populations: against the background of a nonequilibrium population history, accuracy is again much lower. Pavlidis, Jensen, and Stephan (2010) found that *SweepFinder* has very low precision in predicting the target of selection when a sweep takes place during a population bottleneck. This low level of precision is due to the variability in genealogies introduced by the bottleneck, which effectively obscures both the signal of the hitchhiking event and the location of the selected site.

Detecting selection using samples from multiple populations

Genome-wide scans for selection are easy to carry out using data from multiple populations and do not require the sequence of a closely related outgroup species. There are two main approaches to detecting selection from multiple-population samples (Storz 2005): those using differences in allele (or haplotype) frequencies and those based on comparisons of levels of diversity. Most methods take advantage of differences in allele frequencies at individual sites, with the expectation that directional selection in one subpopulation will increase differences, while balancing selection across subpopulations will decrease differences (Chapter 5). One of the reasons these methods have become so popular is that many types of data can be used with them, including genotyping platforms (e.g., Eckert et al. 2010), pooled sequencing (e.g., Turner et al. 2010), and RAD-seq (e.g., Hohenlohe et al. 2010). Methods comparing levels of variation at a locus between multiple populations can also uncover selective sweeps confined to single subpopulations, but are much less reliable when using previously ascertained markers. I discuss these approaches in turn.

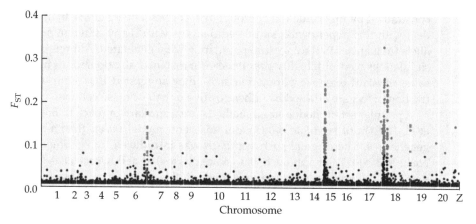

FIGURE 10.4 Genome-wide scan for selection using F_{ST}. Polymorphism data from two populations of the butterfly *Heliconius melpomene* were used to calculate F_{ST}. Significant outlier loci were identified using *BayeScan* (Foll and Gaggiotti 2008). The major peaks of outlier markers on chromosomes 18 and 15 correspond to regions known to control wing-pattern variation between populations. (From Nadeau et al. 2014.)

The most common approach to genome-wide scans for selection involves calculating F_{ST} or F_{ST}-like statistics among populations. Such analyses can be performed pairwise or using all populations at once and are relatively straightforward to carry out and interpret. Viewing the genome-wide spatial distribution of F_{ST} makes it easy to identify extreme values in individual polymorphisms or regions (**FIGURE 10.4**). Outlier approaches using F_{ST} were some of the first genome-wide scans conducted (e.g., Akey et al. 2002) and are still regularly used to identify targets of selection.

The first model-based attempt to detect selection on individual loci with F_{ST} used data from five human populations but only 100 markers (Bowcock et al. 1991). Loci under selection were identified by simulating neutral markers on the tree relating the populations, keeping the average F_{ST} in the simulated data equal to the observed value. Outliers appeared as loci with values of F_{ST} higher or lower than the 95th or 5th percentile, respectively, given their allele frequency (**FIGURE 10.5**). The dependence on allele frequency is key, as the range of F_{ST} is

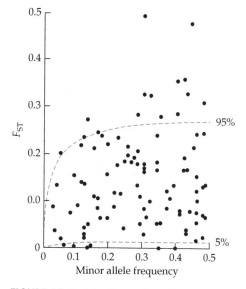

FIGURE 10.5 Identifying F_{ST} outliers. Polymorphism data from 100 loci in five human populations are shown. Curves represent the 95% and 5% cutoffs of F_{ST} values observed in neutral simulations conducted across the population tree, starting from different initial minor allele frequencies. (After Bowcock et al. 1991.)

constrained by this value (Jakobsson, Edge, and Rosenberg 2013). In order to deal with this dependence, simulated datasets with many different starting allele frequencies in the ancestral population were generated. Although these simulations were still highly simplified—for instance, all populations had the same constant effective population size—they suggested that a substantial fraction of loci are affected by either positive or balancing selection.

A popular set of model-based methods use a similar approach to Bowcock et al. (1991) but simulate data under an island model rather than a phylogenetic tree. The original such approach was introduced by Beaumont and Nichols (1996). Their approach was analogous to that of Bowcock et al. (1991), with two differences: an island model with 100 populations was used for the simulations, and F_{ST} outliers were identified as a function of expected heterozygosity rather than allele frequency. The original software running these simulations, *FDIST*, has been parallelized in the newer program, *LOSITAN* (Antao et al. 2008), which automates many of the tedious steps involved. *LOSITAN* can also iteratively remove outlier loci in order to re-run simulations that may better approximate the expected neutral distribution. Further elaborations on this basic model have been made. One of the most significant has been the introduction of a Bayesian approach to identifying outlier loci (Beaumont and Balding 2004; Foll and Gaggiotti 2008; Riebler, Held, and Stephan 2008). The most popular implementation of this model (in the program *BayeScan*; Foll and Gaggiotti 2008) estimates locus-specific selection parameters in order to identify the targets of selection. (This program was used to identify the targets of selection in Figure 10.4.)

One very important issue with using the island model is that it ignores the hierarchical relationships among populations that exist in a phylogenetic tree. This problem is in fact the same one faced by the original Lewontin-Krakauer test (Lewontin and Krakauer 1973; Nei and Maruyama 1975; Robertson 1975) and can lead to many false positives in scans for selection (Meirmans 2012; Vilas, Pérez-Figueroa, and Caballero 2012; Bierne, Roze, and Welch 2013; Fourcade et al. 2013; Lotterhos and Whitlock 2014). One obvious solution is to simulate the phylogenetic relationships among populations (as in Bowcock et al. 1991), but other solutions include using a hierarchical island model (implemented in *Arlequin*; Excoffier, Hofer, and Foll 2009), using only two populations at a time (implemented in *DetSel* [Vitalis, Dawson, and Boursot 2001] and *XP-CLR* [Chen, Patterson, and Reich 2010]), or finding the best-fitting χ^2 distribution of F_{ST} values (implemented in *OutFLANK*; Whitlock and Lotterhos 2015). Note that all of these model-based methods assume that loci are unlinked from one another, which can be a problem when single selective events affect multiple polymorphic sites (see below).

The tests described thus far, with the exception of *DetSel* and *XP-CLR*, are generally carried out by calculating F_{ST} using all populations at once when there are more than two populations. This means that one cannot immediately determine which population(s) might be responsible for significant results and therefore which one contains the signal of selection. F_{ST} can be calculated in a pairwise fashion, and from these pairwise distances one could determine which populations are driving the pattern at outlier loci. Multiple methods have been proposed to do this by explicitly identifying outlier branches on a population tree, allowing researchers to quickly identify which population is the target of selection.

The two most popular methods for doing such a calculation only work for datasets with three populations, largely because there is only one unrooted tree describing these relationships (**FIGURE 10.6**). Given three populations, P_1, P_2, and P_3, we can use pairwise F_{ST} values to estimate population-specific branch lengths for any locus in the genome. If we denote the pairwise F_{ST} value between P_1 and P_2 as F_{12}, that between P_1 and P_3 as F_{13}, and that between P_2 and P_3 as F_{23}, and the branch lengths leading to each population as x_1, x_2, and x_3 (Figure 10.6), then these branch lengths can be estimated by:

$$x_1 = \frac{F_{12} + F_{13} - F_{23}}{2}$$

$$x_2 = \frac{F_{12} + F_{23} - F_{13}}{2} \qquad (10.4)$$

$$x_3 = \frac{F_{13} + F_{23} - F_{12}}{2}$$

This method is used to calculate *locus-specific branch lengths*, or *LSBLs* (Shriver et al. 2004). The same calculations can be made in every window of the genome, and therefore these summary statistics can be used to identify outlier regions specific to individual populations. Similarly, Yi et al. (2010) defined a *population branch statistic (PBS)* that uses the same approach as LSBLs but log-transforms the F_{ST}-values before carrying out the calculations in Equation 10.4.

LSBLs and PBSs are limited to comparisons among three populations. One additional method that can deal with more than three populations builds population trees based on allele frequencies and models change in frequencies via drift (Chapter 9). A background tree can be calculated using the whole genome (similar to the approach used by *SweepFinder*) or with a set of pre-selected markers thought to be unaffected by selection (e.g., Rockman et al. 2003). A likelihood-ratio test can then be conducted by building a tree for every region of the genome and constraining the topology to be the same but allowing branch lengths to vary. Loci under selection will have branch lengths that substantially differ from those of the background tree, resulting in a significant likelihood-ratio test (Rockman et al. 2003). The effects of individual populations can be explored by removing each in turn and recalculating the test.

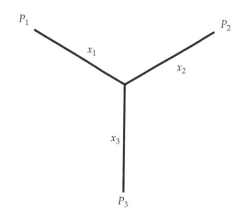

FIGURE 10.6 Locus-specific branch lengths. Given pairwise measures of F_{ST} between three populations, P_1, P_2, and P_3, we can estimate the length of the branches leading to each population: x_1, x_2, and x_3 (Equation 10.4). While one population may have longer branches overall as a result of increased genetic drift, locus-specific outliers can help to identify the targets of selection.

Genome-wide scans using multiple populations can also use levels of variation to identify population-specific sweeps as those regions with low diversity in only a single population. This can be done using π and measures of nucleotide diversity (e.g., Schlenke and Begun 2004; Stajich and Hahn 2005; Oleksyk et al. 2008; Montague et al. 2014), though such an approach has more often been used with microsatellites (e.g., Kohn, Pelz, and Wayne 2000; Harr, Kauer, and Schlötterer 2002; Payseur, Cutter, and Nachman 2002; Kauer, Dieringer, and Schlötterer 2003; Kayser, Brauer, and Stoneking 2003; Storz, Payseur, and Nachman 2004). The comparisons are usually done between pairs of populations in which some form of parent-daughter relationship exists, such that the daughter populations are thought to be the ones undergoing more selection. Examples include African and non-African humans and flies (both of which originated in Africa) and populations of rat that have recently become resistant to rodenticides (Kohn, Pelz, and Wayne 2000).

A popular approach for carrying out such comparisons at microsatellite loci uses the variance in repeat number (V) in each of two populations. Recall that V is an estimate of the population mutation parameter (Equation 3.23) and is therefore a natural statistic with which to measure levels of variation. Using the ratio of the variance in repeat number (RV) in two populations, Schlötterer (2002) proposed taking the log of this value as a test statistic:

$$\ln(RV) = \ln\left(\frac{V_{P1}}{V_{P2}}\right) \tag{10.5}$$

The advantage of using $\ln(RV)$ rather than RV itself is that the log-transformed values are expected to follow a normal distribution when there is no selection. This makes hypothesis testing quite straightforward, and the normality of $\ln(RV)$ is robust under both a wide range of population histories and variation in mutation rates and mutation models (Schlötterer 2002). A related statistic, $\ln(RH)$, uses the ratio of microsatellite heterozygosity in two populations (Kauer, Dieringer, and Schlötterer 2003). This statistic is also expected to be normally distributed when there is no selection, making it easy to identify outlier loci.

CAVEATS ABOUT NON-INDEPENDENCE IN POPULATION GENOMICS

Population genomic datasets, and genome-wide scans for selection in particular, are prone to problems because of the lack of independence among test statistics. A lack of independence results in many issues, from false positive tests to biased parameter estimates. The most widespread problem may simply be that correlation among measures leads to pseudoreplication of patterns present in only a small fraction of observations. These non-independent pieces of evidence can then be interpreted as stronger support for a result than actually exists.

There are multiple causes of non-independence, some biological and some methodological. An obvious methodological cause is the use of sliding windows in genome-wide analyses. In sliding window analyses a summary statistic is calculated for a specific region of the genome, with the same statistic calculated in constant-sized partially overlapping windows moving along the chromosome. Because each window contains some of the same data as the last window, however, the individual measures are non-independent. Using these measures will lead to underestimated standard errors and hence tests that are too liberal (Hahn 2006). At a minimum, non-overlapping windows must be used.

A second cause of non-independence—partially methodological and partially biological—arises from the fact that many different summary statistics in population genetics use the same underlying features of the data. For example, both Tajima's D and F_{ST} are affected by allele frequencies, as is captured in the statistic π. While they are not perfectly correlated, this dependence on a shared variable means that extreme values of both statistics in a genomic region should not be taken as independent support for natural selection. Later in this chapter we will discuss statistically rigorous ways to combine correlated measures, but for now it is important to point out that these sorts of dependencies are common in population genetics and can be quite insidious. A non-obvious example comes from studies that first carry out a divergence-based comparison between species and then conduct population genetic analyses within species. As pointed out by Kern (2009), the selection of candidate regions based on levels of divergence creates an ascertainment bias that affects the derived allele frequency spectrum of polymorphisms. To see why this is, consider the choice of a single sequence from a natural population as a reference genome (the last section of Chapter 6 presents a similar issue). If in comparisons with reference genomes from other species we observe very few nucleotide differences in a particular region, this observation itself implies that the region contains only low-frequency derived polymorphisms, and likely very few of them. Conversely, between-species comparisons that reveal a large number of differences imply a large number of high-frequency derived alleles. Therefore, a skewed frequency spectrum in regions identified first through comparisons between species cannot be used as independent evidence for inferences regarding natural selection, unless a correction for the ascertainment step is applied (Kern 2009).

Other forms of non-independence arise because nearby loci share both genealogical histories and underlying mutation rates. As a consequence, non-overlapping and even non-adjacent nearby loci are likely to have correlated levels of polymorphism, levels of divergence, allele frequency spectra, and other measures of variation (Hahn 2006). Thornton and Jensen (2007) showed how an ascertainment bias can arise when a small number of windows are chosen because they have low levels of diversity, and then flanking windows are subsequently sequenced in order to carry out neutrality tests. The act of selecting only outlier regions as a follow-up induces a bias in these tests because the flanking windows share a genealogical history with the original window and therefore do not represent independent evidence for selection. While this exact procedure is not used very much anymore, similar situations arise even with whole-genome data. The fact that many of our analyses assume that each of our observations is independent means that we may be misled as to the strength of signal in the data. Single selective events can cause very large regions of the genome to have extreme summary statistics (e.g., Tishkoff et al. 2007), and severe demographic events may do the same (e.g., Gray et al. 2009). If we count each gene or window in these large regions as an independent observation, the significant results present in a small subset of the data will be pseudoreplicated—it will not be clear whether a group of outliers represents the same event or multiple events in the same genomic region. As a consequence, there may be more outliers observed than expected by chance (e.g., Akey et al. 2004; Stajich and Hahn 2005). In the next section of the book I discuss newer methods for dealing with spatial correlation in measures of variation.

FUTURE APPROACHES

In this last part of the book I discuss several areas of research in molecular population genetics that are just taking off but seem likely to be very important in the coming years. This is by no means meant to be an exhaustive list—it is simply a short list of four areas that have captured my imagination. The areas reflect the opportunities created by increases in both computational and sequencing power. As the cost of both continues to fall, and as sequencing technologies produce new kinds of data, it is also easy to predict that approaches that do not even exist today will be important in the near future. A discussion of these areas will have to await future editions of the book.

The application of machine learning to population genetics

Machine learning is an active area of research combining methods from computer science and statistics in order to identify patterns in messy data. As population genetics is filled with such data, it seems almost inevitable that the tools from machine learning would be applied here. The basic idea behind the use of machine learning tools is to learn the underlying state or parameter values in our data using algorithmic models. We are familiar with many of the population genetic parameters we would like to learn: N_e, θ, γ, and so on. The underlying states correspond to a similarly wide variety of relevant labels for different loci, including constrained/unconstrained, hard/soft sweeps, and high/medium/low gene flow. Given the stochasticity of the evolutionary process, machine learning methods allow us to make more accurate inferences regarding these values.

Machine learning approaches can be usefully divided into *unsupervised* and *supervised* methods. The distinction between unsupervised and supervised methods will become clearer below when I explain supervised methods, but for now suffice it to say that unsupervised methods are used in cases in which we do not assign the true states or parameters. In Chapter 9 we briefly discussed one popular method that is often used for unsupervised learning—hidden Markov models (see Durbin et al. 1998 for a detailed description). HMMs model a dataset as having been indirectly generated by a Markov chain with varying states, and therefore naturally deal with the correlations among adjoining regions of the genome. If our dataset consists of polymorphism along a chromosome, we imagine that at each position the Markov chain has some unknown (hidden) state. In each hidden state a different set of parameters is used to produce the observed sequence data. As an example, if the states correspond to different T_{MRCA}s (e.g., Li and Durbin 2011), then different numbers of polymorphisms will be "emitted" when the T_{MRCA} is recent versus long ago; if the states correspond to regions experiencing different forms of selection (e.g., Kern and Haussler 2010), different allele frequency spectra will be emitted in each.

One use of HMMs is to segment the genome into regions with different labels (states), and to do so without determining a single window size to use for analysis ahead of time. The first HMM used in biological sequence analysis segmented the genome into regions of differing GC content (Churchill 1989), and HMMs have been applied in phylogenetics for quite a while (e.g., Felsenstein and Churchill 1996; Siepel and Haussler 2004b). In population genetics HMMs have found a number of uses, being employed to identify regions of the genome that are constrained or unconstrained (Schrider and Kern 2014), to label regions of different population ancestry along a chromosome (Falush, Stephens, and Pritchard 2003; Lawson et al. 2012; Brandvain et al. 2014), to uncover regions of higher or lower gene flow between populations (Turner, Hahn, and Nuzhdin 2005; Hofer, Foll, and Excoffier 2012), to identify regions affected by selective sweeps (Boitard, Schlötterer, and Futschik 2009), to determine the form of selection acting on different regions (Kern and Haussler 2010), and to learn demographic histories (discussed in Chapter 9).

Supervised machine learning methods are applied first to training data for which two or more classes of data have known states (labels) in order to learn an optimal classification scheme. In population genetics the training data are often simulated under a variety of known selective and demographic parameters. The methods are then used to classify observed data that are not labeled, thus learning what the underlying states are. In addition to the training step, supervised methods differ from commonly used unsupervised methods in being able to combine many summary statistics of the data, even when they are correlated. Whereas the hidden states of an HMM often only generate one aspect of the data at a time (such as the allele frequency spectrum), supervised methods can be applied to an assembly of measures, each of which captures a slightly different signal in the data. For instance, Pavlidis, Jensen, and Stephan (2010) used a supervised learning approach to show that combining signals from the allele frequency spectrum and linkage disequilibrium resulted in more accurate and powerful detection of selective sweeps than using either statistic alone. The use of many more pieces of data for each region of the genome means that supervised methods can be highly accurate, though the size of the regions considered must be specified *a priori*.

There are many different supervised methods that have been used in population genetics, but almost all have been applied to the problems of detecting regions under positive selection or determining what form of positive selection has been acting (or whether regions are constrained; Schrider and Kern 2015). *Support vector machines (SVMs)* and *boosting* are two methods that have been used to distinguish regions affected by selective sweeps from those that have not been affected (e.g., Pavlidis, Jensen, and Stephan 2010; Lin et al. 2011; Ronen et al. 2013; Pybus et al. 2015). SVMs have further been used to classify regions where sweeps have occurred into those affected by hard sweeps or soft sweeps (Schrider et al. 2015), though even higher accuracy has been achieved in this area using a supervised method known as an *extra-trees classifier* (Schrider and Kern 2016). Finally, *deep learning methods*, based on neural networks, have recently been applied to the task of jointly estimating demography and selection while still determining the form of selection acting on each region of the genome (Sheehan and Song 2016). The application of these and other powerful tools to population genomic data are sure to further transform the way inferences are made in the future.

Ancient DNA and genetic variation through time

One of the most exciting areas opened up by advances in genome sequencing has been the field of ancient DNA, or *paleogenetics*. Sequences from preserved samples were first obtained in the mid-1980s (Higuchi et al. 1984; Pääbo 1985), but the degradation of ancient DNA combined with large amounts of bacterial contamination meant that early studies focused on high-copy-number loci in the mitochondrial or chloroplast genomes. The massive drop in costs that accompanied the first wave of next-generation sequencing technologies meant that large amounts of nuclear sequence could be obtained, even after discarding microbial DNA.

The first large-scale ancient DNA datasets included 40,000-year-old extinct cave bears (Noonan et al. 2005), a woolly mammoth (Poinar et al. 2006), and Neanderthals (Green et al. 2006; Noonan et al. 2006). Whole genomes were soon sequenced for an ancient human (Rasmussen et al. 2010), a Neanderthal (Green et al. 2010), and an ancient hominin from a previously unknown group now referred to as Denisovans (Reich et al. 2010). Genome sequences are currently available from hundreds of ancient humans (reviewed in Slatkin and Racimo 2016)—as well as from many other species (e.g., Ramos-Madrigal et al. 2016)—with more being produced all the time (Shapiro and Hofreiter 2014).

While these ancient sequences are interesting for many different reasons, from a population genetics perspective they present new challenges. The field of *paleopopulation genetics* (Wall and Slatkin 2012) must deal with samples that existed at many different points in time, possibly spread across a geographic region, and without always having clear assignment to known extant or extinct species. Population geneticists have in fact been dealing with such serial samples for some time. Often these datasets have come from viruses sampled within a single individual (e.g., Holmes et al. 1992; Wolinsky et al. 1996), from a small number of samples taken from species of conservation or economic importance (reviewed in Schwarz, Luikart, and Waples 2007), or from mitochondrial sequences in ancient samples (e.g., Hofreiter et al. 2002; Lambert et al. 2002; Orlando et al. 2002; Shapiro et al. 2004). The first methods using such serial samples were able to estimate N_e from allele frequencies taken at two time points (Krimbas and Tsakas 1971; Pamilo and Varvio-Aho 1980; Nei and Tajima 1981b; Waples 1989), the change in N_e from similar data (Williamson and Slatkin 1999; Anderson, Williamson, and Thompson 2000), or the joint history of N_e and migration (Wang and Whitlock 2003). When two or more samples are available, a modification of the standard coalescent model that deals with non-contemporaneous sequences becomes useful (Rodrigo and Felsenstein 1999). There are now a number of genealogically based methods that can be applied to serial samples to estimate changes in N_e over time (e.g., Fu 2001; Berthier et al. 2002; Drummond et al. 2002; Beaumont 2003; Anderson 2005), the generation time of sampled populations (e.g., Rodrigo et al. 1999), and the mutation rate (e.g., Rambaut 2000; Drummond et al. 2002), even after accommodating the DNA damage that occurs with ancient samples (e.g., Rambaut et al. 2009; Racimo, Renaud, and Slatkin 2016). Coalescent simulators also exist that can generate data that come from many different time points (e.g., Anderson et al. 2004; Kern and Schrider 2016).

With the availability of much larger serially sampled datasets has come a focus on detecting the targets of selection over time. General statements about selection in genes known to be involved in important phenotypes can be made from patterns of allelic diversity in ancient samples (e.g., Jaenicke-Despres et al. 2003). But the first method that could estimate $\gamma = 2N_e s$ from temporal datasets and presented a test for selection was provided by Bollback, York, and Nielsen (2008). Application of this method revealed loci under selection during the domestication of horses (Ludwig et al. 2009) but found that a gene thought to be selected in human history had a pattern of allele frequency change consistent with $s = 0$ (Bollback, York, and Nielsen 2008). There are now a number of methods for estimating γ being applied to an ever-expanding set of temporal data (e.g., Malaspinas et al. 2012; Mathieson and McVean 2013; Steinrücken, Bhaskar, and Song 2014; Foll, Shim, and Jensen 2015; Gompert 2016; Jewett, Steinrücken, and Song 2016; Malaspinas 2016; Schraiber, Evans, and Slatkin 2016). Importantly, increased sampling in extant population across seasons or years (e.g., Mueller, Barr, and Ayala 1985; O'Hara 2005; Bergland et al. 2014) means that these methods will be applicable to a wide range of organisms in the near future.

Populations of whole assembled genomes

Within the broad field of population genomics, almost all sequencing and analysis has been carried out using short sequencing reads mapped against a common reference genome (Chapter 2). Whether samples are sequenced individually or in pools, this strategy by necessity means that most of the variation being studied must be present in this reference genome. The current approach to sequencing does not allow us to query nucleotide variants in unassembled regions, in regions that are not present in the individual sequenced to generate the reference (e.g., Kidd et al. 2008), or in regions where non-reference sequences are exceptionally divergent (e.g., Raymond et al. 2005). Even when copy-number variants can be found, most methods only allow us to detect additional copies of sequences found in the reference genome, and this ascertainment bias can affect inferences about the allele frequency spectrum of CNVs (Emerson et al. 2008). In addition, reads from highly repetitive regions often cannot be uniquely mapped, which means that it is difficult to assess variation either in the number of repeats present or in the sequences of these repeats. As a consequence, there are many limitations to current sequencing approaches for studying multiple types of genetic variation, including inversions, insertions, and repeats (Chaisson, Wilson, and Eichler 2015).

Fortunately, next-generation sequencing methods have made it possible to collect populations of whole genomes, each representing a uniquely assembled individual. Single-copy regions not present in the reference, as well as highly diverged haplotypes, have been uncovered by sequencing and assembling a population sample of genomes using standard Illumina technologies (e.g., Gan et al. 2011). However, the most promising approach for understanding more complex variation is to use *long-read* sequencing technologies. These new technologies—sometimes referred to as *third-generation sequencing*—produce contiguous sequences many kilobases in length. At the moment two companies are producing this type of data for general use: Pacific Biosciences ("PacBio") and Oxford Nanopore Technologies. The platforms produced by these companies generate very long sequences that can be used to assemble whole genomes or previously recalcitrant regions of "finished" genomes (e.g., English et al. 2012; Huddleston et al. 2014; Quick, Quinlan, and Loman 2014; Ashton et al. 2015; Chaisson et al. 2015; Jain et al. 2015; Chakraborty et al. 2016) or to discover and genotype CNVs (e.g., Ritz et al. 2014; Rogers et al. 2014). Other approaches use modifications of short-read sequencing technology to produce "synthetic" long reads and have also successfully been able to resolve highly repetitive elements (e.g., McCoy, Taylor, et al. 2014). Regardless of which platforms are used to construct genome assemblies, it is clear that population samples of such assemblies will soon exist for many species.

Samples of assembled genomes will present many challenges not only to population genetics, but also generally to the computational analysis of sequence variation. There are many questions raised when we have multiple assembled genomes, for which there are currently few clear answers. For instance, a reference genome provides a common set of genomic positions to which all researchers can refer, as well as a common set of gene annotations. But when we collect a sample of separately assembled genomes, what coordinate system should we use? If we assign a single genome as the reference, what coordinates should we use to refer to the many megabases of sequence present only in non-reference individuals? One partial solution recently adopted in the representation of the human genome is to include "alternate" reference sequences for several regions that are known to be highly divergent from the reference genome (e.g., MHC). In these regions multiple haplotypes are represented in the reference, which allows reads from individuals carrying the alternate haplotypes (relative to the arbitrary one present in the reference) to be mapped. Another solution is to use a *population reference graph* (Dilthey et al. 2015), a graph structure that contains many possible variants—small and large—present in a population. Such graph structures solve many problems in reference-based mapping (reviewed in Paten et al. 2017). A further alternative solution might appear to be to generate an alignment of all of our genomes and to use positions in the alignment as a shared set of coordinates. However, this solution ignores several important issues with alignments. First, every new genome added will change our alignment, and therefore our coordinates—an unwanted feature in a method intended to allow clear communication about genomic positions. Second, genome alignment is a very challenging computational problem (e.g., Earl et al. 2014), one that is largely avoided by current short-read approaches (which generate the alignment as reads are mapped to the reference). So even after long-read technologies make issues of genome assembly moot—if we imagine that whole chromosomes will come out of sequencing machines one day—this will simply shift the computational burden onto genome alignment.

An additional challenge comes from the diploid nature of eukaryotic samples. Right now the reference genome for most eukaryotic species consists of a haploid representation of the sequence of either a single inbred line, a single diploid individual, or multiple diploid individuals. One reason for using a haploid sequence as the reference is that it is quite hard for assembly methods that have been applied to short sequences to construct a diploid assembly—such assemblies require that every chromosome be represented twice, with haplotypes phased correctly. Some researchers have been able to construct diploid assemblies using clone-based sequencing coupled with shorter reads (e.g., Levy et al. 2007; Kitzman et al. 2011), but long-read sequencing will make this approach much more prevalent (e.g., Chin et al. 2016). This means that for every individual we will get two assemblies—one for each chromosome. As there is no further information contained in a single individual as to how to sort the different chromosomes into haploid complements (i.e., into maternal and paternal sets of chromosomes), it is not clear how we will represent even a single assembly. While advances in sequencing promise many new biological insights, these will almost certainly come with their own logistical and computational issues.

Deep population sequencing

The low cost of DNA sequencing has made it possible to collect over 1,000 genomes from several model species (1001 Genomes Consortium 2016; Lack et al. 2016) and hundreds of genomes from others (e.g., Cook et al. 2017; Zhu, Sherlock, and Petrov 2017). The scientific and medical importance of studies in humans means that we quickly went from 1,000 genomes (1000 Genomes Project Consortium 2012) to 10,000 genomes (Telenti et al. 2016). The increase has been even steeper for human exome sequencing, moving from studies with over 2,400 individuals (Tennessen et al. 2012) to those with over 60,000 individuals (Lek et al. 2016). While these initial efforts often had more modest sample sizes within each global subpopulation, this was not always the case: in Iceland alone 2,636 people were sequenced (Gudbjartsson et al. 2015).

Deep sequencing projects—those with very large sample sizes—present new opportunities as well as challenges for population genetics. As briefly mentioned in Chapter 9, deep sequencing makes it possible to study very rare alleles and therefore to understand very recent demographic history (e.g., Keinan and Clark 2012). Other methods have been developed to infer the strength of selection at loci undergoing selective sweeps using deep population samples (e.g., Messer and Neher 2012). By far the most common use of deep sequencing has been to make inferences about the rate and patterns of mutation (e.g., Messer 2009; Nelson et al. 2012; Schaibley et al. 2013; Harpak, Bhaskar, and Pritchard 2016; Lek et al. 2016; Zhu, Sherlock, and Petrov 2017). Low-frequency variants are the most likely to reflect underlying mutational processes—having not yet been subject to many generations of natural selection—and therefore offer a straightforward way to understand mutation itself.

There are also of course multiple challenges in making inferences from deep samples. While the advantages of this type of data largely come from extremely rare variants, this is also the category of variation most affected by sequencing error. These errors must be taken into account when estimating mutation rates (Messer 2009) or demographic history (Keinan and Clark 2012). We also have to be concerned with recurrent mutation, as the frequency spectrum of very rare alleles will be disproportionately affected by this process (Harpak, Bhaskar, and Pritchard 2016). Most important, however, may be the concern that methods for population genetic inference based on the standard coalescent—which assumes that $n \ll N_e$—will be misleading for sample sizes that approach the effective population size or even the census population size (Wakeley and Takahashi 2003). Fortunately, there are a number of new approaches for dealing with very large sample sizes (e.g., Fu 2006; Bhaskar, Clark, and Song 2014; Chen, Hey, and Chen 2015; Kelleher, Etheridge, and McVean 2016).

As samples grow larger, we also have to be concerned about the underlying population pedigree—that is, the actual familial relationships among individuals. At small sample sizes our coalescent approximations are not affected by this pedigree; however, with large sample sizes our standard expectations may not be correct (Wakeley, King, and Wilton 2016). Knowledge of population pedigrees also offers many interesting opportunities for evolutionary studies. In small, isolated populations, every individual across multiple generations can be included in a pedigree and can be genotyped or sequenced (reviewed in Pemberton 2008). Such pedigrees regularly include thousands of individuals, as in those produced for Soay sheep on the island of St. Kilda ($n = 5,805$; Johnston et al. 2016), for Florida scrub-jays ($n = 3,984$; Chen et al. 2016), or for humans from Iceland ($n = 104,220$; Gudbjartsson et al. 2015). The information carried by pedigrees can be used to make unique types of inferences about the direct fitness advantages of specific genetic variants (e.g., Stefansson et al. 2005; Johnston et al. 2013), meiotic recombination rates (e.g., Johnston et al. 2016), and even the per-generation mutation rate (Kong et al. 2012). As pedigrees grow in size to encompass even whole species, population genetics will transition from being sample-based to truly population-based, with the ability to track the history of every genome in the population.

REFERENCES

1000 Genomes Project Consortium. 2010. A map of human genome variation from population-scale sequencing. *Nature* **467**:1061–1073.

———. 2012. An integrated map of genetic variation from 1,092 human genomes. *Nature* **491**:56–65.

1001 Genomes Consortium. 2016. 1,135 genomes reveal the global pattern of polymorphism in *Arabidopsis thaliana*. *Cell* **166**:481–491.

A

Achaz, G. 2008. Testing for neutrality in samples with sequencing errors. *Genetics* **179**:1409–1424.

———. 2009. Frequency spectrum neutrality tests: one for all and all for one. *Genetics* **183**:249–258.

Adams, A. M., and R. R. Hudson. 2004. Maximum-likelihood estimation of demographic parameters using the frequency spectrum of unlinked single-nucleotide polymorphisms. *Genetics* **168**:1699–1712.

Adams, M. D., S. E. Celniker, R. A. Holt, et al. 2000. The genome sequence of *Drosophila melanogaster*. *Science* **287**:2185–2195.

Aguadé, M., N. Miyashita, and C. H. Langley. 1989. Reduced variation in the *yellow-achaete-scute* region in natural populations of *Drosophila melanogaster*. *Genetics* **122**:607–615.

Akashi, H. 1994. Synonymous codon usage in *Drosophila melanogaster*: natural selection and translational accuracy. *Genetics* **136**:927–935.

———. 1995. Inferring weak selection from patterns of polymorphism and divergence at "silent" sites in *Drosophila* DNA. *Genetics* **139**:1067–1076.

———. 1999. Inferring the fitness effects of DNA mutations from polymorphism and divergence data: statistical power to detect directional selection under stationarity and free recombination. *Genetics* **151**:221–238.

Akashi, H., N. Osada, and T. Ohta. 2012. Weak selection and protein evolution. *Genetics* **192**:15–31.

Akey, J. M., M. A. Eberle, M. J. Rieder, C. S. Carlson, M. D. Shriver, D. A. Nickerson, and L. Kruglyak. 2004. Population history and natural selection shape patterns of genetic variation in 132 genes. *PLoS Biology* **2**:e286.

Akey, J. M., G. Zhang, K. Zhang, L. Jin, and M. D. Shriver. 2002. Interrogating a high-density SNP map for signatures of natural selection. *Genome Research* **12**:1805–1814.

Akey, J. M., K. Zhang, M. Xiong, and L. Jin. 2003. The effect of single nucleotide polymorphism identification strategies on estimates of linkage disequilibrium. *Molecular Biology and Evolution* **20**:232–242.

Alachiotis, N., A. Stamatakis, and P. Pavlidis. 2012. OmegaPlus: a scalable tool for rapid detection of selective sweeps in whole-genome datasets. *Bioinformatics* **28**:2274–2275.

Albert, T. J., M. N. Molla, D. M. Muzny, et al. 2007. Direct selection of human genomic loci by microarray hybridization. *Nature Methods* **4**:903–905.

Albrechtsen, A., F. C. Nielsen, and R. Nielsen. 2010. Ascertainment biases in SNP chips affect measures of population divergence. *Molecular Biology and Evolution* **27**:2534–2547.

Alexander, D. H., J. Novembre, and K. Lange. 2009. Fast model-based estimation of ancestry in unrelated individuals. *Genome Research* **19**:1655–1664.

Alkan, C., B. P. Coe, and E. E. Eichler. 2011. Genome structural variation discovery and genotyping. *Nature Reviews Genetics* **12**:363–376.

Altshuler, D., V. J. Pollara, C. R. Cowles, W. J. Van Etten, J. Baldwin, L. Linton, and E. S. Lander. 2000. An SNP map of the human genome generated by reduced representation shotgun sequencing. *Nature* **407**:513–516.

Anderson, C. N., U. Ramakrishnan, Y. L. Chan, and E. A. Hadly. 2004. Serial SimCoal: a population genetics model for data from multiple populations and points in time. *Bioinformatics* **21**:1733–1734.

Anderson, E. C. 2005. An efficient Monte Carlo method for estimating N_e from temporally spaced samples using a coalescent-based likelihood. *Genetics* **170**:955–967.

Anderson, E. C., E. G. Williamson, and E. A. Thompson. 2000. Monte Carlo evaluation of the likelihood for N_e from temporally spaced samples. *Genetics* **156**:2109–2118.

Andolfatto, P. 2005. Adaptive evolution of non-coding DNA in *Drosophila*. *Nature* **437**:1149–1152.

———. 2007. Hitchhiking effects of recurrent beneficial amino acid substitutions in the *Drosophila melanogaster* genome. *Genome Research* **17**:1755–1762.

———. 2008. Controlling type-I error of the McDonald-Kreitman test in genomewide scans for selection on noncoding DNA. *Genetics* **180**:1767–1771.

Andolfatto, P., D. Davison, D. Erezyilmaz, T. T. Hu, J. Mast, T. Sunayama-Morita, and D. L. Stern. 2011. Multiplexed shotgun genotyping for rapid and efficient genetic mapping. *Genome Research* **21**:610–617.

Andolfatto, P., and M. Nordborg. 1998. The effect of gene conversion on intralocus associations. *Genetics* **148**:1397–1399.

Andolfatto, P., and M. Przeworski. 2000. A genome-wide departure from the standard neutral model in natural populations of *Drosophila*. *Genetics* **156**:257–268.

Andolfatto, P., J. D. Wall, and M. Kreitman. 1999. Unusual haplotype structure at the proximal breakpoint of *In(2L)t* in a natural population of *Drosophila melanogaster*. *Genetics* **153**:1297–1311.

Anisimova, M., J. P. Bielawski, and Z. Yang. 2001. Accuracy and power of the likelihood ratio test in detecting adaptive molecular evolution. *Molecular Biology and Evolution* **18**:1585–1592.

Anisimova, M., R. Nielsen, and Z. Yang. 2003. Effect of recombination on the accuracy of the likelihood method for detecting positive selection at amino acid sites. *Genetics* **164**:1229–1236.

Antao, T., A. Lopes, R. J. Lopes, A. Beja-Pereira, and G. Luikart. 2008. LOSITAN: a workbench to detect molecular adaptation based on a F_{ST}-outlier method. *BMC Bioinformatics* **9**:323.

Aquadro, C. F., and B. D. Greenberg. 1983. Human mitochondrial DNA variation and evolution: analysis of nucleotide sequences from seven individuals. *Genetics* **103**:287–312.

Ardlie, K., S. N. Liu-Cordero, M. A. Eberle, M. Daly, J. Barrett, E. Winchester, E. S. Lander, and L. Kruglyak. 2001. Lower-than-expected linkage disequilibrium between tightly linked markers in humans suggests a role for gene conversion. *American Journal of Human Genetics* **69**:582–589.

Arenas, M., and D. Posada. 2007. Recodon: coalescent simulation of coding DNA sequences with recombination, migration and demography. *BMC Bioinformatics* **8**:458.

Arguello, J. R., Y. Chen, S. Yang, W. Wang, and M. Long. 2006. Origination of an X-linked testes chimeric gene by illegitimate recombination in *Drosophila*. *PLoS Genetics* **2**:e77.

Arnheim, N., M. Krystal, R. Schmickel, G. Wilson, O. Ryder, and E. Zimmer. 1980. Molecular evidence for genetic exchange among ribosomal genes on nonhomologous chromosomes in man and apes. *Proceedings of the National Academy of Sciences* **77**:7323–7327.

Arnheim, N., and C. E. Taylor. 1969. Non-Darwinian evolution: consequences for neutral allelic variation. *Nature* **223**:900–903.

Ashton, P. M., S. Nair, T. Dallman, S. Rubino, W. Rabsch, S. Mwaigwisya, J. Wain, and J. O'Grady. 2015. MinION nanopore sequencing identifies the position and structure of a bacterial antibiotic resistance island. *Nature Biotechnology* **33**:296–300.

Austerlitz, F., B. Jung-Muller, B. Godelle, and P.-H. Gouyon. 1997. Evolution of coalescence times, genetic diversity and structure during colonization. *Theoretical Population Biology* **51**:148–164.

Avise, J. C., J. Shapira, S. W. Daniel, C. F. Aquadro, and R. A. Lansman. 1983. Mitochondrial DNA differentiation during the speciation process in *Peromyscus*. *Molecular Biology and Evolution* **1**:38–56.

B

Bahlo, M., and R. C. Griffiths. 2000. Inference from gene trees in a subdivided population. *Theoretical Population Biology* **57**:79–95.

Baird, N. A., P. D. Etter, T. S. Atwood, M. C. Currey, A. L. Shiver, Z. A. Lewis,

E. U. Selker, W. A. Cresko, and E. A. Johnson. 2008. Rapid SNP discovery and genetic mapping using sequenced RAD markers. *PLoS ONE* 3:e3376.

Balloux, F. 2001. EASYPOP (version 1.7): a computer program for population genetics simulations. *Journal of Heredity* **92**:301–302.

Balloux, F., H. Brunner, N. Lugon-Moulin, J. Hausser, and J. Goudet. 2000. Microsatellites can be misleading: an empirical and simulation study. *Evolution* **54**:1414–1422.

Barbujani, G. 1987. Autocorrelation of gene frequencies under isolation by distance. *Genetics* **117**:777–782.

Barreiro, L. B., G. Laval, H. Quach, E. Patin, and L. Quintana-Murci. 2008. Natural selection has driven population differentiation in modern humans. *Nature Genetics* **40**:340–345.

Barton, N. H. 1998. The effect of hitch-hiking on neutral genealogies. *Genetical Research* **72**:123–133.

Bazin, E., S. Glemin, and N. Galtier. 2006. Population size does not influence mitochondrial genetic diversity in animals. *Science* **312**:570–572.

Beaumont, M. A. 2003. Estimation of population growth or decline in genetically monitored populations. *Genetics* **164**:1139–1160.

Beaumont, M. A., and D. J. Balding. 2004. Identifying adaptive genetic divergence among populations from genome scans. *Molecular Ecology* **13**:969–980.

Beaumont, M., E. M. Barratt, D. Gottelli, A. C. Kitchener, M. J. Daniels, J. K. Pritchard, and M. W. Bruford. 2001. Genetic diversity and introgression in the Scottish wildcat. *Molecular Ecology* **10**:319–336.

Beaumont, M. A., and R. A. Nichols. 1996. Evaluating loci for use in the genetic analysis of population structure. *Proceedings of the Royal Society B: Biological Sciences* **263**:1619–1626.

Beaumont, M. A., W. Zhang, and D. J. Balding. 2002. Approximate Bayesian computation in population genetics. *Genetics* **162**:2025–2035.

Becquet, C., and M. Przeworski. 2007. A new approach to estimate parameters of speciation models with application to apes. *Genome Research* **17**:1505–1519.

———. 2009. Learning about modes of speciation by computational approaches. *Evolution* **63**:2547–2562.

Beerli, P. 2004. Effect of unsampled populations on the estimation of population sizes and migration rates between sampled populations. *Molecular Ecology* **13**:827–836.

———. 2006. Comparison of Bayesian and maximum-likelihood inference of population genetic parameters. *Bioinformatics* **22**:341–345.

Beerli, P., and J. Felsenstein. 1999. Maximum-likelihood estimation of migration rates and effective population numbers in two populations using a coalescent approach. *Genetics* **152**:763–773.

———. 2001. Maximum likelihood estimation of a migration matrix and effective population sizes in *n* subpopulations by using a coalescent approach. *Proceedings of the National Academy of Sciences* **98**:4563–4568.

Begun, D. J., and C. F. Aquadro. 1991. Molecular population genetics of the distal portion of the X chromosome in *Drosophila*: evidence for genetic hitchhiking of the *yellow-achaete* region. *Genetics* **129**:1147–1158.

———. 1992. Levels of naturally occurring DNA polymorphism correlate with recombination rates in *D. melanogaster*. *Nature* **356**:519–520.

Begun, D. J., A. K. Holloway, K. Stevens, et al. 2007. Population genomics: whole-genome analysis of polymorphism and divergence in *Drosophila simulans*. *PLoS Biology* **5**:e310.

Benjamini, Y., and Y. Hochberg. 1995. Controlling the false discovery rate: a practical and powerful approach to multiple testing. *Journal of the Royal Statistical Society: Series B (Statistical Methodology)* **57**:289–300.

Bentley, D. R., S. Balasubramanian, H. P. Swerdlow, G. P. Smith, J. Milton, C. G. Brown, K. P. Hall, D. J. Evers, C. L. Barnes, and H. R. Bignell. 2008. Accurate whole human genome sequencing using reversible terminator chemistry. *Nature* **456**:53–59.

Bergland, A. O., E. L. Behrman, K. R. O'Brien, P. S. Schmidt, and D. A. Petrov. 2014. Genomic evidence of rapid and stable adaptive oscillations over seasonal time scales in *Drosophila*. *PLoS Genetics* **10**:e1004775.

Berry, A. J., J. Ajioka, and M. Kreitman. 1991. Lack of polymorphism on the *Drosophila* fourth chromosome resulting from selection. *Genetics* **129**:1111–1117.

Berthier, P., M. A. Beaumont, J.-M. Cornuet, and G. Luikart. 2002. Likelihood-based estimation of the effective population size using temporal changes in allele frequencies: a genealogical approach. *Genetics* **160**:741–751.

Bertorelle, G., and L. Excoffier. 1998. Inferring admixture proportions from molecular data. *Molecular Biology and Evolution* **15**:1298–1311.

Bhaskar, A., A. G. Clark, and Y. S. Song. 2014. Distortion of genealogical properties when the sample is very large. *Proceedings of the National Academy of Sciences* **111**:2385–2390.

Bhaskar, A., Y. X. R. Wang, and Y. S. Song. 2015. Efficient inference of population size histories and locus-specific mutation rates from large-sample genomic variation data. *Genome Research* **25**:268–279.

Bierne, N. 2010. The distinctive footprints of local hitchhiking in a varied environment and global hitchhiking in a subdivided population. *Evolution* **64**:3254–3272.

Bierne, N., and A. Eyre-Walker. 2004. The genomic rate of adaptive amino acid substitution in *Drosophila*. *Molecular Biology and Evolution* **21**:1350–1360.

Bierne, N., D. Roze, and J. J. Welch. 2013. Pervasive selection or is it …? Why are F_{ST} outliers sometimes so frequent? *Molecular Ecology* **22**:2061–2064.

Bird, A. P. 1980. DNA methylation and the frequency of CpG in animal DNA. *Nucleic Acids Research* **8**:1499–1504.

Birky, C. W., and J. B. Walsh. 1988. Effects of linkage on rates of molecular evolution. *Proceedings of the National Academy of Sciences* **85**:6414–6418.

Boitard, S., C. Schlötterer, and A. Futschik. 2009. Detecting selective sweeps: a new approach based on hidden Markov models. *Genetics* **181**:1567–1578.

Boitard, S., W. Rodríguez, F. Jay, S. Mona, and F. Austerlitz. 2016. Inferring population size history from large samples of genome-wide molecular data: an approximate Bayesian computation approach. *PLoS Genetics* **12**:e1005877.

Bollback, J. P., T. L. York, and R. Nielsen. 2008. Estimation of $2N_e s$ from temporal allele frequency data. *Genetics* **179**:497–502.

Bowcock, A. M., J. R. Kidd, J. L. Mountain, J. M. Hebert, L. Carotenuto, K. K. Kidd, and L. L. Cavalli-Sforza. 1991. Drift, admixture, and selection in human evolution: a study with DNA polymorphisms. *Proceedings of the National Academy of Sciences* **88**:839–843.

Bowcock, A. M., A. Ruizlinares, J. Tomfohrde, E. Minch, J. R. Kidd, and L. L. Cavalli-Sforza. 1994. High resolution of human evolutionary trees with polymorphic microsatellites. *Nature* **368**:455–457.

Boyko, A. R., S. H. Williamson, A. R. Indap, et al. 2008. Assessing the evolutionary impact of amino acid mutations in the human genome. *PLoS Genetics* **4**:e1000083.

Bradburd, G. S., P. L. Ralph, and G. M. Coop. 2016. A spatial framework for understanding population structure and admixture. *PLoS Genetics* **12**:e1005703.

Branca, A., T. D. Paape, P. Zhou, et al. 2011. Whole-genome nucleotide diversity, recombination, and linkage disequilibrium in the model legume *Medicago truncatula*. *Proceedings of the National Academy of Sciences* **108**:E864–E870.

Brandvain, Y., A. M. Kenney, L. Flagel, G. Coop, and A. L. Sweigart. 2014. Speciation and introgression between *Mimulus nasutus* and *Mimulus guttatus*. *PLoS Genetics* **10**:e1004410.

Braverman, J. M., R. R. Hudson, N. L. Kaplan, C. H. Langley, and W. Stephan. 1995. The hitchhiking effect on the site frequency spectrum of DNA polymorphisms. *Genetics* **140**:783–796.

Brown, G. R., G. P. Gill, R. J. Kuntz, C. H. Langley, and D. B. Neale. 2004. Nucleotide diversity and linkage disequilibrium in loblolly pine. *Proceedings of the National Academy of Sciences* **101**:15,255–15,260.

Browning, S. R., and B. L. Browning. 2007. Rapid and accurate haplotype phasing and missing-data inference for whole-genome association studies by use of localized haplotype clustering. *American Journal of Human Genetics* **81**:1084–1097.

———. 2011. Haplotype phasing: existing methods and new developments. *Nature Reviews Genetics* **12**:703–714.

Brues, A. M. 1954. Selection and polymorphism in the A-B-O blood groups. *American Journal of Physical Anthropology* **12**:559–598.

Bryc, K., C. Velez, T. Karafet, A. Moreno-Estrada, A. Reynolds, A. Auton, M. Hammer, C. D. Bustamante, and H. Ostrer. 2010. Genome-wide patterns of population structure and admixture among Hispanic/Latino populations. *Proceedings of the National Academy of Sciences* **107**:8954–8961.

Burgess, R., and Z. Yang. 2008. Estimation of hominoid ancestral population sizes under Bayesian coalescent models incorporating mutation rate variation and sequencing errors. *Molecular Biology and Evolution* **25**:1979–1994.

Burke, M. K., J. P. Dunham, P. Shahrestani, K. R. Thornton, M. R. Rose, and A. D. Long. 2010. Genome-wide analysis of a long-term evolution experiment with *Drosophila*. *Nature* **467**:587–590.

Burnham, K. P., and D. R. Anderson. 2002. *Model Selection and Multimodel Inference: A Practical Information-Theoretic Approach.* New York: Springer.

Bustamante, C. D., A. Fledel-Alon, S. Williamson, et al. 2005. Natural selection on protein-coding genes in the human genome. *Nature* **437**:1153.

Bustamante, C. D., R. Nielsen, S. A. Sawyer, K. M. Olsen, M. D. Purugganan, and D. L. Hartl. 2002. The cost of inbreeding in *Arabidopsis. Nature* **416**:531–534.

C

Caicedo, A. L., S. H. Williamson, R. D. Hernandez, A. Boyko, A. Fledel-Alon, T. L. York, N. R. Polato, K. M. Olsen, R. Nielsen, and S. R. McCouch. 2007. Genome-wide patterns of nucleotide polymorphism in domesticated rice. *PLoS Genetics* **3**:e163.

Calafell, F., F. Roubinet, A. Ramirez-Soriano, N. Saitou, J. Bertranpetit, and A. Blancher. 2008. Evolutionary dynamics of the human *ABO* gene. *Human Genetics* **124**:123–135.

Cannings, C. 1974. The latent roots of certain Markov chains arising in genetics: a new approach, I. Haploid models. *Advances in Applied Probability* **6**:260–290.

Cargill, M., D. Altshuler, J. Ireland, et al. 1999. Characterization of single-nucleotide polymorphisms in coding regions of human genes. *Nature Genetics* **22**:231–238.

Carlson, C. S., D. J. Thomas, M. A. Eberle, J. E. Swanson, R. J. Livingston, M. J. Rieder, and D. A. Nickerson. 2005. Genomic regions exhibiting positive selection identified from dense genotype data. *Genome Research* **15**:1553–1565.

Carneiro, M., F. W. Albert, S. Afonso, R. J. Pereira, H. Burbano, R. Campos, J. Melo-Ferreira, J. A. Blanco-Aguiar, R. Villafuerte, and M. W. Nachman. 2014. The genomic architecture of population divergence between subspecies of the European rabbit. *PLoS Genetics* **10**:e1003519.

Carvajal-Rodriguez, A. 2008. GENOMEPOP: a program to simulate genomes in populations. *BMC Bioinformatics* **9**:223.

Cavalli-Sforza, L. L. 1966. Population structure and human evolution. *Proceedings of the Royal Society B: Biological Sciences* **164**:362–379.

Cavalli-Sforza, L. L., and W. F. Bodmer. 1971. *The Genetics of Human Populations.* San Francisco: W. H. Freeman.

Cavalli-Sforza, L. L., and A. W. Edwards. 1967. Phylogenetic analysis: models and estimation procedures. *American Journal of Human Genetics* **19**:233–257.

Cavalli-Sforza, L. L., P. Menozzi, and A. Piazza. 1993. Demic expansions and human evolution. *Science* **259**:639–646.

Chadeau-Hyam, M., C. J. Hoggart, P. F. O'Reilly, J. C. Whittaker, M. De Iorio, and D. J. Balding. 2008. Fregene: simulation of realistic sequence-level data in populations and ascertained samples. *BMC Bioinformatics* **9**:364.

Chaisson, M. J. P., J. Huddleston, M. Y. Dennis, et al. 2015. Resolving the complexity of the human genome using single-molecule sequencing. *Nature* **517**:608–611.

Chaisson, M. J. P., R. K. Wilson, and E. E. Eichler. 2015. Genetic variation and the *de novo* assembly of human genomes. *Nature Reviews Genetics* **16**:627–640.

Chakraborty, M., J. G. Baldwin-Brown, A. D. Long, and J. J. Emerson. 2016. Contiguous and accurate *de novo* assembly of metazoan genomes with modest long read coverage. *Nucleic Acids Research* **44**:e147.

Chakraborty, R. 1986. Gene admixture in human populations: models and predictions. *American Journal of Physical Anthropology* **29**:1–43.

Chamary, J. V., J. L. Parmley, and L. D. Hurst. 2006. Hearing silence: non-neutral evolution at synonymous sites in mammals. *Nature Reviews Genetics* **7**:98–108.

Chan, A. H., P. A. Jenkins, and Y. S. Song. 2012. Genome-wide fine-scale recombination rate variation in *Drosophila melanogaster. PLoS Genetics* **8**:e1003090.

Chapman, N. H., and E. A. Thompson. 2002. The effect of population history on the lengths of ancestral chromosome segments. *Genetics* **162**:449–458.

Charlesworth, B. 1998. Measures of divergence between populations and the effect of forces that reduce variability. *Molecular Biology and Evolution* **15**:538–543.

———. 2009. Effective population size and patterns of molecular evolution and variation. *Nature Reviews Genetics* **10**:195–205.

———. 2012. The effects of deleterious mutations on evolution at linked sites. *Genetics* **190**:5–22.

Charlesworth, B., M. T. Morgan, and D. Charlesworth. 1993. The effect of deleterious mutations on neutral molecular variation. *Genetics* **134**:1289–1303.

Charlesworth, B., M. Nordborg, and D. Charlesworth. 1997. The effects of local selection, balanced polymorphism and background selection on equilibrium patterns of genetic diversity in subdivided populations. *Genetical Research* **70**:155–174.

Charlesworth, D. 2006. Balancing selection and its effects on sequences in nearby genome regions. *PLoS Genetics* **2**:e64.

Charlesworth, D., B. Charlesworth, and M. T. Morgan. 1995. The pattern of neutral molecular variation under the background selection model. *Genetics* **141**:1619–1632.

Charlesworth, J., and A. Eyre-Walker. 2008. The McDonald-Kreitman test and slightly deleterious mutations. *Molecular Biology and Evolution* **25**:1007–1015.

Chee, M., R. Yang, E. Hubbell, A. Berno, X. C. Huang, D. Stern, J. Winkler, D. J. Lockhart, M. S. Morris, and S. P. Fodor. 1996. Accessing genetic information with high-density DNA arrays. *Science* **274**:610–614.

Chen, F.-C., and W.-H. Li. 2001. Genomic divergences between humans and other hominoids and the effective population size of the common ancestor of humans and chimpanzees. *American Journal of Human Genetics* **68**:444–456.

Chen, G. K., P. Marjoram, and J. D. Wall. 2009. Fast and flexible simulation of DNA sequence data. *Genome Research* **19**:136–142.

Chen, H., J. Hey, and K. Chen. 2015. Inferring very recent population growth rate from population-scale sequencing data: using a large-sample coalescent estimator. *Molecular Biology and Evolution* **32**:2996–3011.

Chen, H., N. Patterson, and D. Reich. 2010. Population differentiation as a test for selective sweeps. *Genome Research* **20**:393–402.

Chen, J.-M., D. N. Cooper, N. Chuzhanova, C. Férec, and G. P. Patrinos. 2007. Gene conversion: mechanisms, evolution and human disease. *Nature Reviews Genetics* **8**:762–775.

Chen, K., M. D. McLellan, L. Ding, M. C. Wendl, Y. Kasai, R. K. Wilson, and E. R. Mardis. 2007. PolyScan: an automatic indel and SNP detection approach to the analysis of human resequencing data. *Genome Research* **17**:659–666.

Chen, N., E. J. Cosgrove, R. Bowman, J. W. Fitzpatrick, and A. G. Clark. 2016. Genomic consequences of population decline in the endangered Florida scrub-jay. *Current Biology* **26**:2974–2979.

Cheng, C., B. J. White, C. Camdem, K. Mockaitis, C. Costantini, M. W. Hahn, and N. J. Besansky. 2012. Ecological genomics of *Anopheles gambiae* along a latitudinal cline in Cameroon: a population resequencing approach. *Genetics* **190**:1417–1432.

Chikhi, L., M. W. Bruford, and M. A. Beaumont. 2001. Estimation of admixture proportions: a likelihood-based approach using Markov chain Monte Carlo. *Genetics* **158**:1347–1362.

Chin, C.-S., P. Peluso, F. J. Sedlazeck, et al. 2016. Phased diploid genome assembly with single molecule real-time sequencing. *Nature Methods* **13**:1050.

Chung, C. S., and N. E. Morton. 1961. Selection at the ABO locus. *American Journal of Human Genetics* **13**:9-27.

Churchill, G. A. 1989. Stochastic models for heterogeneous DNA sequences. *Bulletin of Mathematical Biology* **51**:79–94.

Clark, A. G. 1990. Inference of haplotypes from PCR-amplified samples of diploid populations. *Molecular Biology and Evolution* **7**:111–122.

———. 1997. Neutral behavior of shared polymorphism. *Proceedings of the National Academy of Sciences* **94**:7730–7734.

Clark, A. G., M. J. Hubisz, C. D. Bustamante, S. H. Williamson, and R. Nielsen. 2005. Ascertainment bias in studies of human genome-wide polymorphism. *Genome Research* **15**:1496–1502.

Clark, A. G., and T. S. Whittam. 1992. Sequencing errors and molecular evolutionary analysis. *Molecular Biology and Evolution* **9**:744–752.

Cockerham, C. C. 1969. Variance of gene frequencies. *Evolution* **23**:72–84.

Coffman, A. J., P. H. Hsieh, S. Gravel, and R. N. Gutenkunst. 2016. Computationally efficient composite likelihood statistics for demographic inference. *Molecular Biology and Evolution* **33**:591–593.

Comeron, J. M. 1995. A method for estimating the numbers of synonymous and nonsynonymous substitutions per site. *Journal of Molecular Evolution* **41**:1152–1159.

Comeron, J. M., R. Ratnappan, and S. Bailin. 2012. The many landscapes of recombination in *Drosophila melanogaster*. *PLoS Genetics* **8**:e1002905.

Comeron, J. M., A. Williford, and R. Kliman. 2008. The Hill–Robertson effect: evolutionary consequences of weak selection and linkage in finite populations. *Heredity* **100**:19–31.

Conant, G. C., and K. H. Wolfe. 2008. Turning a hobby into a job: how duplicated genes find new functions. *Nature Reviews Genetics* **9**:938–950.

Cook, D. E., S. Zdraljevic, J. P. Roberts, and E. C. Andersen. 2017. CeNDR, the *Caenorhabditis elegans* natural diversity resource. *Nucleic Acids Research* **45**:D650–D657.

Coop, G., and P. Ralph. 2012. Patterns of neutral diversity under general models of selective sweeps. *Genetics* **192**:205–224.

Coop, G., D. Witonsky, A. Di Rienzo, and J. K. Pritchard. 2010. Using environmental correlations to identify loci underlying local adaptation. *Genetics* **185**:1411–1423.

Corander, J., P. Waldmann, P. Marttinen, and M. J. Sillanpaa. 2004. BAPS 2: enhanced possibilities for the analysis of genetic population structure. *Bioinformatics* **20**:2363–2369.

Corander, J., P. Waldmann, and M. J. Sillanpaa. 2003. Bayesian analysis of genetic differentiation between populations. *Genetics* **163**:367–374.

Corbett-Detig, R. B., D. L. Hartl, and T. B. Sackton. 2015. Natural selection constrains neutral diversity across a wide range of species. *PLoS Biology* **13**:e1002112.

Cornuet, J.-M., S. Piry, G. Luikart, A. Estoup, and M. Solignac. 1999. New methods employing multilocus genotypes to select or exclude populations as origins of individuals. *Genetics* **153**:1989–2000.

Cornuet, J.-M., F. Santos, M. A. Beaumont, C. P. Robert, J.-M. Marin, D. J. Balding, T. Guillemaud, and A. Estoup. 2008. Inferring population history with *DIY ABC*: a user-friendly approach to approximate Bayesian computation. *Bioinformatics* **24**:2713–2719.

Cotton, R. G. H. 2002. Communicating "mutation": modern meanings and connotations. *Human Mutation* **19**:2–3.

Coventry, A., L. M. Bull-Otterson, X. Liu, et al. 2010. Deep resequencing reveals excess rare recent variants consistent with explosive population growth. *Nature Communications* **1**:131.

Crawford, D. L., J. A. Segal, and J. L. Barnett. 1999. Evolutionary analysis of TATA-less proximal promoter function. *Molecular Biology and Evolution* **16**:194–207.

Crisci, J. L., Y.-P. Poh, S. Mahajan, and J. D. Jensen. 2013. The impact of equilibrium assumptions on tests of selection. *Frontiers in Genetics* **4**:235.

Crow, J. F., and M. Kimura. 1970. *An Introduction to Population Genetics Theory*. New York: Harper & Row.

Cruickshank, T. C., and M. W. Hahn. 2014. Reanalysis suggests that genomic islands of speciation are due to reduced diversity, not reduced gene flow. *Molecular Ecology* **23**:3133–3157.

Cutler, D. J., and J. D. Jensen. 2010. To pool, or not to pool? *Genetics* **186**:41–43.

Cutler, D. J., M. E. Zwick, M. M. Carrasquillo, C. T. Yohn, K. P. Tobin, C. Kashuk, D. J. Mathews, N. A. Shah, E. E. Eichler, and J. A. Warrington. 2001. High-throughput variation detection and genotyping using microarrays. *Genome Research* **11**:1913–1925.

Cutter, A. D., and B. A. Payseur. 2013. Genomic signatures of selection at linked sites: unifying the disparity among species. *Nature Reviews Genetics* **14**:262–274.

D

Dawson, K. J., and K. Belkhir. 2001. A Bayesian approach to the identification of panmictic populations and the assignment of individuals. *Genetical Research* **78**:59–77.

De, A., and R. Durrett. 2007. Stepping-stone spatial structure causes slow decay of linkage disequilibrium and shifts the site frequency spectrum. *Genetics* **176**:969–981.

DeGiorgio, M., C. D. Huber, M. J. Hubisz, I. Hellmann, and R. Nielsen. 2016. SweepFinder2: increased sensitivity, robustness and flexibility. *Bioinformatics* **32**:1895–1897.

DeGiorgio, M., K. E. Lohmueller, and R. Nielsen. 2014. A model-based approach for identifying signatures of ancient balancing selection in genetic data. *PLoS Genetics* **10**:e1004561.

DeGiorgio, M., and N. A. Rosenberg. 2013. Geographic sampling scheme as a determinant of the major axis of genetic variation in principal components analysis. *Molecular Biology and Evolution* **30**:480–488.

Degnan, J. H., and N. A. Rosenberg. 2009. Gene tree discordance, phylogenetic inference and the multispecies coalescent. *Trends in Ecology & Evolution* **24**:332–340.

Depaulis, F., and M. Veuille. 1998. Neutrality tests based on the distribution of haplotypes under an infinite-site model. *Molecular Biology and Evolution* **15**:1788–1790.

DePristo, M. A., E. Banks, R. Poplin, et al. 2011. A framework for variation discovery and genotyping using next-generation DNA sequencing data. *Nature Genetics* **43**:491–498.

De Wit, P., M. H. Pespeni, and S. R. Palumbi. 2015. SNP genotyping and population genomics from expressed sequences: current advances and future possibilities. *Molecular Ecology* **24**:2310–2323.

Dilthey, A., C. Cox, Z. Iqbal, M. R. Nelson, and G. McVean. 2015. Improved genome inference in the MHC using a population reference graph. *Nature Genetics* **47**:682–688.

Di Rienzo, A., A. Peterson, J. Garza, A. Valdes, M. Slatkin, and N. Freimer. 1994. Mutational processes of simple-sequence repeat loci in human populations. *Proceedings of the National Academy of Sciences* **91**:3166–3170.

Dobzhansky, T. 1948. Genetics of natural populations. XVI. Altitudinal and seasonal changes produced by natural selection in certain populations of *Drosophila pseudoobscura* and *Drosophila persimilis*. *Genetics* **33**:158–176.

Drmanac, R., A. B. Sparks, M. J. Callow, et al. 2010. Human genome sequencing using unchained base reads on self-assembling DNA nanoarrays. *Science* **327**:78–81.

Drosophila 12 Genomes Consortium. 2007. Evolution of genes and genomes on the *Drosophila* phylogeny. *Nature* **450**:203–218.

Druley, T. E., F. L. Vallania, D. J. Wegner, K. E. Varley, O. L. Knowles, J. A. Bonds, S. W. Robison, S. W. Doniger, A. Hamvas, and F. S. Cole. 2009. Quantification of rare allelic variants from pooled genomic DNA. *Nature Methods* **6**:263–265.

Drummond, A. J., G. K. Nicholls, A. G. Rodrigo, and W. Solomon. 2002. Estimating mutation parameters, population history and genealogy simultaneously from temporally spaced sequence data. *Genetics* **161**:1307–1320.

Drummond, A. J., and A. Rambaut. 2007. BEAST: Bayesian evolutionary analysis by sampling trees. *BMC Evolutionary Biology* **7**:214.

Drummond, A. J., A. Rambaut, B. Shapiro, and O. G. Pybus. 2005. Bayesian coalescent inference of past population dynamics from molecular sequences. *Molecular Biology and Evolution* **22**:1185–1192.

Durand, E. Y., N. Patterson, D. Reich, and M. Slatkin. 2011. Testing for ancient admixture between closely related populations. *Molecular Biology and Evolution* **28**:2239–2252.

Durbin, R., S. R. Eddy, A. Krogh, and G. Mitchison. 1998. *Biological Sequence Analysis: Probabilistic Models of Proteins and Nucleic Acids*. Cambridge: Cambridge University Press.

Dutheil, J. Y. 2017. Hidden Markov models in population genomics. In *Hidden Markov Models: Methods and Protocols*, ed. D. R. Westhead and M. S. Vijayabaskar, pp. 149–164. New York: Springer.

Dutheil, J. Y., G. Ganapathy, A. Hobolth, T. Mailund, M. K. Uyenoyama, and M. H. Schierup. 2009. Ancestral population genomics: the coalescent hidden Markov model approach. *Genetics* **183**:259–274.

Dyer, R. J., and J. D. Nason. 2004. Population graphs: the graph theoretic shape of genetic structure. *Molecular Ecology* **13**:1713–1727.

E

Earl, D., N. Nguyen, G. Hickey, et al. 2014. Alignathon: a competitive assessment of whole-genome alignment methods. *Genome Research* **24**:2077–2089.

Early, A. M., J. R. Arguello, M. Cardoso-Moreira, S. Gottipati, J. K. Grenier, and A. G. Clark. 2017. Survey of global genetic diversity within the *Drosophila* immune system. *Genetics* **205**:353–366.

Eckert, A. J., J. van Heerwaarden, J. L. Wegrzyn, C. D. Nelson, J. Ross-Ibarra, S. C. González-Martínez, and D. B. Neale. 2010. Patterns of population structure and environmental associations to aridity across the range of loblolly pine (*Pinus taeda* L., Pinaceae). *Genetics* **185**:969–982.

Edwards, S. V., and P. Beerli. 2000. Perspective: gene divergence, population divergence, and the variance in coalescence time in phylogeographic studies. *Evolution* **54**:1839–1854.

Ellegren, H. 2004. Microsatellites: simple sequences with complex evolution. *Nature Reviews Genetics* **5**:435–445.

Ellegren, H., L. Smeds, R. Burri, et al. 2012. The genomic landscape of species divergence in *Ficedula* flycatchers. *Nature* **491**:756–760.

Elshire, R. J., J. C. Glaubitz, Q. Sun, J. A. Poland, K. Kawamoto, E. S. Buckler, and S. E. Mitchell. 2011. A robust, simple genotyping-by-sequencing (GBS) approach for high diversity species. *PLoS ONE* **6**:e19379.

Emerson, J. J., M. Cardoso-Moreira, J. O. Borevitz, and M. Long. 2008. Natural selection shapes genome-wide patterns of copy-number polymorphism in *Drosophila melanogaster*. *Science* **320**:1629–1631.

Enard, D., L. Cai, C. Gwennap, and D. A. Petrov. 2016. Viruses are a dominant driver of protein adaptation in mammals. *eLife* **5**:e12469.

Engelhardt, B. E., and M. Stephens. 2010. Analysis of population structure: a unifying framework and novel methods based on sparse factor analysis. *PLoS Genetics* **6**:e1001117.

English, A. C., S. Richards, Y. Han, et al. 2012. Mind the gap: upgrading genomes with Pacific Biosciences RS long-read sequencing technology. *PLoS ONE* **7**:e47768.

Epperson, B. K. 2003. *Geographical Genetics*. Princeton, NJ: Princeton University Press.

Epperson, B. K., and T. Li. 1996. Measurement of genetic structure within populations using Moran's spatial autocorrelation statistics. *Proceedings of the National Academy of Sciences* **93**:10,528–10,532.

Eriksson, A., and A. Manica. 2012. Effect of ancient population structure on the degree of polymorphism shared between modern human populations and ancient hominins. *Proceedings of the National Academy of Sciences* **109**:13,956–13,960.

Evanno, G., S. Regnaut, and J. Goudet. 2005. Detecting the number of clusters of individuals using the software STRUCTURE: a simulation study. *Molecular Ecology* **14**:2611–2620.

Ewens, W. J. 1972. The sampling theory of selectively neutral alleles. *Theoretical Population Biology* **3**:87–112.

———. 1973. Testing for increased mutation rate for neutral alleles. *Theoretical Population Biology* **4**:251–258.

———. 1974. A note on the sampling theory for infinite alleles and infinite sites models. *Theoretical Population Biology* **6**:143–148.

———. 2004. *Mathematical Population Genetics. I. Theoretical Introduction*. New York: Springer-Verlag.

Ewing, B., and P. Green. 1998. Base-calling of automated sequencer traces using *phred*. II. Error probabilities. *Genome Research* **8**:186–194.

Ewing, B., L. Hillier, M. C. Wendl, and P. Green. 1998. Base-calling of automated sequencer traces using *phred*. I. Accuracy assessment. *Genome Research* **8**:175–185.

Ewing, G., and J. Hermisson. 2010. MSMS: a coalescent simulation program including

recombination, demographic structure and selection at a single locus. *Bioinformatics* **26**:2064–2065.

Ewing, G. B., and J. D. Jensen. 2016. The consequences of not accounting for background selection in demographic inference. *Molecular Ecology* **25**:135–141.

Excoffier, L. 2007. Analysis of population subdivision. In *Handbook of Statistical Genetics*, vol. 2, ed. D. J. Balding, M. Bishop, and C. Cannings, pp. 980–1020. West Sussex, England: Wiley.

Excoffier, L., I. Dupanloup, E. Huerta-Sánchez, V. C. Sousa, and M. Foll. 2013. Robust demographic inference from genomic and SNP data. *PLoS Genetics* **9**:e1003905.

Excoffier, L., T. Hofer, and M. Foll. 2009. Detecting loci under selection in a hierarchically structured population. *Heredity* **103**:285–298.

Excoffier, L., and M. Foll. 2011. fastsimcoal: a continuous-time coalescent simulator of genomic diversity under arbitrarily complex evolutionary scenarios. *Bioinformatics* **27**:1332–1334.

Excoffier, L., and H. E. Lischer. 2010. Arlequin suite ver 3.5: a new series of programs to perform population genetics analyses under Linux and Windows. *Molecular Ecology Resources* **10**:564–567.

Excoffier, L., J. Novembre, and S. Schneider. 2000. SIMCOAL: A general coalescent program for the simulation of molecular data in interconnected populations with arbitrary demography. *Journal of Heredity* **91**:506–509.

Excoffier, L., and M. Slatkin. 1995. Maximum-likelihood estimation of molecular haplotype frequencies in a diploid population. *Molecular Biology and Evolution* **12**:921–927.

Excoffier, L., P. E. Smouse, and J. M. Quattro. 1992. Analysis of molecular variance inferred from metric distances among DNA haplotypes: application to human mitochondrial DNA restriction data. *Genetics* **131**:479–491.

Eyre-Walker, A. 2002. Changing effective population size and the McDonald-Kreitman test. *Genetics* **162**:2017–2024.

Eyre-Walker, A., and P. D. Keightley. 2009. Estimating the rate of adaptive molecular evolution in the presence of slightly deleterious mutations and population size change. *Molecular Biology and Evolution* **26**:2097–2108.

Eyre-Walker, A., P. D. Keightley, N. G. C. Smith, and D. Gaffney. 2002. Quantifying the slightly deleterious mutation model of molecular evolution. *Molecular Biology and Evolution* **19**:2142–2149.

F

Falush, D., M. Stephens, and J. K. Pritchard. 2003. Inference of population structure using multilocus genotype data: linked loci and correlated allele frequencies. *Genetics* **164**:1567–1587.

Fay, J. C., and C.-I. Wu. 1999. A human population bottleneck can account for the discordance between patterns of mitochondrial versus nuclear DNA variation. *Molecular Biology and Evolution* **16**:1003–1005.

———. 2000. Hitchhiking under positive Darwinian selection. *Genetics* **155**:1405–1413.

———. 2001. The neutral theory in the genomic era. *Current Opinion in Genetics & Development* **11**:642–646.

Fay, J. C., G. J. Wyckoff, and C.-I. Wu. 2001. Positive and negative selection on the human genome. *Genetics* **158**:1227–1234.

———. 2002. Testing the neutral theory of molecular evolution with genomic data from *Drosophila*. *Nature* **415**:1024–1026.

Fearnhead, P., and P. Donnelly. 2001. Estimating recombination rates from population genetic data. *Genetics* **159**:1299–1318.

Feder, J. L., X. Xie, J. Rull, S. Velez, A. Forbes, B. Leung, H. Dambroski, K. E. Filchak, and M. Aluja. 2005. Mayr, Dobzhansky, and Bush and the complexities of sympatric speciation in *Rhagoletis*. *Proceedings of the National Academy of Sciences* **102**:6573–6580.

Feldman, M. W., and F. B. Christiansen. 1974. Effect of population subdivision on two loci without selection. *Genetical Research* **24**:151–162.

Felsenstein, J. 1974. The evolutionary advantage of recombination. *Genetics* **78**:737–756.

———. 1976. The theoretical population genetics of variable selection and migration. *Annual Review of Genetics* **10**:253–280.

———. 1981a. Evolutionary trees from DNA sequences: a maximum likelihood approach. *Journal of Molecular Evolution* **17**:368–376.

———. 1981b. Evolutionary trees from gene frequencies and quantitative characters: finding maximum likelihood estimates. *Evolution* **35**:1229–1242.

———. 1989. PHYLIP—Phylogeny Inference Package (version 3.2). *Cladistics* **5**:164–166.

———. 1992. Estimating effective population size from samples of sequences: inefficiency of pairwise and segregating sites as compared to phylogenetic estimates. *Genetical Research* **59**:139–147.

Felsenstein, J., and G. A. Churchill. 1996. A hidden Markov model approach to variation among sites in rate of evolution. *Molecular Biology and Evolution* **13**:93–104.

Ferrer-Admetlla, A., M. Liang, T. Korneliussen, and R. Nielsen. 2014. On detecting incomplete soft or hard selective sweeps using haplotype structure. *Molecular Biology and Evolution* **31**:1275–1291.

Ferretti, L., E. Raineri, and S. Ramos-Onsins. 2012. Neutrality tests for sequences with missing data. *Genetics* **191**:1397–1401.

Fisher, R. A. 1930a. The distribution of gene ratios for rare mutations. *Proceedings of the Royal Society of Edinburgh* **50**:205–220.

———. 1930b. *The Genetical Theory of Natural Selection*. Oxford: Clarendon Press.

Fogelqvist, J., A. Niittyvuopio, J. Agren, O. Savolainen, and M. Lascoux. 2010. Cryptic population genetic structure: the number of inferred clusters depends on sample size. *Molecular Ecology Resources* **10**:314–323.

Foll, M., M. A. Beaumont, and O. Gaggiotti. 2008. An approximate Bayesian computation approach to overcome biases that arise when using amplified fragment length polymorphism markers to study population structure. *Genetics* **179**:927–939.

Foll, M., and O. Gaggiotti. 2008. A genome-scan method to identify selected loci appropriate for both dominant and codominant markers: a Bayesian perspective. *Genetics* **180**:977–993.

Foll, M., H. Shim, and J. D. Jensen. 2015. WFABC: a Wright–Fisher ABC-based approach for inferring effective population sizes and selection coefficients from time-sampled data. *Molecular Ecology Resources* **15**:87–98.

Fontaine, M. C., J. B. Pease, A. Steele, et al. 2015. Extensive introgression in a malaria vector species complex revealed by phylogenomics. *Science* **347**:1258524.

Ford, M. J., and C. F. Aquadro. 1996. Selection on X-linked genes during speciation in the *Drosophila athabasca* complex. *Genetics* **144**:689–703.

Fourcade, Y., A. Chaput-Bardy, J. Secondi, C. Fleurant, and C. Lemaire. 2013. Is local selection so widespread in river organisms? Fractal geometry of river networks leads to high bias in outlier detection. *Molecular Ecology* **22**:2065–2073.

Fournier-Level, A., A. Korte, M. D. Cooper, M. Nordborg, J. Schmitt, and A. M. Wilczek. 2011. A map of local adaptation in *Arabidopsis thaliana. Science* **334**:86–89.

Foxe, J. P., and S. I. Wright. 2009. Signature of diversifying selection on members of the pentatricopeptide repeat protein family in *Arabidopsis lyrata. Genetics* **183**:663–672.

François, O., S. Ancelet, and G. Guillot. 2006. Bayesian clustering using hidden Markov random fields in spatial population genetics. *Genetics* **174**:805–816.

François, O., M. Currat, N. Ray, E. Han, L. Excoffier, and J. Novembre. 2010. Principal component analysis under population genetic models of range expansion and admixture. *Molecular Biology and Evolution* **27**:1257–1268.

Frichot, E., S. D. Schoville, G. Bouchard, and O. François. 2013. Testing for associations between loci and environmental gradients using latent factor mixed models. *Molecular Biology and Evolution* **30**:1687–1699.

Frisse, L., R. R. Hudson, A. Bartoszewicz, J. D. Wall, J. Donfack, and A. Di Rienzo. 2001. Gene conversion and different population histories may explain the contrast between polymorphism and linkage disequilibrium levels. *American Journal of Human Genetics* **69**:831–843.

Fu, Y.-X. 1995. Statistical properties of segregating sites. *Theoretical Population Biology* **48**:172–197.

———. 1996. New statistical tests of neutrality for DNA samples from a population. *Genetics* **143**:557–570.

———. 1997. Statistical tests of neutrality of mutations against population growth, hitchhiking and background selection. *Genetics* **147**:915–925.

———. 2001. Estimating mutation rate and generation time from longitudinal samples of DNA sequences. *Molecular Biology and Evolution* **18**:620–626.

———. 2006. Exact coalescent for the Wright-Fisher model. *Theoretical Population Biology* **69**:385–394.

Fu, Y.-X., and W.-H. Li. 1993a. Statistical tests of neutrality of mutations. *Genetics* **133**:693–709.

———. 1993b. Maximum likelihood estimation of population parameters. *Genetics* **134**:1261–1270.

Fumagalli, M., M. Sironi, U. Pozzoli, A. Ferrer-Admettla, L. Pattini, and R. Nielsen. 2011. Signatures of environmental genetic adaptation pinpoint pathogens as the main selective pressure through human evolution. *PLoS Genetics* **7**:e1002355.

Futschik, A., and C. Schlötterer. 2010. The next generation of molecular markers from massively parallel sequencing of pooled DNA samples. *Genetics* **186**:207–218.

G

Galtier, N., G. Piganeau, D. Mouchiroud, and L. Duret. 2001. GC-content evolution in mammalian genomes: the biased gene conversion hypothesis. *Genetics* **159**:907–911.

Gan, X., O. Stegle, J. Behr, et al. 2011. Multiple reference genomes and transcriptomes for *Arabidopsis thaliana. Nature* **477**:419–423.

Gao, F., and A. Keinan. 2016. Inference of super-exponential human population growth via efficient computation of the site frequency spectrum for generalized models. *Genetics* **202**:235–245.

Gao, H., K. Bryc, and C. D. Bustamante. 2011. On identifying the optimal number of population clusters via the deviance information criterion. *PLoS ONE* **6**:e21014.

Gao, H., S. Williamson, and C. D. Bustamante. 2007. A Markov chain Monte Carlo approach for joint inference of population structure and inbreeding rates from multilocus genotype data. *Genetics* **176**:1635–1651.

Gao, Z., M. Przeworski, and G. Sella. 2015. Footprints of ancient-balanced polymorphisms in genetic variation data from closely related species. *Evolution* **69**:431–446.

Garrigan, D. 2009. Composite likelihood estimation of demographic parameters. *BMC Genetics* **10**:72.

Garud, N. R., P. W. Messer, E. O. Buzbas, and D. A. Petrov. 2015. Recent selective sweeps in North American *Drosophila melanogaster* show signatures of soft sweeps. *PLoS Genetics* **11**:e1005004.

Gattepaille, L., T. Günther, and M. Jakobsson. 2016. Inferring past effective population size from distributions of coalescent times. *Genetics* **204**:1191–1206.

Gattepaille, L. M., M. Jakobsson, and M. G. B. Blum. 2013. Inferring population size changes with sequence and SNP data:

lessons from human bottlenecks. *Heredity* 110:409–419.

Gayral, P., J. Melo-Ferreira, S. Glémin, et al. 2013. Reference-free population genomics from next-generation transcriptome data and the vertebrate–invertebrate gap. *PLoS Genetics* 9:e1003457.

Gazave, E., L. Ma, D. Chang, et al. 2014. Neutral genomic regions refine models of recent rapid human population growth. *Proceedings of the National Academy of Sciences* 111:757–762.

Geneva, A. J., C. A. Muirhead, S. B. Kingan, and D. Garrigan. 2015. A new method to scan genomes for introgression in a secondary contact model. *PLoS ONE* 10:e0118621.

Gibson, G., and S. V. Muse. 2009. *A Primer of Genome Science.* Sunderland, MA: Sinauer Associates.

Gillespie, J. H. 1991. *The Causes of Molecular Evolution.* Oxford: Oxford University Press.

———. 2000. Genetic drift in an infinite population: the pseudohitchhiking model. *Genetics* 155:909–919.

———. 2001. Is the population size of a species relevant to its evolution? *Evolution* 55:2161–2169.

———. 2004. *Population Genetics: A Concise Guide.* Baltimore: Johns Hopkins University Press.

Gillespie, J. H., and C. H. Langley. 1979. Are evolutionary rates really variable? *Journal of Molecular Evolution* 13:27–34.

Glinka, S., L. Ometto, S. Mousset, W. Stephan, and D. De Lorenzo. 2003. Demography and natural selection have shaped genetic variation in *Drosophila melanogaster*: a multi-locus approach. *Genetics* 165:1269–1278.

Gnirke, A., A. Melnikov, J. Maguire, et al. 2009. Solution hybrid selection with ultra-long oligonucleotides for massively parallel targeted sequencing. *Nature Biotechnology* 27:182–189.

Golding, G. B. 1997. The effect of purifying selection on genealogies. *Institute for Mathematics and Its Applications* 87:271.

Goldman, N., and Z. Yang. 1994. A codon-based model of nucleotide substitution for protein-coding DNA sequences. *Molecular Biology and Evolution* 11:725–736.

Goldstein, D. B., A. R. Linares, L. L. Cavalli-Sforza, and M. W. Feldman. 1995. An evaluation of genetic distances for use with microsatellite loci. *Genetics* 139:463–471.

Goldstein, D. B., and C. Schlötterer. 1999. *Microsatellites: Evolution and Applications.* Oxford: Oxford University Press.

Gompert, Z. 2016. Bayesian inference of selection in a heterogeneous environment from genetic time-series data. *Molecular Ecology* 25:121–134.

González, J., K. Lenkov, M. Lipatov, J. M. Macpherson, and D. A. Petrov. 2008. High rate of recent transposable element-induced adaptation in *Drosophila melanogaster*. *PLoS Biology* 6:e251.

Goudet, J., M. Raymond, T. de-Meeus, and F. Rousset. 1996. Testing differentiation in diploid populations. *Genetics* 144:1933–1940.

Gravel, S., B. M. Henn, R. N. Gutenkunst, A. R. Indap, G. T. Marth, A. G. Clark, F. Yu, R. A. Gibbs, T. G. Project, and C. D. Bustamante. 2011. Demographic history and rare allele sharing among human populations. *Proceedings of the National Academy of Sciences* 108:11,983–11,988.

Gray, M. M., J. M. Granka, C. D. Bustamante, N. B. Sutter, A. R. Boyko, L. Zhu, E. A. Ostrander, and R. K. Wayne. 2009. Linkage disequilibrium and demographic history of wild and domestic canids. *Genetics* 181:1493–1505.

Green, R. E., J. Krause, A. W. Briggs, et al. 2010. A draft sequence of the Neandertal genome. *Science* 328:710–722.

Green, R. E., J. Krause, S. E. Ptak, et al. 2006. Analysis of one million base pairs of Neanderthal DNA. *Nature* 444:330–336.

Griffiths, R. C. 1981. Neutral two-locus multiple allele models with recombination. *Theoretical Population Biology* 19:169–186.

Griffiths, R. C., and P. Marjoram. 1996. Ancestral inference from samples of DNA sequences with recombination. *Journal of Computational Biology* 3:479–502.

Griffiths, R. C., and S. Tavaré. 1998. The age of a mutation in a general coalescent tree. *Communications in Statistics: Stochastic Models* 14:273–295.

Gronau, I., M. J. Hubisz, B. Gulko, C. G. Danko, and A. Siepel. 2011. Bayesian inference of ancient human demography from individual genome sequences. *Nature Genetics* 43:1031–1034.

Gudbjartsson, D. F., H. Helgason, S. A. Gudjonsson, et al. 2015. Large-scale whole-genome sequencing of the Icelandic population. *Nature Genetics* 47:435–444.

Guillaume, F., and J. Rougemont. 2006. Nemo: an evolutionary and population genetics programming framework. *Bioinformatics* **22**:2556–2557.

Guillot, G., A. Estoup, F. Mortier, and J. F. Cosson. 2005. A spatial statistical model for landscape genetics. *Genetics* **170**:1261–1280.

Guillot, G., and M. Foll. 2009. Correcting for ascertainment bias in the inference of population structure. *Bioinformatics* **25**:552–554.

Guillot, G., and F. Rousset. 2013. Dismantling the Mantel tests. *Methods in Ecology and Evolution* **4**:336–344.

Günther, T., and G. Coop. 2013. Robust identification of local adaptation from allele frequencies. *Genetics* **195**:205–220.

Guryev, V., B. M. G. Smits, J. van de Belt, M. Verheul, N. Hubner, and E. Cuppen. 2006. Haplotype block structure is conserved across mammals. *PLoS Genetics* **2**:e121.

Gutenkunst, R. N., R. D. Hernandez, S. H. Williamson, and C. D. Bustamante. 2009. Inferring the joint demographic history of multiple populations from multidimensional SNP frequency data. *PLoS Genetics* **5**:e1000695.

H

Haasl, R. J., and B. A. Payseur. 2010. The number of alleles at a microsatellite defines the allele frequency spectrum and facilitates fast accurate estimation of θ. *Molecular Biology and Evolution* **27**:2702–2715.

———. 2016. Fifteen years of genomewide scans for selection: trends, lessons and unaddressed genetic sources of complication. *Molecular Ecology* **25**:5–23.

Hacia, J. G., L. C. Brody, M. S. Chee, S. P. Fodor, and F. S. Collins. 1996. Detection of heterozygous mutations in *BRCA1* using high density oligonucleotide arrays and two-colour fluorescence analysis. *Nature Genetics* **14**:441–447.

Hahn, M. W. 2006. Accurate inference and estimation in population genomics. *Molecular Biology and Evolution* **23**:911–918.

———. 2007. Detecting natural selection on *cis*-regulatory DNA. *Genetica* **129**:7–18.

———. 2008. Toward a selection theory of molecular evolution. *Evolution* **62**:255–265.

———. 2009. Distinguishing among evolutionary models for the maintenance of gene duplicates. *Journal of Heredity* **100**:605–617.

Hahn, M. W., M. D. Rausher, and C. W. Cunningham. 2002. Distinguishing between selection and population expansion in an experimental lineage of bacteriophage T7. *Genetics* **161**:11–20.

Hahn, M. W., M. V. Rockman, N. Soranzo, D. B. Goldstein, and G. A. Wray. 2004. Population genetic and phylogenetic evidence for positive selection on regulatory mutations at the *Factor VII* locus in humans. *Genetics* **167**:867–877.

Haldane, J. B. S. 1927. A mathematical theory of natural and artificial selection, part v: selection and mutation. *Proceedings of the Cambridge Philosophical Society* **23**:838–844.

Haller, B. C., and P. W. Messer. 2017. SLiM 2: flexible, interactive forward genetic simulations. *Molecular Biology and Evolution* **34**:230–240.

Hanchard, N. A., K. A. Rockett, C. Spencer, G. Coop, M. Pinder, M. Jallow, M. Kimber, G. McVean, R. Mott, and D. P. Kwiatkowski. 2006. Screening for recently selected alleles by analysis of human haplotype similarity. *American Journal of Human Genetics* **78**:153–159.

Hancock, A. M., B. Brachi, N. Faure, M. W. Horton, L. B. Jarymowycz, F. G. Sperone, C. Toomajian, F. Roux, and J. Bergelson. 2011. Adaptation to climate across the *Arabidopsis thaliana* genome. *Science* **334**:83–86.

Hancock, A. M., D. B. Witonsky, A. S. Gordon, G. Eshel, J. K. Pritchard, G. Coop, and A. Di Rienzo. 2008. Adaptations to climate in candidate genes for common metabolic disorders. *PLoS Genetics* **4**:e32.

Hardy, G. H. 1908. Mendelian proportions in a mixed population. *Science* **28**:49–50.

Hardy, O. J., and X. Vekemans. 1999. Isolation by distance in a continuous population: reconciliation between spatial autocorrelation analysis and population genetics models. *Heredity* **83**:145–154.

Harpak, A., A. Bhaskar, and J. K. Pritchard. 2016. Mutation rate variation is a primary determinant of the distribution of allele frequencies in humans. *PLoS Genetics* **12**:e1006489.

Harr, B., M. Kauer, and C. Schlötterer. 2002. Hitchhiking mapping: a population-based fine-mapping strategy for adaptive mutations in *Drosophila melanogaster*. *Proceedings of the National Academy of Sciences* **99**:12,949–12,954.

Harrigan, R. J., M. E. Mazza, and M. D. Sorenson. 2008. Computation vs. cloning: evaluation of two methods for haplotype determination. *Molecular Ecology Resources* **8**:1239–1248.

Harris, H. 1966. Enzyme polymorphisms in man. *Proceedings of the Royal Society B: Biological Sciences* **164**:298–310.

Harris, K., and R. Nielsen. 2013. Inferring demographic history from a spectrum of shared haplotype lengths. *PLoS Genetics* **9**:e1003521.

———. 2014. Error-prone polymerase activity causes multinucleotide mutations in humans. *Genome Research* **24**:1445–1454.

Harrison, R. G. 1991. Molecular changes at speciation. *Annual Review in Ecology and Systematics* **22**:281–308.

Hasegawa, M., H. Kishino, and T. A. Yano. 1985. Dating of the human-ape splitting by a molecular clock of mitochondrial DNA. *Journal of Molecular Evolution* **22**:160–174.

Hedenfalk, I., D. Duggan, Y. D. Chen, et al. 2001. Gene expression profiles in hereditary breast cancer. *New England Journal of Medicine* **344**:539–548.

Hedrick, P. W. 1998. Balancing selection and MHC. *Genetica* **104**:207–214.

———. 2005. A standardized genetic differentiation measure. *Evolution* **59**:1633–1638.

Hein, J., M. H. Schierup, and C. Wiuf. 2005. *Gene Genealogies, Variation and Evolution: A Primer in Coalescent Theory*. Oxford: Oxford University Press.

Heled, J., and A. J. Drummond. 2008. Bayesian inference of population size history from multiple loci. *BMC Evolutionary Biology* **8**:289.

Heliconius Genome Consortium. 2012. Butterfly genome reveals promiscuous exchange of mimicry adaptations among species. *Nature* **487**:94–98.

Hellenthal, G., J. K. Pritchard, and M. Stephens. 2006. The effects of genotype-dependent recombination, and transmission asymmetry, on linkage disequilibrium. *Genetics* **172**:2001–2005.

Hengen, P. N. 1995. Fidelity of DNA polymerases for PCR. *Trends in Biochemical Sciences* **20**:324–325.

Hermisson, J., and P. S. Pennings. 2005. Soft sweeps: molecular population genetics of adaptation from standing genetic variation. *Genetics* **169**:2335–2352.

Hernandez, R. D. 2008. A flexible forward simulator for populations subject to selection and demography. *Bioinformatics* **24**:2786–2787.

Hey, J. 2010. Isolation with migration models for more than two populations. *Molecular Biology and Evolution* **27**:905–920.

Hey, J., Y. Chung, and A. Sethuraman. 2015. On the occurrence of false positives in tests of migration under an isolation-with-migration model. *Molecular Ecology* **24**:5078–5083.

Hey, J., and E. Harris. 1999. Population bottlenecks and patterns of human polymorphism. *Molecular Biology and Evolution* **16**:1423–1426.

Hey, J., and R. Nielsen. 2004. Multilocus methods for estimating population sizes, migration rates and divergence time, with applications to the divergence of *Drosophila pseudoobscura* and *D. persimilis*. *Genetics* **167**:747–760.

———. 2007. Integration within the Felsenstein equation for improved Markov chain Monte Carlo methods in population genetics. *Proceedings of the National Academy of Sciences* **104**:2785–2790.

Hey, J., and J. Wakeley. 1997. A coalescent estimator of the population recombination rate. *Genetics* **145**:833–846.

Hickerson, M. J., E. Stahl, and N. Takebayashi. 2007. msBayes: pipeline for testing comparative phylogeographic histories using hierarchical approximate Bayesian computation. *BMC Bioinformatics* **8**:268.

Higuchi, R., B. Bowman, M. Freiberger, O. A. Ryder, and A. C. Wilson. 1984. DNA sequences from the quagga, an extinct member of the horse family. *Nature* **312**:282–284.

Hill, R. E., and N. D. Hastie. 1987. Accelerated evolution in reactive centre regions of serine protease inhibitors. *Nature* **326**:96–99.

Hill, W. G., and A. Robertson. 1966. The effect of linkage on limits to artificial selection. *Genetical Research* **8**:269–294.

———. 1968. Linkage disequilibrium in finite populations. *Theoretical and Applied Genetics* **38**:226–231.

Hill, W. G., and B. S. Weir. 1988. Variances and covariances of squared linkage disequilibria in finite populations. *Theoretical Population Biology* **33**:54–78.

Hilliker, A. J., G. Harauz, A. G. Reaume, M. Gray, S. H. Clark, and A. Chovnick. 1994. Meiotic gene conversion tract length distribution within the *rosy* locus of *Drosophila melanogaster*. *Genetics* **137**:1019–1024.

Hinds, D. A., L. L. Stuve, G. B. Nilsen, E. Halperin, E. Eskin, D. G. Ballinger, K. A. Frazer, and D. R. Cox. 2005. Whole-genome patterns of common DNA variation in three human populations. *Science* **307**:1072–1079.

Ho, S. Y. W., and B. Shapiro. 2011. Skyline-plot methods for estimating demographic history from nucleotide sequences. *Molecular Ecology Resources* **11**:423–434.

Hoban, S., J. L. Kelley, K. E. Lotterhos, M. F. Antolin, G. Bradburd, D. B. Lowry, M. L. Poss, L. K. Reed, A. Storfer, and M. C. Whitlock. 2016. Finding the genomic basis of local adaptation: pitfalls, practical solutions, and future directions. *American Naturalist* **188**:379–397.

Hobolth, A., O. F. Christensen, T. Mailund, and M. H. Schierup. 2007. Genomic relationships and speciation times of human, chimpanzee, and gorilla inferred from a coalescent hidden Markov model. *PLoS Genetics* **3**:e7.

Hobolth, A., J. Y. Dutheil, J. Hawks, M. H. Schierup, and T. Mailund. 2011. Incomplete lineage sorting patterns among human, chimpanzee, and orangutan suggest recent orangutan speciation and widespread selection. *Genome Research* **21**:349–356.

Hodges, E., Z. Xuan, V. Balija, et al. 2007. Genome-wide *in situ* exon capture for selective resequencing. *Nature Genetics* **39**:1522–1527.

Hodgkinson, A., and A. Eyre-Walker. 2010. Human triallelic sites: evidence for a new mutational mechanism? *Genetics* **184**:233–241.

———. 2011. Variation in the mutation rate across mammalian genomes. *Nature Reviews Genetics* **12**:756–766.

Hoekstra, H. E., and J. A. Coyne. 2007. The locus of evolution: evo devo and the genetics of adaptation. *Evolution* **61**:995–1016.

Hofer, T., M. Foll, and L. Excoffier. 2012. Evolutionary forces shaping genomic islands of population differentiation in humans. *BMC Genomics* **13**:107.

Hofreiter, M., C. Capelli, M. Krings, L. Waits, N. Conard, S. Münzel, G. Rabeder, D. Nagel, M. Paunovic, and G. Jambresić. 2002. Ancient DNA analyses reveal high mitochondrial DNA sequence diversity and parallel morphological evolution of late Pleistocene cave bears. *Molecular Biology and Evolution* **19**:1244–1250.

Hoggart, C. J., M. Chadeau-Hyam, T. G. Clark, R. Lampariello, J. C. Whittaker, M. De Iorio, and D. J. Balding. 2007. Sequence-level population simulations over large genomic regions. *Genetics* **177**:1725–1731.

Hoggart, C. J., E. J. Parra, M. D. Shriver, C. Bonilla, R. A. Kittles, D. G. Clayton, and P. M. McKeigue. 2003. Control of confounding of genetic associations in stratified populations. *American Journal of Human Genetics* **72**:1492–1504.

Hoggart, C. J., M. D. Shriver, R. A. Kittles, D. G. Clayton, and P. M. McKeigue. 2004. Design and analysis of admixture mapping studies. *American Journal of Human Genetics* **74**:965–978.

Hohenlohe, P. A., S. Bassham, P. D. Etter, N. Stiffler, E. A. Johnson, and W. A. Cresko. 2010. Population genomics of parallel adaptation in threespine stickleback using sequenced RAD tags. *PLoS Genetics* **6**:e1000862.

Holloway, A. K., M. K. N. Lawniczak, J. G. Mezey, D. J. Begun, and C. D. Jones. 2007. Adaptive gene expression divergence inferred from population genomics. *PLoS Genetics* **3**:e187.

Holmes, E. C., L. Q. Zhang, P. Simmonds, C. A. Ludlam, and A. Brown. 1992. Convergent and divergent sequence evolution in the surface envelope glycoprotein of human immunodeficiency virus type 1 within a single infected patient. *Proceedings of the National Academy of Sciences* **89**:4835–4839.

Holsinger, K. E., and B. S. Weir. 2009. Genetics in geographically structured populations: defining, estimating and interpreting F_{ST}. *Nature Reviews Genetics* **10**:639–650.

Holt, R. A., G. M. Subramanian, A. Halpern, et al. 2002. The genome sequence of the malaria mosquito *Anopheles gambiae*. *Science* **298**:129–149.

House, G. L., and M. W. Hahn. 2017. Evaluating methods to visualize patterns of genetic differentiation on a landscape. *Molecular Ecology Resources* doi:10.1111/1755-0998.12747.

Hu, X., J. Yuan, Y. Shi, et al. 2012. pIRS: profile-based Illumina pair-end reads simulator. *Bioinformatics* **28**:1533–1535.

Huang, X., Q. Feng, Q. Qian, et al. 2009. High-throughput genotyping by whole-genome resequencing. *Genome Research* **19**:1068–1076.

Hubbell, S. P. 2001. *The Unified Neutral Theory of Biodiversity and Biogeography*. Princeton, NJ: Princeton University Press.

Hubby, J. L., and R. C. Lewontin. 1966. A molecular approach to the study of genic heterozygosity in natural populations. I. The number of alleles at different loci in *Drosophila pseudoobscura*. Genetics **54**:577.

Huber, C. D., M. DeGiorgio, I. Hellmann, and R. Nielsen. 2016. Detecting recent selective sweeps while controlling for mutation rate and background selection. *Molecular Ecology* **25**:142–156.

Hubisz, M. J., D. Falush, M. Stephens, and J. K. Pritchard. 2009. Inferring weak population structure with the assistance of sample group information. *Molecular Ecology Resources* **9**:1322–1332.

Huddleston, J., S. Ranade, M. Malig, et al. 2014. Reconstructing complex regions of genomes using long-read sequencing technology. *Genome Research* **24**:688–696.

Hudson, R. R. 1983a. Testing the constant-rate neutral allele model with protein sequence data. *Evolution* **37**:203–217.

———. 1983b. Properties of a neutral allele model with intragenic recombination. *Theoretical Population Biology* **23**:183–201.

———. 1987. Estimating the recombination parameter of a finite population model without selection. *Genetical Research* **50**:245–250.

———. 1990. Gene genealogies and the coalescent process. In *Oxford Surveys in Evolutionary Biology*, ed. D. Futuyma and J. Antonovics, pp. 1–44. Oxford: Oxford University Press.

———. 1992. Gene trees, species trees and the segregation of ancestral alleles. *Genetics* **131**:509–513.

———. 1993. The how and why of generating gene genealogies. In *Mechanisms of Molecular Evolution*, ed. N. Takahata and A. G. Clark, pp. 23–36. Sunderland, MA: Sinauer Associates.

———. 2000. A new statistic for detecting genetic differentiation. *Genetics* **155**:2011–2014.

———. 2001. Two-locus sampling distributions and their application. *Genetics* **159**:1805–1817.

———. 2002. Generating samples under a Wright-Fisher neutral model of genetic variation. *Bioinformatics* **18**:337–338.

Hudson, R. R., K. Bailey, D. Skarecky, J. Kwiatowski, and F. J. Ayala. 1994. Evidence for positive selection in the superoxide dismutase (*Sod*) region of *Drosophila melanogaster*. *Genetics* **136**:1329–1340.

Hudson, R. R., D. D. Boos, and N. L. Kaplan. 1992. A statistical test for detecting geographic subdivision. *Molecular Biology and Evolution* **9**:138–151.

Hudson, R. R., and J. A. Coyne. 2002. Mathematical consequences of the genealogical species concept. *Evolution* **56**:1557–1565.

Hudson, R. R., and N. L. Kaplan. 1985. Statistical properties of the number of recombination events in the history of a sample of DNA sequences. *Genetics* **111**:147–164.

———. 1988. The coalescent process in models with selection and recombination. *Genetics* **120**:831–840.

———. 1994. Gene trees with background selection. In *Non-neutral Evolution: Theories and Molecular Data*, ed. B. Golding, pp. 140–153. New York: Chapman & Hall.

Hudson, R. R., M. Kreitman, and M. Aguadé. 1987. A test of neutral molecular evolution based on nucleotide data. *Genetics* **116**:153–159.

Hudson, R. R., M. Slatkin, and W. P. Maddison. 1992. Estimation of levels of gene flow from DNA sequence data. *Genetics* **132**:583–589.

Huelsenbeck, J. P., and P. Andolfatto. 2007. Inference of population structure under a Dirichlet process model. *Genetics* **175**:1787–1802.

Hughes, A. L., and M. Nei. 1988. Pattern of nucleotide substitution at major histocompatibility complex class I loci reveals overdominant selection. *Nature* **335**:167–170.

Huson, D. H., T. Klöpper, P. J. Lockhart, and M. A. Steel. 2005. Reconstruction of reticulate networks from gene trees. In *Research in Computational Molecular Biology*, pp. 233–249. Springer.

I

Ikemura, T. 1981. Correlation between the abundance of *Escherichia coli* transfer RNAs and the occurrence of the respective codons in its protein genes. *Journal of Molecular Biology* **146**:1–21.

Ina, Y. 1995. New methods for estimating the numbers of synonymous and nonsynonymous substitutions. *Journal of Molecular Evolution* **40**:190–226.

Ingvarsson, P. K. 2004. Population subdivision and the Hudson-Kreitman-Aguade test: testing for deviations from the neutral model in organelle genomes. *Genetical Research* **83**:31–39.

Innan, H., and H. Watanabe. 2006. The effect of gene flow on the coalescent time in the human-chimpanzee ancestral population. *Molecular Biology and Evolution* **23**:1040–1047.

Innan, H., K. Zhang, P. Marjoram, S. Tavaré, and N. A. Rosenberg. 2005. Statistical tests of the coalescent model based on the haplotype frequency distribution and the number of segregating sites. *Genetics* **169**:1763–1777.

International HapMap Consortium. 2005. A haplotype map of the human genome. *Nature* **437**:1299–1320.

J

Jaenicke-Despres, V., E. S. Buckler, B. D. Smith, M. T. P. Gilbert, A. Cooper, J. Doebley, and S. Pääbo. 2003. Early allelic selection in maize as revealed by ancient DNA. *Science* **302**:1206–1208.

Jain, M., I. T. Fiddes, K. H. Miga, H. E. Olsen, B. Paten, and M. Akeson. 2015. Improved data analysis for the MinION nanopore sequencer. *Nature Methods* **12**:351–356.

Jakobsson, M., M. D. Edge, and N. A. Rosenberg. 2013. The relationship between F_{ST} and the frequency of the most frequent allele. *Genetics* **193**:515–528.

Janes, J. K., J. M. Miller, J. R. Dupuis, R. M. Malenfant, J. C. Gorrell, C. I. Cullingham, and R. L. Andrew. 2017. The $K = 2$ conundrum. *Molecular Ecology* **26**:3594–3602.

Jeffreys, A. J., L. Kauppi, and R. Neumann. 2001. Intensely punctate meiotic recombination in the class II region of the major histocompatibility complex. *Nature Genetics* **29**:217–222.

Jenkins, D. L., C. A. Ortori, and J. F. Y. Brookfield. 1995. A test for adaptive change in DNA sequences controlling transcription. *Proceedings of the Royal Society B: Biological Sciences* **261**:203–207.

Jensen, J. D., Y. Kim, V. B. DuMont, C. F. Aquadro, and C. D. Bustamante. 2005. Distinguishing between selective sweeps and demography using DNA polymorphism data. *Genetics* **170**:1401–1410.

Jensen, J. D., K. R. Thornton, C. D. Bustamante, and C. F. Aquadro. 2007. On the utility of linkage disequilibrium as a statistic for identifying targets of positive selection in nonequilibrium populations. *Genetics* **176**:2371–2379.

Jensen, J. L., A. J. Bohonak, and S. T. Kelley. 2005. Isolation by distance, web service. *BMC Genetics* **6**:13.

Jeong, S., M. Rebeiz, P. Andolfatto, T. Werner, J. True, and S. B. Carroll. 2008. The evolution of gene regulation underlies a morphological difference between two *Drosophila* sister species. *Cell* **132**:783–793.

Jewett, E. M., M. Steinrücken, and Y. S. Song. 2016. The effects of population size histories on estimates of selection coefficients from time-series genetic data. *Molecular Biology and Evolution* **33**:3002–3027.

Jobling, M. A., M. E. Hurles, and C. Tyler-Smith. 2004. *Human Evolutionary Genetics: Origins, Peoples & Disease*. New York: Garland Science.

Johnson, P. L. F., and M. Slatkin. 2008. Accounting for bias from sequencing error in population genetic estimates. *Molecular Biology and Evolution* **25**:199–206.

Johnston, S. E., C. Bérénos, J. Slate, and J. M. Pemberton. 2016. Conserved genetic architecture underlying individual recombination rate variation in a wild population of Soay sheep (*Ovis aries*). *Genetics* **203**:583–598.

Johnston, S. E., J. Gratten, C. Berenos, J. G. Pilkington, T. H. Clutton-Brock, J. M. Pemberton, and J. Slate. 2013. Life history trade-offs at a single locus maintain sexually selected genetic variation. *Nature* **502**:93–95.

Joly, S., P. A. McLenachan, and P. J. Lockhart. 2009. A statistical approach for distinguishing hybridization and incomplete lineage sorting. *American Naturalist* **174**:E54–E70.

Jones, C. D., and D. J. Begun. 2005. Parallel evolution of chimeric fusion genes. *Proceedings of the National Academy of Sciences* **102**:11,373–11,378.

Jones, F. C., M. G. Grabherr, Y. F. Chan, et al. 2012. The genomic basis of adaptive evolution in threespine sticklebacks. *Nature* **484**:55–61.

Jones, M. R., and J. M. Good. 2016. Targeted capture in evolutionary and ecological genomics. *Molecular Ecology* **25**:185–202.

Jukes, T. H., and C. R. Cantor. 1969. Evolution of protein molecules. In *Mammalian Protein Metabolism*, ed. H. N. Munro, pp. 21–32. New York: Academic Press.

K

Kaplan, N., and R. R. Hudson. 1985. The use of sample genealogies for studying a selectively neutral *m*-loci model with recombination. *Theoretical Population Biology* **28**:382–396.

Kaplan, N. L., R. R. Hudson, and C. H. Langley. 1989. The "hitchhiking effect" revisited. *Genetics* **123**:887–899.

Karasov, T., P. W. Messer, and D. A. Petrov. 2010. Evidence that adaptation in *Drosophila* is not limited by mutation at single sites. *PLoS Genetics* **6**:e1000924.

Karsten, K. B., L. N. Andriamandimbiarisoa, S. F. Fox, and C. J. Raxworthy. 2008. A unique life history among tetrapods: an annual chameleon living mostly as an egg. *Proceedings of the National Academy of Sciences* **105**:8980–8984.

Kauer, M., D. Dieringer, and C. Schlötterer. 2003. A microsatellite variability screen for positive selection associated with the "out of Africa" habitat expansion of *Drosophila melanogaster*. *Genetics* **165**:1137–1148.

Kaur, T., and M. V. Rockman. 2014. Crossover heterogeneity in the absence of hotspots in *Caenorhabditis elegans*. Genetics **196**:137–148.

Kayser, M., S. Brauer, and M. Stoneking. 2003. A genome scan to detect candidate regions influenced by local natural selection in human populations. *Molecular Biology and Evolution* **20**:893–900.

Keightley, P. D., and A. Eyre-Walker. 2007. Joint inference of the distribution of fitness effects of deleterious mutations and population demography based on nucleotide polymorphism frequencies. *Genetics* **177**:2251–2261.

———. 2010. What can we learn about the distribution of fitness effects of new mutations from DNA sequence data? *Philosophical Transactions of the Royal Society B: Biological Sciences* **365**:1187–1193.

Keinan, A., and A. G. Clark. 2012. Recent explosive human population growth has resulted in an excess of rare genetic variants. *Science* **336**:740–743.

Keinan, A., J. C. Mullikin, N. Patterson, and D. Reich. 2007. Measurement of the human allele frequency spectrum demonstrates greater genetic drift in East Asians than in Europeans. *Nature Genetics* **39**:1251–1255.

———. 2009. Accelerated genetic drift on chromosome X during the human dispersal out of Africa. *Nature Genetics* **41**:66–70.

Kelleher, J., A. M. Etheridge, and G. McVean. 2016. Efficient coalescent simulation and genealogical analysis for large sample sizes. *PLoS Computational Biology* **12**:e1004842.

Kelley, J. L., J. Madeoy, J. C. Calhoun, W. Swanson, and J. M. Akey. 2006. Genomic signatures of positive selection in humans and the limits of outlier approaches. *Genome Research* **16**:980–989.

Kelly, J. K. 1997. A test of neutrality based on interlocus associations. *Genetics* **146**:1197–1206.

Kern, A. D. 2009. Correcting the site frequency spectrum for divergence-based ascertainment. *PLoS ONE* **4**:e5152.

Kern, A. D., and D. Haussler. 2010. A population genetic hidden Markov model for detecting genomic regions under selection. *Molecular Biology and Evolution* **27**:1673–1685.

Kern, A. D., and D. R. Schrider. 2016. Discoal: flexible coalescent simulations with selection. *Bioinformatics* **32**:3839–3841.

Khaitovich, P., S. Pääbo, and G. Weiss. 2005. Toward a neutral evolutionary model of gene expression. *Genetics* **170**:929–939.

Kidd, J. M., G. M. Cooper, W. F. Donahue, et al. 2008. Mapping and sequencing of structural variation from eight human genomes. *Nature* **453**:56–64.

Kim, S., V. Plagnol, T. T. Hu, C. Toomajian, R. M. Clark, S. Ossowski, J. R. Ecker, D. Weigel, and M. Nordborg. 2007. Recombination and linkage disequilibrium in *Arabidopsis thaliana*. *Nature Genetics* **39**:1151–1155.

Kim, Y., and R. Nielsen. 2004. Linkage disequilibrium as a signature of selective sweeps. *Genetics* **167**:1513–1524.

Kim, Y., and W. Stephan. 2000. Joint effects of genetic hitchhiking and background selection on neutral variation. *Genetics* **155**:1415–1427.

———. 2002. Detecting a local signature of genetic hitchhiking along a recombining chromosome. *Genetics* **160**:765–777.

Kimmel, M., and R. Chakraborty. 1996. Measures of variation at DNA repeat loci under a general stepwise mutation model. *Theoretical Population Biology* **50**:345–367.

Kimura, M. 1953. "Stepping stone" model of population. In *Annual Report*, ed. National Institute of Genetics, pp. 62–63. Japan.

———. 1955. Solution of a process of random genetic drift with a continuous model. *Proceedings of the National Academy of Sciences* **41**:144–150.

———. 1957. Some problems of stochastic processes in genetics. *Annals of Mathematical Statistics* **28**:882–901.

———. 1968. Evolutionary rate at the molecular level. *Nature* **217**:624–626.

———. 1969. The number of heterozygous nucleotide sites maintained in a finite population due to steady flux of mutations. *Genetics* **61**:893–903.

———. 1970. The length of time required for a selectively neutral mutant to reach fixation through random frequency drift in a finite population. *Genetical Research* **15**:131–133.

———. 1971. Theoretical foundation of population genetics at the molecular level. *Theoretical Population Biology* **2**:174–208.

———. 1977. Preponderance of synonymous changes as evidence for the neutral theory of molecular evolution. *Nature* **267**:275–276.

———. 1980. A simple method for estimating evolutionary rates of base substitutions through comparative studies of nucleotide sequences. *Journal of Molecular Evolution* **16**:111–120.

———. 1981. Possibility of extensive neutral evolution under stabilizing selection with special reference to nonrandom usage of synonymous codons. *Proceedings of the National Academy of Sciences* **78**:5773–5777.

———. 1983. *The Neutral Theory of Molecular Evolution*. Cambridge: Cambridge University Press.

Kimura, M., and J. F. Crow. 1964. The number of alleles that can be maintained in a finite population. *Genetics* **49**:725–738.

Kimura, M., and T. Ohta. 1969. The average number of generations until fixation of a mutant gene in a finite population. *Genetics* **61**:763.

———. 1971. Protein polymorphism as a phase of molecular evolution. *Nature* **229**:467–469.

Kimura, M., and G. H. Weiss. 1964. The stepping stone model of population structure and the decrease of genetic correlation with distance. *Genetics* **49**:561–576.

King, J. L., and T. H. Jukes. 1969. Non-Darwinian evolution. *Science* **164**:788–798.

Kingman, J. F. C. 1982a. The coalescent. *Stochastic Processes and Their Applications* **13**:235–248.

———. 1982b. On the genealogy of large populations. *Journal of Applied Probability* **19**:27–43.

———. 1982c. Exchangeability and the evolution of large populations. In Exchangeability in Probability and Statistics, ed. G. Koch and F. Spizzichino, pp. 97–112. Amsterdam: North-Holland.

Kitzman, J. O., A. P. MacKenzie, A. Adey, et al. 2011. Haplotype-resolved genome sequencing of a Gujarati Indian individual. *Nature Biotechnology* **29**:59–63.

Kliman, R. M., P. Andolfatto, J. A. Coyne, F. Depaulis, M. Kreitman, A. J. Berry, J. McCarter, J. Wakeley, and J. Hey. 2000. The population genetics of the origin and divergence of the *Drosophila simulans* complex species. *Genetics* **156**:1913–1931.

Knudsen, B., and M. M. Miyamoto. 2009. Accurate and fast methods to estimate the population mutation rate from error prone sequences. *BMC Bioinformatics* **10**:247.

Kofler, R., A. J. Betancourt, and C. Schlötterer. 2012. Sequencing of pooled DNA samples (pool-seq) uncovers complex dynamics of transposable element insertions in *Drosophila melanogaster*. *PLoS Genetics* **8**:e1002487.

Kohn, M. H., S. Fang, and C.-I. Wu. 2004. Inference of positive and negative selection on the 5′ regulatory regions of *Drosophila* genes. *Molecular Biology and Evolution* **21**:374–383.

Kohn, M. H., H.-J. Pelz, and R. K. Wayne. 2000. Natural selection mapping of the warfarin-resistance gene. *Proceedings of the National Academy of Sciences* **97**:7911–7915.

Kojima, K.-I., and H. E. Schaffer. 1967. Survival process of linked mutant genes. *Evolution* **21**:518–531.

Kolaczkowski, B., A. D. Kern, A. K. Holloway, and D. J. Begun. 2011. Genomic differentiation between temperate and tropical Australian populations of *Drosophila melanogaster*. *Genetics* **187**:245–260.

Kong, A., M. L. Frigge, G. Masson, et al. 2012. Rate of *de novo* mutations and the importance of father's age to disease risk. *Nature* **488**:471–475.

Kong, A., G. Masson, M. L. Frigge, et al. 2008. Detection of sharing by descent, long-range phasing and haplotype imputation. *Nature Genetics* **40**:1068–1075.

Koopman, W. J. M., Y. H. Li, E. Coart, E. V. De Weg, B. Vosman, I. Roldan-Ruiz, and M. J. M. Smulders. 2007. Linked vs. unlinked markers: multilocus microsatellite haplotype-sharing as a tool to estimate gene flow and introgression. *Molecular Ecology* **16**:243–256.

Korneliussen, T. S., A. Albrechtsen, and R. Nielsen. 2014. ANGSD: analysis of next generation sequencing data. *BMC Bioinformatics* **15**:356.

Kreitman, M. 1983. Nucleotide polymorphism at the alcohol dehydrogenase locus of *Drosophila melanogaster*. *Nature* **304**:412–417.

———. 1996. The neutral theory is dead. Long live the neutral theory. *Bioessays* **18**:678–683.

Kreitman, M. E., and M. Aguadé. 1986. Excess polymorphism at the *Adh* locus in *Drosophila melanogaster*. *Genetics* **114**:93–110.

Kreitman, M., and R. R. Hudson. 1991. Inferring the evolutionary histories of the *Adh* and *Adh-dup* loci in *Drosophila melanogaster* from patterns of polymorphism and divergence. *Genetics* 127:565–582.

Krimbas, C. B., and S. Tsakas. 1971. The genetics of *Dacus oleae*. V. Changes of esterase polymorphism in a natural population following insecticide control—selection or drift? *Evolution* 25:454–460.

Kruglyak, L. 1999. Prospects for whole-genome linkage disequilibrium mapping of common disease genes. *Nature Genetics* 22:139–144.

Kubatko, L. S. 2009. Identifying hybridization events in the presence of coalescence via model selection. *Systematic Biology* 58:478–488.

Kuhner, M. K., P. Beerli, J. Yamato, and J. Felsenstein. 2000. Usefulness of single nucleotide polymorphism data for estimating population parameters. *Genetics* 156:439–447.

Kuhner, M. K., J. Yamato, and J. Felsenstein. 1995. Estimating effective population size and mutation rate from sequence data using Metropolis-Hastings sampling. *Genetics* 140:1421–1430.

———. 2000. Maximum likelihood estimation of recombination rates from population data. *Genetics* 156:1393–1401.

L

Lack, J. B., J. D. Lange, A. D. Tang, R. B. Corbett-Detig, and J. E. Pool. 2016. A thousand fly genomes: an expanded *Drosophila* genome nexus. *Molecular Biology and Evolution* 33:3308–3313.

Lambert, B. W., J. D. Terwilliger, and K. M. Weiss. 2008. ForSim: a tool for exploring the genetic architecture of complex traits with controlled truth. *Bioinformatics* 24:1821–1822.

Lambert, D. M., P. A. Ritchie, C. D. Millar, B. Holland, A. J. Drummond, and C. Baroni. 2002. Rates of evolution in ancient DNA from Adélie penguins. *Science* 295:2270–2273.

Lande, R. 1976. Natural selection and random genetic drift in phenotypic evolution. *Evolution* 30:314–334.

Lander, E. S., L. M. Linton, B. Birren, et al. 2001. Initial sequencing and analysis of the human genome. *Nature* 409:860–921.

Lander, E. S., and M. S. Waterman. 1988. Genomic mapping by fingerprinting random clones: a mathematical analysis. *Genomics* 2:231–239.

Langley, C. H., and J. F. Crow. 1974. The direction of linkage disequilibrium. *Genetics* 78:937–941.

Langley, C. H., B. P. Lazzaro, W. Phillips, E. Heikkinen, and J. M. Braverman. 2000. Linkage disequilibria and the site frequency spectra in the *su(s)* and *su(w^a)* regions of the *Drosophila melanogaster* X chromosome. *Genetics* 156:1837–1852.

Langley, C. H., K. Stevens, C. Cardeno, et al. 2012. Genomic variation in natural populations of *Drosophila melanogaster*. *Genetics* 192:533–598.

Langley, C. H., Y. N. Tobari, and K.-I. Kojima. 1974. Linkage disequilibrium in natural populations of *Drosophila melanogaster*. *Genetics* 78:921–936.

Langmead, B., C. Trapnell, M. Pop, and S. Salzberg. 2009. Ultrafast and memory-efficient alignment of short DNA sequences to the human genome. *Genome Biology* 10:R25.

Lansing, J. S., and M. P. Cox. 2011. The domain of the replicators: selection, neutrality, and cultural evolution. *Current Anthropology* 52:105–125.

Lathrop, G. M. 1982. Evolutionary trees and admixture: phylogenetic inference when some populations are hybridized. *Annals of Human Genetics* 46:245–255.

Laval, G., and L. Excoffier. 2004. SIMCOAL 2.0: a program to simulate genomic diversity over large recombining regions in a subdivided population with a complex history. *Bioinformatics* 20:2485–2487.

Lawrie, D. S., P. W. Messer, R. Hershberg, and D. A. Petrov. 2013. Strong purifying selection at synonymous sites in *D. melanogaster*. *PLoS Genetics* 9:e1003527.

Lawson, D. J., G. Hellenthal, S. Myers, and D. Falush. 2012. Inference of population structure using dense haplotype data. *PLoS Genetics* 8:e1002453.

Le, S. Q., and R. Durbin. 2011. SNP detection and genotyping from low-coverage sequencing data on multiple diploid samples. *Genome Research* 21:952–960.

Leffler, E. M., K. Bullaughey, D. R. Matute, W. K. Meyer, L. Ségurel, A. Venkat, P. Andolfatto, and M. Przeworski. 2012. Revisiting an old riddle: what determines genetic diversity levels within species? *PLoS Biology* 10:e1001388.

Leffler, E. M., Z. Gao, S. Pfeifer, et al. 2013. Multiple instances of ancient balancing selection shared between humans and chimpanzees. *Science* **339**:1578–1582.

Lek, M., K. J. Karczewski, E. V. Minikel, et al. 2016. Analysis of protein-coding genetic variation in 60,706 humans. *Nature* **536**:285–291.

Leman, S. C., Y. G. Chen, J. E. Stajich, M. A. F. Noor, and M. K. Uyenoyama. 2005. Likelihoods from summary statistics: recent divergence between species. *Genetics* **171**:1419–1436.

Levy, S., G. Sutton, P. C. Ng, et al. 2007. The diploid genome sequence of an individual human. *PLoS Biology* **5**:e254.

Lewontin, R. C. 1964. The interaction of selection and linkage. I. General considerations; heterotic models. *Genetics* **49**:49–67.

———. 1974. *The Genetic Basis for Evolutionary Change*. New York: Columbia University Press.

———. 1988. On measures of gametic disequilibrium. *Genetics* **120**:849–852.

Lewontin, R. C., and J. L. Hubby. 1966. A molecular approach to the study of genic heterozygosity in natural populations. II. Amount of variation and degree of heterozygosity in natural populations of *Drosophila pseudoobscura*. *Genetics* **54**:595–609.

Lewontin, R. C., and K.-I. Kojima. 1960. The evolutionary dynamics of complex polymorphisms. *Evolution* **14**:458–472.

Lewontin, R. C., and J. Krakauer. 1973. Distribution of gene frequency as a test of the theory of selective neutrality of polymorphisms. *Genetics* **74**:175–195.

Li, H. 2011. A statistical framework for SNP calling, mutation discovery, association mapping and population genetical parameter estimation from sequencing data. *Bioinformatics* **27**:2987–2993.

Li, H., and R. Durbin. 2010. Fast and accurate short read alignment with Burrows–Wheeler transform. *Bioinformatics* **25**:1754–1760.

———. 2011. Inference of human population history from individual whole-genome sequences. *Nature* **475**:493–496.

Li, H., B. Handsaker, A. Wysoker, T. Fennell, J. Ruan, N. Homer, G. Marth, G. Abecasis, R. Durbin, and 1000 Genome Project Data Processing Subgroup. 2009. The sequence alignment/map format and SAMtools. *Bioinformatics* **25**:2078–2079.

Li, H., J. Ruan, and R. Durbin. 2008. Mapping short DNA sequencing reads and calling variants using mapping quality scores. *Genome Research* **18**:1851–1858.

Li, H., and W. Stephan. 2006. Inferring the demographic history and rate of adaptive substitution in *Drosophila*. *PLoS Genetics* **2**:e166.

Li, N., and M. Stephens. 2003. Modeling linkage disequilibrium and identifying recombination hotspots using single-nucleotide polymorphism data. *Genetics* **165**:2213–2233.

Li, R., Y. Li, K. Kristiansen, and J. Wang. 2008. SOAP: short oligonucleotide alignment program. *Bioinformatics* **24**:713–714.

Li, W.-H. 1977. Distribution of nucleotide differences between two randomly chosen cistrons in a finite population. *Genetics* **85**:331–337.

———. 1993. Unbiased estimation of the rates of synonymous and nonsynonymous substitution. *Journal of Molecular Evolution* **36**:96–99.

———. 1997. *Molecular Evolution*. Sunderland, MA: Sinauer Associates.

Li, W.-H., C.-I. Wu, and C.-C. Luo. 1985. A new method for estimating synonymous and nonsynonymous rates of nucleotide substitution considering the relative likelihood of nucleotide and codon changes. *Molecular Biology and Evolution* **2**:150–174.

Li, Y. F., J. C. Costello, A. K. Holloway, and M. W. Hahn. 2008. "Reverse ecology" and the power of population genomics. *Evolution* **62**:2984–2994.

Li, Y., H. Zheng, R. Luo, et al. 2011. Structural variation in two human genomes mapped at single-nucleotide resolution by whole genome *de novo* assembly. *Nature Biotechnology* **29**:723–730.

Liang, L. M., S. Zollner, and G. R. Abecasis. 2007. GENOME: a rapid coalescent-based whole genome simulator. *Bioinformatics* **23**:1565–1567.

Librado, P., and J. Rozas. 2009. DnaSP v5: a software for comprehensive analysis of DNA polymorphism data. *Bioinformatics* **25**:1451–1452.

Lin, K., H. Li, C. Schlötterer, and A. Futschik. 2011. Distinguishing positive selection from neutral evolution: boosting the performance of summary statistics. *Genetics* **187**:229–244.

Lipson, M., P.-R. Loh, A. Levin, D. Reich, N. Patterson, and B. Berger. 2013. Efficient moment-based inference of admixture parameters and sources of gene flow. *Molecular Biology and Evolution* **30**:1788–1802.

Liti, G., D. M. Carter, A. M. Moses, et al. 2009. Population genomics of domestic and wild yeasts. *Nature* **458**:337–341.

Liu, K. J., J. Dai, K. Truong, Y. Song, M. H. Kohn, and L. Nakhleh. 2014. An HMM-based comparative genomic framework for detecting introgression in eukaryotes. *PLoS Computational Biology* 10:e1003649.

Liu, X., and Y.-X. Fu. 2015. Exploring population size changes using SNP frequency spectra. *Nature Genetics* 47:555–559.

Loewe, L., and B. Charlesworth. 2007. Background selection in single genes may explain patterns of codon bias. *Genetics* 175:1381–1393.

Loewe, L., B. Charlesworth, C. Bartolome, and V. Noel. 2006. Estimating selection on nonsynonymous mutations. *Genetics* 172:1079–1092.

Loh, P.-R., M. Lipson, N. Patterson, P. Moorjani, J. K. Pickrell, D. Reich, and B. Berger. 2013. Inferring admixture histories of human populations using linkage disequilibrium. *Genetics* 193:1233–1254.

Lohmueller, K. E., C. D. Bustamante, and A. G. Clark. 2009. Methods for human demographic inference using haplotype patterns from genomewide single-nucleotide polymorphism data. *Genetics* 182:217–231.

Lohse, K., R. Harrison, and N. H. Barton. 2011. A general method for calculating likelihoods under the coalescent process. *Genetics* 189:977–987.

Loman, N. J., R. V. Misra, T. J. Dallman, C. Constantinidou, S. E. Gharbia, J. Wain, and M. J. Pallen. 2012. Performance comparison of benchtop high-throughput sequencing platforms. *Nature Biotechnology* 30:434–439.

Lopes, J. S., D. Balding, and M. A. Beaumont. 2009. PopABC: a program to infer historical demographic parameters. *Bioinformatics* 25:2747–2749.

Lotterhos, K. E., and M. C. Whitlock. 2014. Evaluation of demographic history and neutral parameterization on the performance of F_{ST} outlier tests. *Molecular Ecology* 23:2178–2192.

Lourenço, J. M., S. Glémin, and N. Galtier. 2013. The rate of molecular adaptation in a changing environment. *Molecular Biology and Evolution* 30:1292–1301.

Luca, F., R. R. Hudson, D. B. Witonsky, and A. Di Rienzo. 2011. A reduced representation approach to population genetic analyses and applications to human evolution. *Genome Research* 21:1087–1098.

Ludwig, A., M. Pruvost, M. Reissmann, et al. 2009. Coat color variation at the beginning of horse domestication. *Science* 324:485–485.

Ludwig, M. Z., and M. Kreitman. 1995. Evolutionary dynamics of the enhancer region of *even-skipped* in Drosophila. *Molecular Biology and Evolution* 12:1002–1011.

Lukić, S., and J. Hey. 2012. Demographic inference using spectral methods on SNP data, with an analysis of the human out-of-Africa expansion. *Genetics* 192:619–639.

Lunter, G., and M. Goodson. 2011. Stampy: a statistical algorithm for sensitive and fast mapping of Illumina sequence reads. *Genome Research* 21:936–939.

Lunter, G., C. P. Ponting, and J. Hein. 2006. Genome-wide identification of human functional DNA using a neutral indel model. *PLoS Computational Biology* 2:e5.

Lynch, M. 2006. The origins of eukaryotic gene structure. *Molecular Biology and Evolution* 23:450–468.

———. 2008. Estimation of nucleotide diversity, disequilibrium coefficients, and mutation rates from high-coverage genome-sequencing projects. *Molecular Biology and Evolution* 25:2409–2419.

———. 2010. Evolution of the mutation rate. *Trends in Genetics* 26:345–352.

Lynch, M., and T. J. Crease. 1990. The analysis of population survey data on DNA sequence variation. *Molecular Biology and Evolution* 7:377–394.

Lynch, M., D. Bost, S. Wilson, T. Maruki, and S. Harrison. 2014. Population-genetic inference from pooled-sequencing data. *Genome Biology and Evolution* 6:1210–1218.

Lynch, M., and W. G. Hill. 1986. Phenotypic evolution by neutral mutation. *Evolution* 40:915–935.

M

Macdonald, S. J., and A. D. Long. 2005. Prospects for identifying functional variation across the genome. *Proceedings of the National Academy of Sciences* 102:6614–6621.

Machado, C. A., R. M. Kliman, J. A. Markert, and J. Hey. 2002. Inferring the history of speciation from multilocus DNA sequence data: the case of *Drosophila pseudoobscura* and close relatives. *Molecular Biology and Evolution* 19:472–488.

Mackay, T. F., S. Richards, E. A. Stone, et al. 2012. The *Drosophila melanogaster* genetic reference panel. *Nature* 482:173–178.

Macpherson, J. M., G. Sella, J. C. Davis, and D. A. Petrov. 2007. Genomewide spatial correspondence between nonsynonymous divergence and neutral polymorphism reveals extensive adaptation in *Drosophila*. *Genetics* **177**:2083–2099.

Maddison, W. P. 1997. Gene trees in species trees. *Systematic Biology* **46**:523–536.

Mailund, T., J. Y. Dutheil, A. Hobolth, G. Lunter, and M. H. Schierup. 2011. Estimating divergence time and ancestral effective population size of Bornean and Sumatran orangutan subspecies using a coalescent hidden Markov model. *PLoS Genetics* **7**:e1001319.

Mailund, T., A. E. Halager, M. Westergaard, J. Y. Dutheil, K. Munch, L. N. Andersen, G. Lunter, K. Prüfer, A. Scally, and A. Hobolth. 2012. A new isolation with migration model along complete genomes infers very different divergence processes among closely related great ape species. *PLoS Genetics* **8**:e1003125.

Mailund, T., M. H. Schierup, C. N. S. Pedersen, P. J. M. Mechlenborg, J. N. Madsen, and L. Schauser. 2005. CoaSim: a flexible environment for simulating genetic data under coalescent models. *BMC Bioinformatics* **6**:252.

Malaspinas, A.-S. 2016. Methods to characterize selective sweeps using time serial samples: an ancient DNA perspective. *Molecular Ecology* **25**:24–41.

Malaspinas, A.-S., O. Malaspinas, S. N. Evans, and M. Slatkin. 2012. Estimating allele age and selection coefficient from time-serial data. *Genetics* **192**:599–607.

Malécot, G. 1948. *Les Mathématiques de l'Hérédité*. Paris: Masson & Cie.

Manel, S., M. K. Schwartz, G. Luikart, and P. Taberlet. 2003. Landscape genetics: combining landscape ecology and population genetics. *Trends in Ecology & Evolution* **18**:189–197.

Mantel, N. 1967. The detection of disease clustering and a generalized regression approach. *Cancer Research* **27**:209–220.

Manzano-Winkler, B., S. E. McGaugh, and M. A. Noor. 2013. How hot are Drosophila hotspots? Examining recombination rate variation and associations with nucleotide diversity, divergence, and maternal age in *Drosophila pseudoobscura*. *PLoS ONE* **8**:e71582.

Marais, G. 2003. Biased gene conversion: implications for genome and sex evolution. *Trends in Genetics* **19**:330–338.

Marchini, J., D. Cutler, N. Patterson, et al. 2006. A comparison of phasing algorithms for trios and unrelated individuals. *American Journal of Human Genetics* **78**:437–450.

Margulies, M., M. Egholm, W. E. Altman, S. Attiya, J. S. Bader, L. A. Bemben, J. Berka, M. S. Braverman, Y.-J. Chen, and Z. Chen. 2005. Genome sequencing in microfabricated high-density picolitre reactors. *Nature* **437**:376–380.

Marjoram, P., and J. D. Wall. 2006. Fast "coalescent" simulation. *BMC Genetics* **7**:16.

Markovtsova, L., P. Marjoram, and S. Tavaré. 2001. On a test of Depaulis and Veuille. *Molecular Biology and Evolution* **18**:1132–1133.

Marth, G. T., E. Czabarka, J. Murvai, and S. T. Sherry. 2004. The allele frequency spectrum in genome-wide human variation data reveals signals of differential demographic history in three large world populations. *Genetics* **166**:351–372.

Martin, S. H., K. K. Dasmahapatra, N. J. Nadeau, C. Salazar, J. R. Walters, F. Simpson, M. Blaxter, A. Manica, J. Mallet, and C. D. Jiggins. 2013. Genome-wide evidence for speciation with gene flow in *Heliconius* butterflies. *Genome Research* **23**:1817–1828.

Maruyama, T., and P. A. Fuerst. 1985. Population bottlenecks and nonequilibrium models in population genetics. II. Number of alleles in a small population that was formed by a recent bottleneck. *Genetics* **111**:675–689.

Mathew, L. A., and J. D. Jensen. 2015. Evaluating the ability of the pairwise joint site frequency spectrum to co-estimate selection and demography. *Frontiers in Genetics* **6**:268.

Mathieson, I., and G. McVean. 2013. Estimating selection coefficients in spatially structured populations from time series data of allele frequencies. *Genetics* **193**:973–984.

Maynard Smith, J. 2002. Equations of life. In *It Must Be Beautiful: Great Equations of Modern Science*, 193–211. ed. G. Farmelo. London: Granta.

Maynard Smith, J., and J. Haigh. 1974. The hitch-hiking effect of a favorable gene. *Genetical Research* **23**:23–35.

McCarroll, S. A., F. G. Kuruvilla, J. M. Korn, et al. 2008. Integrated detection and population-genetic analysis of SNPs and copy number variation. *Nature Genetics* **40**:1166–1174.

McCoy, R. C., N. R. Garud, J. L. Kelley, C. L. Boggs, and D. A. Petrov. 2014. Genomic inference accurately predicts the timing and severity of a recent bottleneck in a nonmodel insect population. *Molecular Ecology* **23**:136–150.

McCoy, R. C., R. W. Taylor, T. A. Blauwkamp, J. L. Kelley, M. Kertesz, D. Pushkarev, D. A. Petrov, and A.-S. Fiston-Lavier. 2014. Illumina TruSeq synthetic long-reads empower *de novo* assembly and resolve complex, highly-repetitive transposable elements. *PLoS ONE* **9**:e106689.

McDonald, J. H., and M. Kreitman. 1991. Adaptive protein evolution at the *Adh* locus in *Drosophila*. *Nature* **351**:652–654.

McGaugh, S. E., C. S. Heil, B. Manzano-Winkler, L. Loewe, S. Goldstein, T. L. Himmel, and M. A. Noor. 2012. Recombination modulates how selection affects linked sites in *Drosophila*. *PLoS Biology* **10**:e1001422.

McKenna, A., M. Hanna, E. Banks, et al. 2010. The genome analysis toolkit: a MapReduce framework for analyzing next-generation DNA sequencing data. *Genome Research* **20**:1297-1303.

McKernan, K. J., H. E. Peckham, G. L. Costa, S. F. McLaughlin, Y. Fu, E. F. Tsung, C. R. Clouser, C. Duncan, J. K. Ichikawa, and C. C. Lee. 2009. Sequence and structural variation in a human genome uncovered by short-read, massively parallel ligation sequencing using two-base encoding. *Genome Research* **19**:1527–1541.

McTaggart, S. J., D. J. Obbard, C. Conlon, and T. J. Little. 2012. Immune genes undergo more adaptive evolution than non-immune system genes in *Daphnia pulex*. *BMC Evolutionary Biology* **12**:63.

McVean, G. A. T. 2002. A genealogical interpretation of linkage disequilibrium. *Genetics* **162**:987–991.

———. 2007. The structure of linkage disequilibrium around a selective sweep. *Genetics* **175**:1395–1406.

———. 2009. A genealogical interpretation of principal components analysis. *PLoS Genetics* **5**:e1000686.

McVean, G., P. Awadalla, and P. Fearnhead. 2002. A coalescent-based method for detecting and estimating recombination from gene sequences. *Genetics* **160**:1231–1241.

McVean, G. A. T., and N. J. Cardin. 2005. Approximating the coalescent with recombination. *Philosophical Transactions of the Royal Society B: Biological Sciences* **360**:1387–1393.

McVean, G. A. T., S. R. Myers, S. Hunt, P. Deloukas, D. R. Bentley, and P. Donnelly. 2004. The fine-scale structure of recombination rate variation in the human genome. *Science* **304**:581–584.

McVicker, G., D. Gordon, C. Davis, and P. Green. 2009. Widespread genomic signatures of natural selection in hominid evolution. *PLoS Genetics* **5**:e1000471.

Meacham, F., D. Boffelli, J. Dhahbi, D. I. K. Martin, M. Singer, and L. Pachter. 2011. Identification and correction of systematic error in high-throughput sequence data. *BMC Bioinformatics* **12**:451.

Meiklejohn, C. D., Y. Kim, D. L. Hartl, and J. Parsch. 2004. Identification of a locus under complex positive selection in *Drosophila simulans* by haplotype mapping and composite-likelihood estimation. *Genetics* **168**:265–279.

Meirmans, P. G. 2012. The trouble with isolation by distance. *Molecular Ecology* **21**:2839–2846.

Menozzi, P., A. Piazza, and L. Cavalli-Sforza. 1978. Synthetic maps of human gene frequencies in Europeans. *Science* **201**:786–792.

Messer, P. W. 2009. Measuring the rates of spontaneous mutation from deep and large-scale polymorphism data. *Genetics* **182**:1219.

———. 2013. SLiM: Simulating evolution with selection and linkage. *Genetics* **194**:1037–1039.

Messer, P. W., and R. A. Neher. 2012. Estimating the strength of selective sweeps from deep population diversity data. *Genetics* **191**:593–605.

Messier, W., and C.-B. Stewart. 1997. Episodic adaptive evolution of primate lysozymes. *Nature* **385**:151–154.

Meyer, D., and G. Thomson. 2001. How selection shapes variation of the human major histocompatibility complex: a review. *Annals of Human Genetics* **65**:1–26.

Michalakis, Y., and L. Excoffier. 1996. A generic estimation of population subdivision using distances between alleles with special reference for microsatellite loci. *Genetics* **142**:1061–1064.

Mikkelsen, T. S., L. W. Hillier, E. E. Eichler, et al. 2005. Initial sequence of the chimpanzee genome and comparison with the human genome. *Nature* **437**:69–87.

Miller, D. E., S. Takeo, K. Nandanan, et al. 2012. A whole-chromosome analysis of meiotic recombination in *Drosophila melanogaster*. *G3: Genes | Genomes | Genetics* **2**:249–260.

Minin, V. N., E. W. Bloomquist, and M. A. Suchard. 2008. Smooth skyride through a rough skyline: Bayesian coalescent-based inference of population dynamics. *Molecular Biology and Evolution* **25**:1459–1471.

Miyata, T., and T. Yasunaga. 1980. Molecular evolution of messenger RNA: a method for estimating evolutionary rates of synonymous and amino acid substitutions from homologous nucleotide sequences and its application. *Journal of Molecular Evolution* **16**:23–36.

Miyata, T., T. Yasunaga, Y. Yamawakikataoka, M. Obata, and T. Honjo. 1980. Nucleotide sequence divergence of mouse immunoglobulin γ_1 and γ_{2b} chain genes and the hypothesis of intervening sequence-mediated domain transfer. *Proceedings of the National Academy of Sciences* **77**:2143–2147.

Montague, M. J., G. Li, B. Gandolfi, et al. 2014. Comparative analysis of the domestic cat genome reveals genetic signatures underlying feline biology and domestication. *Proceedings of the National Academy of Sciences* **111**:17,230–17,235.

Moore, R. C., and M. D. Purugganan. 2003. The early stages of duplicate gene evolution. *Proceedings of the National Academy of Sciences* **100**:15,682–15,687.

Moorjani, P., N. Patterson, J. N. Hirschhorn, A. Keinan, L. Hao, G. Atzmon, E. Burns, H. Ostrer, A. L. Price, and D. Reich. 2011. The history of African gene flow into southern Europeans, Levantines, and Jews. *PLoS Genetics* **7**:e1001373.

Moran, P. A. P. 1958. Random processes in genetics. *Proceedings of the Cambridge Philosophical Society* **54**:60–71.

———. 1975. Wandering distributions and the electrophoretic profile. *Theoretical Population Biology* **8**:318–330.

Mousset, S., L. Brazier, M.-L. Cariou, F. Chartois, F. Depaulis, and M. Veuille. 2003. Evidence of a high rate of selective sweeps in African *Drosophila melanogaster*. *Genetics* **163**:599–609.

Moyle, L. C. 2006. Correlates of genetic differentiation and isolation by distance in 17 congeneric *Silene* species. *Molecular Ecology* **15**:1067–1081.

Mueller, L. D., L. G. Barr, and F. J. Ayala. 1985. Natural selection vs. random drift: evidence from temporal variation in allele frequencies in nature. *Genetics* **111**:517–554.

Munch, K., K. Nam, M. H. Schierup, and T. Mailund. 2016. Selective sweeps across twenty millions years of primate evolution. *Molecular Biology and Evolution* **33**:3065–3074.

Murray, M. C., and M. P. Hare. 2006. A genomic scan for divergent selection in a secondary contact zone between Atlantic and Gulf of Mexico oysters, *Crassostrea virginica*. *Molecular Ecology* **15**:4229–4242.

Murrell, B., J. O. Wertheim, S. Moola, T. Weighill, K. Scheffler, and S. L. Kosakovsky Pond. 2012. Detecting individual sites subject to episodic diversifying selection. *PLoS Genetics* **8**:e1002764.

Muse, S. V., and B. S. Gaut. 1994. A likelihood approach for comparing synonymous and nonsynonymous nucleotide substitution rates with application to the chloroplast genome. *Molecular Biology and Evolution* **11**:715–724.

Myers, S., C. Fefferman, and N. Patterson. 2008. Can one learn history from the allelic spectrum? *Theoretical Population Biology* **73**:342–348.

Myers, S. R., and R. C. Griffiths. 2003. Bounds on the minimum number of recombination events in a sample history. *Genetics* **163**:375–394.

N

Nachman, M. W., S. N. Boyer, and C. F. Aquadro. 1994. Nonneutral evolution at the mitochondrial NADH dehydrogenase subunit 3 gene in mice. *Proceedings of the National Academy of Sciences* **91**:6364–6368.

Nachman, M. W., and B. A. Payseur. 2012. Recombination rate variation and speciation: theoretical predictions and empirical results from rabbits and mice. *Philosophical Transactions of the Royal Society B: Biological Sciences* **367**:409–421.

Nadeau, N. J., M. Ruiz, P. Salazar, B. Counterman, J. A. Medina, H. Ortiz-Zuazaga, A. Morrison, W. O. McMillan, C. D. Jiggins, and R. Papa. 2014. Population genomics of parallel hybrid zones in the mimetic butterflies, *H. melpomene* and *H. erato*. *Genome Research* **24**:1316–1333.

Naduvilezhath, L., L. E. Rose, and D. Metzler. 2011. Jaatha: a fast composite-likelihood approach to estimate demographic parameters. *Molecular Ecology* **20**:2709–2723.

Nee, S., E. C. Holmes, A. Rambaut, and P. H. Harvey. 1995. Inferring population history from molecular phylogenies. *Philosophical Transactions of the Royal Society B: Biological Sciences* **349**:25–31.

Neher, R. A. 2013. Genetic draft, selective interference, and population genetics of rapid adaptation. *Annual Review of Ecology, Evolution, and Systematics* **44**:195–215.

Nei, M. 1972. Genetic distance between populations. *American Naturalist* **106**:283–292.

———. Analysis of gene diversity in subdivided populations. *Proceedings of the National Academy of Sciences* **70**:3321–3323.

———. 1982. Evolution of human races at the gene level. In *Human Genetics, Part A: The Unfolding Genome*, ed. B. Bonne-Tamir, pp. 167–181. New York: Alan R. Liss.

———. 1986. Stochastic errors in DNA evolution and molecular phylogeny. In *Evolutionary Perspectives and the New Genetics*, ed. H. Gershowitz, D. L. Rucknagel, and R. E. Tashian, pp. 133–147. New York: Alan R. Liss.

———. 1987. *Molecular Evolutionary Genetics*. New York: Columbia University Press.

Nei, M., and A. Chakravarti. 1977. Drift variances of F_{ST} and G_{ST} statistics obtained from a finite number of isolated populations. *Theoretical Population Biology* **11**:307–325.

Nei, M., and T. Gojobori. 1986. Simple methods for estimating the numbers of synonymous and nonsynonymous nucleotide substitutions. *Molecular Biology and Evolution* **3**:418–426.

Nei, M., and S. Kumar. 2000. *Molecular Evolution and Phylogenetics*. Oxford: Oxford University Press.

Nei, M., and W. H. Li. 1973. Linkage disequilibrium in subdivided populations. *Genetics* **75**:213–219.

———. 1979. Mathematical model for studying genetic variation in terms of restriction endonucleases. *Proceedings of the National Academy of Sciences* **76**:5269–5273.

Nei, M., and T. Maruyama. 1975. Lewontin-Krakauer test for neutral genes. *Genetics* **80**:395.

Nei, M., T. Maruyama, and R. Chakraborty. 1975. The bottleneck effect and genetic variability in populations. *Evolution* **29**:1–10.

Nei, M., and F. Tajima. 1981a. DNA polymorphism detectable by restriction endonucleases. *Genetics* **97**:145–163.

———. 1981b. Genetic drift and estimation of effective population size. *Genetics* **98**:625–640.

Neigel, J. E. 2002. Is F_{ST} obsolete? *Conservation Genetics* **3**:167–173.

Nelson, M. R., D. Wegmann, M. G. Ehm, et al. 2012. An abundance of rare functional variants in 202 drug target genes sequenced in 14,002 people. *Science* **337**:100–104.

Neuhauser, C., and S. M. Krone. 1997. The genealogy of samples in models with selection. *Genetics* **145**:519–534.

Nickerson, D. A., V. O. Tobe, and S. L. Taylor. 1997. PolyPhred: automating the detection and genotyping of single nucleotide substitutions using fluorescence-based resequencing. *Nucleic Acids Research* **25**:2745–2751.

Nielsen, E. E., M. M. Hansen, D. E. Ruzzante, D. Meldrup, and P. Gronkjaer. 2003. Evidence of a hybrid-zone in Atlantic cod (*Gadus morhua*) in the Baltic and the Danish Belt Sea revealed by individual admixture analysis. *Molecular Ecology* **12**:1497–1508.

Nielsen, R. 1997. A likelihood approach to population samples of microsatellite alleles. *Genetics* **146**:711–716.

———. 2000. Estimation of population parameters and recombination rates from single nucleotide polymorphisms. *Genetics* **154**:931–942.

———. 2005. Molecular signatures of natural selection. *Annual Review of Genetics* **39**:197–218.

Nielsen, R., and M. A. Beaumont. 2009. Statistical inferences in phylogeography. *Molecular Ecology* **18**:1034–1047.

Nielsen, R., M. J. Hubisz, and A. G. Clark. 2004. Reconstituting the frequency spectrum of ascertained single-nucleotide polymorphism data. *Genetics* **168**:2373–2382.

Nielsen, R., T. Korneliussen, A. Albrechtsen, Y. Li, and J. Wang. 2012. SNP calling, genotype calling, and sample allele frequency estimation from new-generation sequencing data. *PLoS ONE* **7**:e37558.

Nielsen, R., and J. Signorovitch. 2003. Correcting for ascertainment biases when analyzing SNP data: applications to the estimation of linkage disequilibrium. *Theoretical Population Biology* **63**:245–255.

Nielsen, R., and J. Wakeley. 2001. Distinguishing migration from isolation: a Markov chain Monte Carlo approach. *Genetics* **158**:885–896.

Nielsen, R., S. Williamson, Y. Kim, M. J. Hubisz, A. G. Clark, and C. Bustamante. 2005. Genomic scans for selective sweeps using SNP data. *Genome Research* **15**:1566–1575.

Nikiforov, T. T., R. R. Rendie, P. Goelet, Y.-H. Rogers, M. L. Kotewicz, S. Anderson, G. L. Trainor, and M. R. Knapp. 1994. Genetic bit analysis: a solid phase method for typing single nucleotide polymorphisms. *Nucleic Acids Research* **22**:4167–4175.

Noonan, J. P., G. Coop, S. Kudaravalli, et al. 2006. Sequencing and analysis of Neanderthal genomic DNA. *Science* **314**:1113–1118.

Noonan, J. P., M. Hofreiter, D. Smith, J. R. Priest, N. Rohland, G. Rabeder, J. Krause, J. C. Detter, S. Pääbo, and E. M. Rubin. 2005. Genomic sequencing of Pleistocene cave bears. *Science* **309**:597–599.

Noor, M. A. F., and S. M. Bennett. 2009. Islands of speciation or mirages in the desert? Examining the role of restricted recombination in maintaining species. *Heredity* **103**:439–444.

Nordborg, M., T. T. Hu, Y. Ishino, et al. 2005. The pattern of polymorphism in *Arabidopsis thaliana*. *PLoS Biology* **3**:e196.

Nosil, P., D. J. Funk, and D. Ortiz-Barrientos. 2009. Divergent selection and heterogeneous genomic divergence. *Molecular Ecology* **18**:375–402.

Novembre, J., T. Johnson, K. Bryc, et al. 2008. Genes mirror geography within Europe. *Nature* **456**:98–101.

Novembre, J., and M. Stephens. 2008. Interpreting principal component analyses of spatial population genetic variation. *Nature Genetics* **40**:646–649.

O

Obbard, D. J., J. J. Welch, K.-W. Kim, and F. M. Jiggins. 2009. Quantifying adaptive evolution in the *Drosophila* immune system. *PLoS Genetics* **5**:e1000698.

O'Hara, R. B. 2005. Comparing the effects of genetic drift and fluctuating selection on genotype frequency changes in the scarlet tiger moth. *Proceedings of the Royal Society B: Biological Sciences* **272**:211–217.

Ohta, T. 1972a. Evolutionary rate of cistrons and DNA divergence. *Journal of Molecular Evolution* **1**:150–157.

———. 1972b. Population size and rate of evolution. *Journal of Molecular Evolution* **1**:305–314.

———. 1992. The nearly neutral theory of molecular evolution. *Annual Review of Ecology and Systematics* **23**:263–286.

Ohta, T., and M. Kimura. 1971. Linkage disequilibrum between two segregating nucleotide sites under the steady flux of mutations in a finite population. *Genetics* **68**:571–580.

———. 1973. A model of mutation appropriate to estimate the number of electrophoretically detectable alleles in a finite population. *Genetical Research* **22**:201–204.

Okou, D. T., K. M. Steinberg, C. Middle, D. J. Cutler, T. J. Albert, and M. E. Zwick. 2007. Microarray-based genomic selection for high-throughput resequencing. *Nature Methods* **4**:907–909.

Oleksyk, T. K., K. Zhao, F. M. De La Vega, D. A. Gilbert, S. J. O'Brien, and M. W. Smith. 2008. Identifying selected regions from heterozygosity and divergence using a light-coverage genomic dataset from two human populations. *PLoS ONE* **3**:e1712.

Ometto, L., S. Glinka, D. De Lorenzo, and W. Stephan. 2005. Inferring the effects of demography and selection on *Drosophila melanogaster* populations from a chromosome-wide scan of DNA variation. *Molecular Biology and Evolution* **22**:2119–2130.

O'Reilly, P. F., E. Birney, and D. J. Balding. 2008. Confounding between recombination and selection, and the Ped/Pop method for detecting selection. *Genome Research* **18**:1304–1313.

Orlando, L., D. Bonjean, H. Bocherens, A. Thenot, A. Argant, M. Otte, and C. Hänni. 2002. Ancient DNA and the population genetics of cave bears (*Ursus spelaeus*) through space and time. *Molecular Biology and Evolution* **19**:1920–1933.

Orozco-terWengel, P., J. Corander, and C. Schlötterer. 2011. Genealogical lineage sorting leads to significant, but incorrect Bayesian multilocus inference of population structure. *Molecular Ecology* **20**:1108–1121.

P

Pääbo, S. 1985. Molecular cloning of ancient Egyptian mummy DNA. *Nature* **314**:644–645.

Padhukasahasram, B., P. Marjoram, J. D. Wall, C. D. Bustamante, and M. Nordborg. 2008. Exploring population genetic models with recombination using efficient forward-time simulations. *Genetics* **178**:2417–2427.

Padhukasahasram, B., J. D. Wall, P. Marjoram, and M. Nordborg. 2006. Estimating recombination rates from single-nucleotide polymorphisms using summary statistics. *Genetics* **174**:1517–1528.

Paetkau, D., W. Calvert, I. Stirling, and C. Strobeck. 1995. Microsatellite analysis of population structure in Canadian polar bears. *Molecular Ecology* **4**:347–354.

Palacios, J. A., J. Wakeley, and S. Ramachandran. 2015. Bayesian nonparametric inference of population size changes from sequential genealogies. *Genetics* **201**:281–304.

Palamara, P. F. 2016. ARGON: fast, whole-genome simulation of the discrete time Wright-Fisher process. *Bioinformatics* **32**:3032–3034.

Palamara, P. F., T. Lencz, A. Darvasi, and I. Pe'er. 2012. Length distributions of identity by descent reveal fine-scale demographic history. *American Journal of Human Genetics* **91**:809–822.

Pamilo, P., and N. O. Bianchi. 1993. Evolution of the *Zfx* and *Zfy* genes: rates and inter-dependence between the genes. *Molecular Biology and Evolution* **10**:271–281.

Pamilo, P., and M. Nei. 1988. Relationships between gene trees and species trees. *Molecular Biology and Evolution* **5**:568–583.

Pamilo, P., and S. L. Varvio-Aho. 1980. On the estimation of population size from allele frequency changes. *Genetics* **95**:1055–1057.

Parchman, T. L., K. S. Geist, J. A. Grahnen, C. W. Benkman, and C. A. Buerkle. 2010. Transcriptome sequencing in an ecologically important tree species: assembly, annotation, and marker discovery. *BMC Genomics* **11**:180.

Pardo-Diaz, C., C. Salazar, and C. D. Jiggins. 2015. Towards the identification of the loci of adaptive evolution. *Methods in Ecology and Evolution* **6**:445–464.

Parsch, J., C. D. Meiklejohn, and D. L. Hartl. 2001. Patterns of DNA sequence variation suggest the recent action of positive selection in the *janus-ocnus* region of *Drosophila simulans*. *Genetics* **159**:647–657.

Parsch, J., Z. Zhang, and J. F. Baines. 2009. The influence of demography and weak selection on the McDonald-Kreitman test: an empirical study in *Drosophila*. *Molecular Biology and Evolution* **26**:691–698.

Paten, B., A. M. Novak, J. M. Eizenga, and E. Garrison. 2017. Genome graphs and the evolution of genome inference. *Genome Research* **27**:665–676.

Patterson, N., N. Hattangadi, B. Lane, et al. 2004. Methods for high-density admixture mapping of disease genes. *American Journal of Human Genetics* **74**:979–1000.

Patterson, N., P. Moorjani, Y. Luo, S. Mallick, N. Rohland, Y. Zhan, T. Genschoreck, T. Webster, and D. Reich. 2012. Ancient admixture in human history. *Genetics* **192**:1065–1093.

Patterson, N., A. L. Price, and D. Reich. 2006. Population structure and eigenanalysis. *PLoS Genetics* **2**:e190.

Paul, J. S., M. Steinrücken, and Y. S. Song. 2011. An accurate sequentially Markov conditional sampling distribution for the coalescent with recombination. *Genetics* **187**:1115–1128.

Pavlidis, P., S. Hutter, and W. Stephan. 2008. A population genomic approach to map recent positive selection in model species. *Molecular Ecology* **17**:3585–3598.

Pavlidis, P., J. D. Jensen, and W. Stephan. 2010. Searching for footprints of positive selection in whole-genome SNP data from nonequilibrium populations. *Genetics* **185**:907–922.

Pavlidis, P., J. D. Jensen, W. Stephan, and A. Stamatakis. 2012. A critical assessment of storytelling: gene ontology categories and the importance of validating genomic scans. *Molecular Biology and Evolution* **29**:3237–3248.

Pavlidis, P., S. Laurent, and W. Stephan. 2010. msABC: a modification of Hudson's ms to facilitate multi-locus ABC analysis. *Molecular Ecology Resources* **10**:723–727.

Pavlidis, P., D. Živković, A. Stamatakis, and N. Alachiotis. 2013. SweeD: likelihood-based detection of selective sweeps in thousands of genomes. *Molecular Biology and Evolution* **30**:2224–2234.

Payseur, B. A., A. D. Cutter, and M. W. Nachman. 2002. Searching for evidence of positive selection in the human genome using patterns of microsatellite variability. *Molecular Biology and Evolution* **19**:1143–1153.

Pease, J. B., D. C. Haak, M. W. Hahn, and L. C. Moyle. 2016. Phylogenomics reveals three sources of adaptive variation during a rapid radiation. *PLoS Biology* **14**:e1002379.

Pease, J. B., and M. W. Hahn. 2013. More accurate phylogenies inferred from low-recombination regions in the presence of incomplete lineage sorting. *Evolution* **67**:2376–2384.

———. 2015. Detection and polarization of introgression in a five-taxon phylogeny. *Systematic Biology* **64**:651–662.

Pemberton, J. M. 2008. Wild pedigrees: the way forward. *Proceedings of the Royal Society B: Biological Sciences* **275**:613–621.

Peng, B., and M. Kimmel. 2005. simuPOP: a forward-time population genetics simulation environment. *Bioinformatics* **21**:3686–3687.

Pennings, P. S., and J. Hermisson. 2006a. Soft sweeps II: molecular population genetics of adaptation from recurrent mutation or migration. *Molecular Biology and Evolution* 23:1076–1084.

———. 2006b. Soft sweeps III: the signature of positive selection from recurrent mutation. *PLoS Genetics* 2:e186.

Perry, G. H., N. J. Dominy, K. G. Claw et al. 2007. Diet and the evolution of human amylase gene copy number variation. *Nature Genetics* 39:1256–1260.

Peter, B. M. 2016. Admixture, population structure, and *F*-statistics. *Genetics* 202:1485–1501.

Peter, B. M., and M. Slatkin. 2013. Detecting range expansions from genetic data. *Evolution* 67:3274–3289.

Peter, B. M., D. Wegmann, and L. Excoffier. 2010. Distinguishing between population bottleneck and population subdivision by a Bayesian model choice procedure. *Molecular Ecology* 19:4648–4660.

Peterson, B. K., J. N. Weber, E. H. Kay, H. S. Fisher, and H. E. Hoekstra. 2012. Double digest RADseq: an inexpensive method for *de novo* SNP discovery and genotyping in model and non-model species. *PLoS ONE* 7:e37135.

Petkova, D., J. Novembre, and M. Stephens. 2016. Visualizing spatial population structure with estimated effective migration surfaces. *Nature Genetics* 48:94–100.

Petrov, D. A., E. R. Lozovskaya, and D. L. Hartl. 1996. High intrinsic rate of DNA loss in *Drosophila*. *Nature* 384:346–349.

Pickrell, J. K., and J. K. Pritchard. 2012. Inference of population splits and mixtures from genome-wide allele frequency data. *PLoS Genetics* 8:e1002967.

Pinho, C., and J. Hey. 2010. Divergence with gene flow: models and data. *Annual Review of Ecology, Evolution, and Systematics* 41:215–230.

Plagnol, V., and J. D. Wall. 2006. Possible ancestral structure in human populations. *PLoS Genetics* 2:e105.

Plotkin, J. B., and G. Kudla. 2011. Synonymous but not the same: the causes and consequences of codon bias. *Nature Reviews Genetics* 12:32–42.

Pluzhnikov, A., and P. Donnelly. 1996. Optimal sequencing strategies for surveying molecular genetic diversity. *Genetics* 144:1247–1262.

Podlaha, O., and J. Zhang. 2003. Positive selection on protein-length in the evolution of

a primate sperm ion channel. *Proceedings of the National Academy of Sciences* 100:12,241–12,246.

Poinar, H. N., C. Schwarz, J. Qi, et al. 2006. Metagenomics to paleogenomics: large-scale sequencing of mammoth DNA. *Science* 311:392–394.

Polanski, A., and M. Kimmel. 2003. New explicit expressions for relative frequencies of single-nucleotide polymorphisms with application to statistical inference on population growth. *Genetics* 165:427–436.

Pond, S. L. K., S. D. W. Frost, and S. V. Muse. 2005. HyPhy: hypothesis testing using phylogenies. *Bioinformatics* 21:676–679.

Pond, S. K., and S. V. Muse. 2005. Site-to-site variation of synonymous substitution rates. *Molecular Biology and Evolution* 22:2375–2385.

Pool, J. E., and R. Nielsen. 2008. The impact of founder events on chromosomal variability in multiply mating species. *Molecular Biology and Evolution* 25:1728–1736.

———. 2009. Inference of historical changes in migration rate from the lengths of migrant tracts. *Genetics* 181:711–719.

Posada, D., and K. A. Crandall. 2001. Evaluation of methods for detecting recombination from DNA sequences: computer simulations. *Proceedings of the National Academy of Sciences* 98:13,757–13,762.

Prakash, S., R. Lewontin, and J. L. Hubby. 1969. A molecular approach to the study of genic heterozygosity in natural populations IV. Patterns of genic variation in central, marginal and isolated populations of *Drosophila pseudoobscura*. *Genetics* 61:841–858.

Presgraves, D. C. 2005. Recombination enhances protein adaptation in *Drosophila melanogaster*. *Current Biology* 15:1651–1656.

Price, A. L., N. J. Patterson, R. M. Plenge, M. E. Weinblatt, N. A. Shadick, and D. Reich. 2006. Principal components analysis corrects for stratification in genome-wide association studies. *Nature Genetics* 38:904–909.

Pritchard, J. K., M. Stephens, and P. Donnelly. 2000. Inference of population structure using multilocus genotype data. *Genetics* 155:945–959.

Prout, T. 1973. Population genetics of marine pelecypods. III. Epistasis between functionally related isoenzymes of *Mytilus edulis* (appendix). *Genetics* 73:487–496.

Prüfer, K., K. Munch, I. Hellmann, et al. 2012. The bonobo genome compared with the chimpanzee and human genomes. *Nature* **486**:527–531.

Przeworski, M. 2002. The signature of positive selection at randomly chosen loci. *Genetics* **160**:1179–1189.

Przeworski, M., B. Charlesworth, and J. D. Wall. 1999. Genealogies and weak purifying selection. *Molecular Biology and Evolution* **16**:246–252.

Przeworski, M., G. Coop, and J. D. Wall. 2005. The signature of positive selection on standing genetic variation. *Evolution* **59**:2312–2323.

Ptak, S. E., and M. Przeworski. 2002. Evidence for population growth in humans is confounded by fine-scale population structure. *Trends in Genetics* **18**:559–563.

Pybus, M., P. Luisi, G. M. Dall'Olio, M. Uzkudun, H. Laayouni, J. Bertranpetit, and J. Engelken. 2015. Hierarchical boosting: a machine-learning framework to detect and classify hard selective sweeps in human populations. *Bioinformatics* **31**:3946–3952.

Pybus, O. G., A. Rambaut, and P. H. Harvey. 2000. An integrated framework for the inference of viral population history from reconstructed genealogies. *Genetics* **155**:1429–1437.

Pyhäjärvi, T., M. B. Hufford, S. Mezmouk, and J. Ross-Ibarra. 2013. Complex patterns of local adaptation in teosinte. *Genome Biology and Evolution* **5**:1594–1609.

Q

Quick, J., A. R. Quinlan, and N. J. Loman. 2014. A reference bacterial genome dataset generated on the MinION™ portable single-molecule nanopore sequencer. *GigaScience* **3**:22.

R

Racimo, F., G. Renaud, and M. Slatkin. 2016. Joint estimation of contamination, error and demography for nuclear DNA from ancient humans. *PLoS Genetics* **12**:e1005972.

Ramachandran, S., O. Deshpande, C. C. Roseman, N. A. Rosenberg, M. W. Feldman, and L. L. Cavalli-Sforza. 2005. Support from the relationship of genetic and geographic distance in human populations for a serial founder effect originating in Africa. *Proceedings of the National Academy of Sciences* **102**:15,942–15,947.

Rambaut, A. 2000. Estimating the rate of molecular evolution: incorporating non-contemporaneous sequences into maximum likelihood phylogenies. *Bioinformatics* **16**:395–399.

Rambaut, A., S. Y. Ho, A. J. Drummond, and B. Shapiro. 2009. Accommodating the effect of ancient DNA damage on inferences of demographic histories. *Molecular Biology and Evolution* **26**:245–248.

Ramírez-Soriano, A., S. E. Ramos-Onsins, J. Rozas, F. Calafell, and A. Navarro. 2008. Statistical power analysis of neutrality tests under demographic expansions, contractions and bottlenecks with recombination. *Genetics* **179**:555–567.

Ramos-Madrigal, J., B. D. Smith, J. V. Moreno-Mayar, S. Gopalakrishnan, J. Ross-Ibarra, M. T. P. Gilbert, and N. Wales. 2016. Genome sequence of a 5,310-year-old maize cob provides insights into the early stages of maize domestication. *Current Biology* **26**:3195–3201.

Ramos-Onsins, S. E., and J. Rozas. 2002. Statistical properties of new neutrality tests against population growth. *Molecular Biology and Evolution* **19**:2092–2100.

Rand, D. M., and L. M. Kann. 1996. Excess amino acid polymorphism in mitochondrial DNA: contrasts among genes from *Drosophila*, mice, and humans. *Molecular Biology and Evolution* **13**:735–748.

Rannala, B., and J. L. Mountain. 1997. Detecting immigration by using multilocus genotypes. *Proceedings of the National Academy of Sciences* **94**:9197–9201.

Rannala, B., and Z. Yang. 2003. Bayes estimation of species divergence times and ancestral population sizes using DNA sequences from multiple loci. *Genetics* **164**:1645–1656.

Rasmussen, M. D., M. J. Hubisz, I. Gronau, and A. Siepel. 2014. Genome-wide inference of ancestral recombination graphs. *PLoS Genetics* **10**:e1004342.

Rasmussen, M., Y. Li, S. Lindgreen, et al. 2010. Ancient human genome sequence of an extinct Palaeo-Eskimo. *Nature* **463**:757–762.

Rat Genome Sequencing Project Consortium. 2004. Genome sequence of the Brown Norway rat yields insights into mammalian evolution. *Nature* **428**:493–521.

Raymond, C. K., A. Kas, M. Paddock, et al. 2005. Ancient haplotypes of the HLA Class II region. *Genome Research* **15**:1250–1257.

Reich, D. E., M. Cargill, S. Bolk et al. 2001. Linkage disequilibrium in the human genome. *Nature* **411**:199–204.

Reich, D., R. E. Green, M. Kircher, et al. 2010. Genetic history of an archaic hominin group from Denisova cave in Siberia. *Nature* **468**:1053–1060.

Reich, D., K. Thangaraj, N. Patterson, A. L. Price, and L. Singh. 2009. Reconstructing Indian population history. *Nature* **461**:489–494.

Resch, A. M., L. Carmel, L. Marino-Ramirez, A. Y. Ogurtsov, S. A. Shabalina, I. B. Rogozin, and E. V. Koonin. 2007. Widespread positive selection in synonymous sites of mammalian genes. *Molecular Biology and Evolution* **24**: 1821–1831.

Rice, W. R. 1989. Analyzing tables of statistical tests. *Evolution* **43**:223–225.

Riebler, A., L. Held, and W. Stephan. 2008. Bayesian variable selection for detecting adaptive genomic differences among populations. *Genetics* **178**:1817–1829.

Ritz, A., A. Bashir, S. Sindi, D. Hsu, I. Hajirasouliha, and B. J. Raphael. 2014. Characterization of structural variants with single molecule and hybrid sequencing approaches. *Bioinformatics* **30**:3458–3466.

Robbins, R. B. 1918. Some applications of mathematics to breeding problems III. *Genetics* **3**:375–389.

Robertson, A. 1975. Remarks on the Lewontin-Krakauer test. *Genetics* **80**:396.

Rockman, M. V., M. W. Hahn, N. Soranzo, D. B. Goldstein, and G. A. Wray. 2003. Positive selection on a human-specific transcription factor binding site regulating *IL4* expression. *Current Biology* **13**: 2118–2123.

Rockman, M. V., M. W. Hahn, N. Soranzo, D. A. Loisel, D. B. Goldstein, and G. A. Wray. 2004. Positive selection on *MMP3* regulation has shaped heart disease risk. *Current Biology* **14**:1531–1539.

Rodrigo, A. G., and J. Felsenstein. 1999. Coalescent approaches to HIV population genetics. In *The Evolution of HIV*, ed. K. Crandall, pp. 233–271. Baltimore: Johns Hopkins University Press.

Rodrigo, A. G., E. G. Shpaer, E. L. Delwart, A. K. N. Iversen, M. V. Gallo, J. Brojatsch, M. S. Hirsch, B. D. Walker, and J. I. Mullins. 1999. Coalescent estimates of HIV-1 generation time *in vivo. Proceedings of the National Academy of Sciences* **96**:2187–2191.

Rogers, R. L., J. M. Cridland, L. Shao, T. T. Hu, P. Andolfatto, and K. R. Thornton. 2014. Landscape of standing variation for tandem duplications in *Drosophila yakuba* and *Drosophila simulans. Molecular Biology and Evolution* **31**:1750–1766.

Rogers, A. R., and H. Harpending. 1992. Population growth makes waves in the distribution of pairwise genetic differences. *Molecular Biology and Evolution* **9**:552–569.

Romiguier, J., P. Gayral, M. Ballenghien, et al. 2014. Comparative population genomics in animals uncovers the determinants of genetic diversity. *Nature* **515**:261–263.

Ronen, R., N. Udpa, E. Halperin, and V. Bafna. 2013. Learning natural selection from the site frequency spectrum. *Genetics* **195**:181–193.

Rosenberg, N. A. 2003. The shapes of neutral gene genealogies in two species: Probabilities of monophyly, paraphyly, and polyphyly in a coalescent model. *Evolution* **57**:1465–1477.

Rosenberg, N. A., S. Mahajan, S. Ramachandran, C. Zhao, J. K. Pritchard, and M. W. Feldman. 2005. Clines, clusters, and the effect of study design on the inference of human population structure. *PLoS Genetics* **1**:e70.

Rosenberg, N. A., J. K. Pritchard, J. L. Weber, H. M. Cann, K. K. Kidd, L. A. Zhivotovsky, and M. W. Feldman. 2002. Genetic structure of human populations. *Science* **298**:2381–2385.

Rosenblum, E. B., and J. Novembre. 2007. Ascertainment bias in spatially structured populations: a case study in the eastern fence lizard. *Journal of Heredity* **98**:331–336.

Rosenzweig, B. K., J. B. Pease, N. J. Besansky, and M. W. Hahn. 2016. Powerful methods for detecting introgressed regions from population genomic data. *Molecular Ecology* **25**:2387–2397.

Rousset, F. 1997. Genetic differentiation and estimation of gene flow from *F*-statistics under isolation by distance. *Genetics* **145**:1219–1228.

———. 2000. Genetic differentiation between individuals. *Journal of Evolutionary Biology* **13**:58–62.

———. 2007. Inferences from spatial population genetics. In *Handbook of Statistical Genetics*, vol. 2, ed. D. J. Balding, M. Bishop, and C. Cannings, pp. 945–979. West Sussex, England: John Wiley & Sons.

Roux, C., C. Fraïsse, J. Romiguier, Y. Anciaux, N. Galtier, and N. Bierne. 2016. Shedding light on the grey zone of speciation along a continuum of genomic divergence. *PLoS Biology* **14**:e2000234.

RoyChoudhury, A., and M. Stephens. 2007. Fast and accurate estimation of the population-scaled mutation rate, θ, from microsatellite genotype data. *Genetics* **176**: 1363–1366.

Rozas, J., M. Gullaud, G. Blandin, and M. Aguadé. 2001. DNA variation at the *rp49* gene region of *Drosophila simulans*: Evolutionary inferences from an unusual haplotype structure. *Genetics* **158**:1147–1155.

Rubinsztein, D. C., W. Amos, J. Leggo, S. Goodburn, S. Jain, S.-H. Li, R. L. Margolis, C. A. Ross, and M. A. Ferguson-Smith. 1995. Microsatellite evolution—evidence for directionality and variation in rate between species. *Nature Genetics* **10**:337–343.

Ruvolo, M. 1997. Molecular phylogeny of the hominoids: inferences from multiple independent DNA sequence data sets. *Molecular Biology and Evolution* **14**:248–265.

S

Sabeti, P. C., D. E. Reich, J. M. Higgins, et al. 2002. Detecting recent positive selection in the human genome from haplotype structure. *Nature* **419**:832–837.

Sabeti, P. C., P. Varilly, B. Fry, et al. 2007. Genome-wide detection and characterization of positive selection in human populations. *Nature* **449**:913–918.

Sackton, T. B., R. J. Kulathinal, C. M. Bergman, A. R. Quinlan, E. B. Dopman, M. Carneiro, G. T. Marth, D. L. Hartl, and A. G. Clark. 2009. Population genomic inferences from sparse high-throughput sequencing of two populations of *Drosophila melanogaster*. *Genome Biology and Evolution* **1**:449–465.

Sackton, T. B., B. P. Lazzaro, T. A. Schlenke, J. D. Evans, D. Hultmark, and A. G. Clark. 2007. Dynamic evolution of the innate immune system in *Drosophila*. *Nature Genetics* **39**:1461–1468.

Saiki, R. K., S. Scharf, F. Faloona, K. B. Mullis, G. T. Horn, H. A. Erlich, and N. Arnheim. 1985. Enzymatic amplification of β-globin genomic sequences and restriction site analysis for diagnosis of sickle cell anemia. *Science* **230**:1350–1354.

Sanger, F., S. Nickleu, and A. R. Coulson. 1977. DNA sequencing with chain-terminating inhibitors. *Proceedings of the National Academy of Sciences* **74**:5463–5467.

Sankararaman, S., G. Kimmel, E. Halperin, and M. I. Jordan. 2008. On the inference of ancestries in admixed populations. *Genome Research* **18**:668–675.

Sawyer, S. A. 1989. Statistical tests for detecting gene conversion. *Molecular Biology and Evolution* **6**:526–538.

Sawyer, S. A., D. E. Dykhuizen, and D. L. Hartl. 1987. Confidence interval for the number of selectively neutral amino acid polymorphisms. *Proceedings of the National Academy of Sciences* **84**:6225–6228.

Sawyer, S. A., and D. L. Hartl. 1992. Population genetics of polymorphism and divergence. *Genetics* **132**:1161–1176.

Scally, A., J. Y. Dutheil, L. W. Hillier, et al. 2012. Insights into hominid evolution from the gorilla genome sequence. *Nature* **483**:169–175.

Schaeffer, S. W. 2002. Molecular population genetics of sequence length diversity in the *Adh* region of *Drosophila pseudoobscura*. *Genetical Research* **80**:163–175.

Schaibley, V. M., M. Zawistowski, D. Wegmann, M. G. Ehm, M. R. Nelson, P. L. St. Jean, G. R. Abecasis, J. Novembre, S. Zöllner, and J. Z. Li. 2013. The influence of genomic context on mutation patterns in the human genome inferred from rare variants. *Genome Research* **23**:1974–1984.

Scheet, P., and M. Stephens. 2006. A fast and flexible statistical model for large-scale population genotype data: applications to inferring missing genotypes and haplotypic phase. *American Journal of Human Genetics* **78**:629–644.

Scherer, S., and R. W. Davis. 1980. Recombination of dispersed repeated DNA sequences in yeast. *Science* **209**:1380–1384.

Schierup, M. H., and J. Hein. 2000. Consequences of recombination on traditional phylogenetic analysis. *Genetics* **156**:879–891.

Schiffels, S., and R. Durbin. 2014. Inferring human population size and separation history from multiple genome sequences. *Nature Genetics* **46**:919–925.

Schirmer, M., R. D'Amore, U. Z. Ijaz, N. Hall, and C. Quince. 2016. Illumina error profiles: resolving fine-scale variation in metagenomic sequencing data. *BMC Bioinformatics* **17**:125.

Schlenke, T. A., and D. J. Begun. 2004. Strong selective sweep associated with a transposon insertion in *Drosophila simulans*. *Proceedings of the National Academy of Sciences* **101**:1626–1631.

Schlötterer, C. 2002. A microsatellite-based multilocus screen for the identification of local selective sweeps. *Genetics* **160**:753–763.

Schlötterer, C., R. Tobler, R. Kofler, and
V. Nolte. 2014. Sequencing pools of
individuals—mining genome-wide polymorphism data without big funding. *Nature
Reviews Genetics* **15**:749–763.

Schmid, K., and Z. Yang. 2008. The
trouble with sliding windows and the
selective pressure in BRCA1. *PLoS ONE*
3:e3746.

Schraiber, J. G., and J. M. Akey. 2015. Methods
and models for unravelling human evolutionary history. *Nature Reviews Genetics*
16:727–740.

Schraiber, J. G., S. N. Evans, and M. Slatkin.
2016. Bayesian inference of natural selection
from allele frequency time series. *Genetics*
203:493–511.

Schrider, D. R., D. J. Begun, and M. W. Hahn.
2013. Detecting highly differentiated copy-
number variants from pooled population sequencing. *Pacific Symposium on Biocomputing*
18:344–355.

Schrider, D. R., and M. W. Hahn. 2010. Gene
copy-number polymorphism in nature.
*Proceedings of the Royal Society B: Biological
Sciences* **277**:3213–3221.

Schrider, D. R., and A. D. Kern. 2014.
Discovering functional DNA elements using
population genomic information: a proof
of concept using human mtDNA. *Genome
Biology and Evolution* **6**:1542–1548.

———. 2015. Inferring selective constraint
from population genomic data suggests
recent regulatory turnover in the human
brain. *Genome Biology and Evolution*
7:3511–3528.

———. 2016. S/HIC: robust identification of
soft and hard sweeps using machine learning. *PLoS Genetics* **12**:e1005928.

Schrider, D. R., J. N. Hourmozdi, and
M. W. Hahn. 2011. Pervasive multinu-
cleotide mutational events in eukaryotes.
Current Biology **21**:1051–1054.

Schrider, D. R., F. K. Mendes, M. W. Hahn,
and A. D. Kern. 2015. Soft shoulders ahead:
spurious signatures of soft and partial selective sweeps result from linked hard sweeps.
Genetics **200**:267–284.

Schrider, D. R., A. G. Shanku, and A. D. Kern.
2016. Effects of linked selective sweeps on
demographic inference and model selection.
Genetics **204**:1207–1223.

Schwartz, M. K., G. Luikart, and R. S. Waples.
2007. Genetic monitoring as a promising
tool for conservation and management.
Trends in Ecology & Evolution **22**:25–33.

Scornavacca, C., and N. Galtier. 2017.
Incomplete lineage sorting in mammalian phylogenomics. *Systematic Biology*
66:112–120.

Ségurel, L., E. E. Thompson, T. Flutre, et al.
2012. The ABO blood group is a trans-
species polymorphism in primates.
Proceedings of the National Academy of Sciences
109:18,493–18,498.

Sella, G., D. A. Petrov, M. Przeworski, and
P. Andolfatto. 2009. Pervasive natural selection in the *Drosophila* genome. *PLoS Genetics*
5:e1000495.

Serre, D., and S. Pääbo. 2004. Evidence for
gradients of human genetic diversity within
and among continents. *Genome Research*
14:1679–1685.

Shapiro, B., A. J. Drummond, A. Rambaut, et al.
2004. Rise and fall of the Beringian steppe
bison. *Science* **306**:1561–1565.

Shapiro, B., and M. Hofreiter. 2014. A paleoge-
nomic perspective on evolution and gene
function: new insights from ancient DNA.
Science **343**:1236573.

Shapiro, J. A., W. Huang, C. Zhang, et al. 2007.
Adaptive genic evolution in the *Drosophila*
genomes. *Proceedings of the National Academy
of Sciences* **104**:2271–2276.

Sheehan, S., K. Harris, and Y. S. Song. 2013.
Estimating variable effective population sizes from multiple genomes: a
sequentially Markov conditional sampling distribution approach. *Genetics*
194:647–662.

Sheehan, S., and Y. S. Song. 2016. Deep learning for population genetic inference. *PLoS
Computational Biology* **12**:e1004845.

Shriver, M. D., G. C. Kennedy, E. J. Parra,
H. A. Lawson, V. Sonpar, J. Huang,
J. M. Akey, and K. W. Jones. 2004. The
genomic distribution of population
substructure in four populations using
8,525 autosomal SNPs. *Human Genomics*
1:274–286.

Šidàk, Z. 1967. Rectangular confidence regions
for the means of multivariate normal distributions. *Journal of the American Statistical
Association* **62**:626–633.

Siepel, A., and D. Haussler. 2004a. Phylogenetic
estimation of context-dependent substitution rates by maximum likelihood.
Molecular Biology and Evolution **21**:468–488.

———. 2004b. Combining phylogenetic and
hidden Markov models in biosequence
analysis. *Journal of Computational Biology*
11:413–428.

Simonsen, K. L., G. A. Churchill, and C. F. Aquadro. 1995. Properties of statistical tests of neutrality for DNA polymorphism data. *Genetics* **141**:413–429.

Singh, N. D., C. F. Aquadro, and A. G. Clark. 2009. Estimation of fine-scale recombination intensity variation in the *white–echinus* interval of *D. melanogaster*. *Journal of Molecular Evolution* **69**:42–53.

Sinnock, P., and C. F. Sing. 1972. Analysis of multilocus genetic systems in Tecumseh, Michigan. II. Consideration of the correlation between nonalleles in gametes. *American Journal of Human Genetics* **24**:393–415.

Sjödin, P., I. Kaj, S. Krone, M. Lascoux, and M. Nordborg. 2005. On the meaning and existence of an effective population size. *Genetics* **169**:1061–1070.

Slatkin, M. 1985. Rare alleles as indicators of gene flow. *Evolution* **39**:53–65.

———. 1991. Inbreeding coefficients and coalescence times. *Genetical Research* **58**:167–175.

———. 1993. Isolation by distance in equilibrium and non-equilibrium populations. *Evolution* **47**:264–279.

———. 1994a. An exact test for neutrality based on the Ewens sampling distribution. *Genetical Research* **64**:71–74.

———. 1994b. Linkage disequilibrium in growing and stable populations. *Genetics* **137**:331–336.

———. 1995. A measure of population subdivision based on microsatellite allele frequencies. *Genetics* **139**:457–462.

———. 1996. A correction to the exact test based on the Ewens sampling distribution. *Genetical Research* **68**:259–260.

Slatkin, M., and H. E. Arter. 1991. Spatial autocorrelation methods in population genetics. *American Naturalist* **138**:499–517.

Slatkin, M., and G. Bertorelle. 2001. The use of intraallelic variability for testing neutrality and estimating population growth rate. *Genetics* **158**:865–874.

Slatkin, M., and R. R. Hudson. 1991. Pairwise comparisons of mitochondrial DNA sequences in stable and exponentially growing populations. *Genetics* **129**:555–562.

Slatkin, M., and W. P. Maddison. 1989. A cladistic measure of gene flow inferred from the phylogenies of alleles. *Genetics* **123**:603–613.

Slatkin, M., and J. L. Pollack. 2008. Subdivision in an ancestral species creates asymmetry in gene trees. *Molecular Biology and Evolution* **25**:2241–2246.

Slatkin, M., and F. Racimo. 2016. Ancient DNA and human history. *Proceedings of the National Academy of Sciences* **113**:6380–6387.

Smith, N. G. C., and A. Eyre-Walker. 2002. Adaptive protein evolution in *Drosophila*. *Nature* **415**:1022–1024.

Smouse, P. E., J. C. Long, and R. R. Sokal. 1986. Multiple regression and correlation extensions of the Mantel test of matrix correspondence. *Systematic Zoology* **35**:627–632.

Smouse, P. E., and R. Peakall. 1999. Spatial autocorrelation analysis of individual multiallele and multilocus genetic structure. *Heredity* **82**:561–573.

Snyder, M. W., A. Adey, J. O. Kitzman, and J. Shendure, 2015. Haplotype-resolved genome sequencing: experimental methods and applications. *Nature Reviews Genetics* **16**:344–358.

Sokal, R. R. 1979. Testing statistical significance of geographic variation patterns. *Systematic Zoology* **28**:227–232.

Sokal, R. R., and N. L. Oden. 1978. Spatial autocorrelation in biology 1: methodology. *Biological Journal of the Linnean Society* **10**:199–228.

Solís-Lemus, C., and C. Ané. 2016. Inferring phylogenetic networks with maximum pseudolikelihood under incomplete lineage sorting. *PLoS Genetics* **12**:e1005896.

Song, Y. S., and J. Hein. 2005. Constructing minimal ancestral recombination graphs. *Journal of Computational Biology* **12**:147–169.

Sousa, V. C., M. Fritz, M. A. Beaumont, and L. Chikhi. 2009. Approximate Bayesian computation without summary statistics: the case of admixture. *Genetics* **181**:1507–1519.

Sousa, V. C., A. Grelaud, and J. Hey. 2011. On the nonidentifiability of migration time estimates in isolation with migration models. *Molecular Ecology* **20**:3956–3962.

Spencer, C. C. A., and G. Coop. 2004. SelSim: a program to simulate population genetic data with natural selection and recombination. *Bioinformatics* **20**:3673–3675.

Stajich, J. E., and M. W. Hahn. 2005. Disentangling the effects of demography and selection in human history. *Molecular Biology and Evolution* **22**:63–73.

Stefansson, H., A. Helgason, G. Thorleifsson, et al. 2005. A common inversion under selection in Europeans. *Nature Genetics* **37**:129–137.

Steinrücken, M., A. Bhaskar, and Y. S. Song. 2014. A novel spectral method for inferring general diploid selection from time series genetic data. *Annals of Applied Statistics* 8:2203–2222.

Stephan, W. 2010. Genetic hitchhiking versus background selection: the controversy and its implications. *Philosophical Transactions of the Royal Society B: Biological Sciences* 365:1245–1253.

Stephens, J. C. 1986. On the frequency of un-detectable recombination events. *Genetics* 112:923–926.

Stephens, M., and P. Donnelly. 2003. A comparison of Bayesian methods for haplotype reconstruction from population genotype data. *American Journal of Human Genetics* 73:1162–1169.

Stephens, M., J. S. Sloan, P. D. Robertson, P. Scheet, and D. A. Nickerson. 2006. Automating sequence-based detection and genotyping of SNPs from diploid samples. *Nature Genetics* 38:375–381.

Stephens, M., N. J. Smith, and P. Donnelly. 2001. A new statistical method for hap-lotype reconstruction from population data. *American Journal of Human Genetics* 68:978–989.

Stern, D. L., and V. Orgogozo. 2008. The loci of evolution: how predictable is genetic evolution? *Evolution* 62:2155–2177.

Stinchcombe, J. R., and H. E. Hoekstra. 2008. Combining population genomics and quantitative genetics: finding the genes underlying ecologically important traits. *Heredity* 100:158–170.

Stoletzki, N., and A. Eyre-Walker. 2011. Estimation of the neutrality index. *Molecular Biology and Evolution* 28:63–70.

Storey, J. D. 2002. A direct approach to false discovery rates. *Journal of the Royal Statistical Society: Series B (Statistical Methodology)* 64:479–498.

Storey, J. D., and R. Tibshirani. 2003. Statistical significance for genomewide studies. *Proceedings of the National Academy of Sciences* 100:9440–9445.

Storz, J. F. 2005. Using genome scans of DNA polymorphism to infer adaptive population divergence. *Molecular Ecology* 14:671–688.

Storz, J. F., B. A. Payseur, and M. W. Nachman. 2004. Genome scans of DNA variability in humans reveal evidence for selective sweeps outside of Africa. *Molecular Biology and Evolution* 21:1800–1811.

Strasburg, J. L., and L. H. Rieseberg. 2010. How robust are "isolation with migration" analyses to violations of the IM model? A simulation study. *Molecular Biology and Evolution* 27:297–310.

———. 2011. Interpreting the estimated timing of migration events between hybridizing species. *Molecular Ecology* 20:2353–2366.

Strobeck, C. 1987. Average number of nucleotide differences in a sample from a single subpopulation: a test for population subdivision. *Genetics* 117:149–153.

Sun, J. X., A. Helgason, G. Masson, S. S. Ebenesersdóttir, H. Li, S. Mallick, S. Gnerre, N. Patterson, A. Kong, and D. Reich. 2012. A direct characterization of human mutation based on microsatellites. *Nature Genetics* 44:1161–1165.

Sved, J. A. 1971. Linkage disequilibrium and homozygosity of chromosome segments in finite populations. *Theoretical Population Biology* 2:125–141.

Szpiech, Z. A., and R. D. Hernandez. 2014. Selscan: an efficient multi-threaded program to perform EHH-based scans for positive selection. *Molecular Biology and Evolution* 31:2824–2827.

T

Tachida, H. 2000. DNA evolution under weak selection. *Gene* 261:3–9.

Tajima, F. 1983. Evolutionary relationship of DNA sequences in finite populations. *Genetics* 105:437–460.

———. 1989a. Statistical method for testing the neutral mutation hypothesis by DNA polymorphism. *Genetics* 123:585–595.

———. 1989b. The effect of change in population size on DNA polymorphism. *Genetics* 123:597–601.

———. 1996. The amount of DNA polymorphism maintained in a finite population when the neutral mutation rate varies among sites. *Genetics* 143:1457–1465.

Takahasi, K. R., and H. Innan. 2008. The direction of linkage disequilibrium: a new measure based on the ancestral-derived status of segregating alleles. *Genetics* 179:1705–1712.

Takahata, N. 1986. An attempt to estimate the effective size of the ancestral species common to two extant species from which homologous genes are sequenced. *Genetical Research* 48:187–190.

Takahata, N., and M. Nei. 1985. Gene genealogy and variance of interpopulational nucleotide differences. *Genetics* **110**:325–344.

Takahata, N., Y. Satta, and J. Klein. 1995. Divergence time and population size in the lineage leading to modern humans. *Theoretical Population Biology* **48**:198–221.

Tamura, K., and M. Nei. 1993. Estimation of the number of nucleotide substitutions in the control region of mitochondrial DNA in humans and chimpanzees. *Molecular Biology and Evolution* **10**:512–526.

Tang, H., M. Coram, P. Wang, X. F. Zhu, and N. Risch. 2006. Reconstructing genetic ancestry blocks in admixed individuals. *American Journal of Human Genetics* **79**:1–12.

Tang, H., J. Peng, P. Wang, and N. J. Risch. 2005. Estimation of individual admixture: analytical and study design considerations. *Genetic Epidemiology* **28**:289–301.

Tang, K., K. R. Thornton, and M. Stoneking. 2007. A new approach for using genome scans to detect recent positive selection in the human genome. *PLoS Biology* **5**:e171.

Tavaré, S. 1984. Line-of-descent and genealogical processes, and their applications in population genetics models. *Theoretical Population Biology* **26**:119–164.

Tavaré, S., D. J. Balding, R. C. Griffiths, and P. Donnelly. 1997. Inferring coalescence times from DNA sequence data. *Genetics* **145**:505–518.

Telenti, A., L. C. T. Pierce, W. H. Biggs, et al. 2016. Deep sequencing of 10,000 human genomes. *Proceedings of the National Academy of Sciences* **113**:11,901–11,906.

Templeton, A. R. 1996. Contingency tests of neutrality using intra/interspecific gene trees: the rejection of neutrality for the evolution of the mitochondrial cytochrome oxidase II gene in the hominoid primates. *Genetics* **144**:1263–1270.

Tennessen, J. A., A. W. Bigham, T. D. O'Connor, et al. 2012. Evolution and functional impact of rare coding variation from deep sequencing of human exomes. *Science* **337**:64–69.

Terhorst, J., J. A. Kamm, and Y. S. Song. 2017. Robust and scalable inference of population history from hundreds of unphased whole genomes. *Nature Genetics* **49**:303–309.

Terhorst, J., and Y. S. Song. 2015. Fundamental limits on the accuracy of demographic inference based on the sample frequency spectrum. *Proceedings of the National Academy of Sciences* **112**:7677–7682.

Teshima, K. M., G. Coop, and M. Przeworski. 2006. How reliable are empirical genomic scans for selective sweeps? *Genome Research* **16**:702–712.

Teshima, K. M., and H. Innan. 2009. mbs: modifying Hudson's ms software to generate samples of DNA sequences with a biallelic site under selection. *BMC Bioinformatics* **10**:166.

Than, C., D. Ruths, and L. Nakhleh. 2008. PhyloNet: a software package for analyzing and reconstructing reticulate evolutionary relationships. *BMC Bioinformatics* **9**:322.

Thompson, E. A. 1973. The method of minimum evolution. *Annals of Human Genetics* **36**:333–340.

Thornton, K. 2005. Recombination and the properties of Tajima's *D* in the context of approximate-likelihood calculation. *Genetics* **171**:2143–2148.

Thornton, K. R. 2009. Automating approximate Bayesian computation by local linear regression. *BMC Genetics* **10**:35.

———. 2014. A C++ template library for efficient forward-time population genetic simulation of large populations. *Genetics* **198**:157–166.

Thornton, K. R., and J. D. Jensen. 2007. Controlling the false-positive rate in multilocus genome scans for selection. *Genetics* **175**:737–750.

Thornton, K., and M. Long. 2005. Excess of amino acid substitutions relative to polymorphism between X-linked duplications in *Drosophila melanogaster*. *Molecular Biology and Evolution* **22**:273–284.

Thorvaldsdóttir, H., J. T. Robinson, and J. P. Mesirov. 2012. Integrative Genomics Viewer (IGV): high-performance genomics data visualization and exploration. *Briefings in Bioinformatics* **14**:178–192.

Tishkoff, S. A., E. Dietzsch, W. Speed et al. 1996. Global patterns of linkage disequilibrium at the CD4 locus and modern human origins. *Science* **271**:1380–1387.

Tishkoff, S. A., F. A. Reed, A. Ranciaro, et al. 2007. Convergent adaptation of human lactase persistence in Africa and Europe. *Nature Genetics* **39**:31–40.

Toomajian, C., R. S. Ajioka, L. B. Jorde, J. P. Kushner, and M. Kreitman. 2003. A method for detecting recent selection in the human genome from allele age estimates. *Genetics* **165**:287–297.

Turner, T. L., E. C. Bourne, E. J. Von Wettberg, T. T. Hu, and S. V. Nuzhdin. 2010. Population resequencing reveals local adaptation of *Arabidopsis lyrata* to serpentine soils. *Nature Genetics* **42**:260–263.

Turner, T. L., M. W. Hahn, and S. V. Nuzhdin. 2005. Genomic islands of speciation in *Anopheles gambiae*. *PLoS Biology* **3**:e285.

Turner, T. L., M. T. Levine, M. L. Eckert, and D. J. Begun. 2008. Genomic analysis of adaptive differentiation in *Drosophila melanogaster*. *Genetics* **179**:455–473.

V

Valdes, A. M., M. Slatkin, and N. B. Freimer. 1993. Allele frequencies at microsatellite loci: the stepwise mutation model revisited. *Genetics* **133**:737–749.

Van Tassell, C. P., T. P. Smith, L. K. Matukumalli, J. F. Taylor, R. D. Schnabel, C. T. Lawley, C. D. Haudenschild, S. S. Moore, W. C. Warren, and T. S. Sonstegard. 2008. SNP discovery and allele frequency estimation by deep sequencing of reduced representation libraries. *Nature Methods* **5**:247–252.

Vera, J. C., C. W. Wheat, H. W. Fescemyer, M. J. Frilander, D. L. Crawford, I. Hanski, and J. H. Marden. 2008. Rapid transcriptome characterization for a nonmodel organism using 454 pyrosequencing. *Molecular Ecology* **17**:1636–1647.

Vilas, A., A. Pérez-Figueroa, and A. Caballero. 2012. A simulation study on the performance of differentiation-based methods to detect selected loci using linked neutral markers. *Journal of Evolutionary Biology* **25**:1364–1376.

Villa-Angulo, R., L. K. Matukumalli, C. A. Gill, J. Choi, C. P. Van Tassell, and J. J. Grefenstette. 2009. High-resolution haplotype block structure in the cattle genome. *BMC Genetics* **10**:19.

Vitalis, R., K. Dawson, and P. Boursot. 2001. Interpretation of variation across marker loci as evidence of selection. *Genetics* **158**:1811–1823.

Voight, B. F., A. M. Adams, L. A. Frisse, Y. Qian, R. R. Hudson, and A. Di Rienzo. 2005. Interrogating multiple aspects of variation in a full resequencing data set to infer human population size changes. *Proceedings of the National Academy of Sciences* **102**:18,508–18,513.

Voight, B. F., S. Kudaravalli, X. Wen, and J. K. Pritchard. 2006. A map of recent positive selection in the human genome. *PLoS Biology* **4**:e72.

W

Wahlund, S. 1928. Zusammensetzung von Population und Korrelationserscheinung vom Standpunkt der Vererbungslehre aus betrachtet. *Hereditas* **11**:65–106.

Wakeley, J. 1996a. The variance of pairwise nucleotide differences in two populations with migration. *Theoretical Population Biology* **49**:39–57.

———. 1996b. Distinguishing migration from isolation using the variance of pairwise differences. *Theoretical Population Biology* **49**:369–386.

———. 1997. Using the variance of pairwise differences to estimate the recombination rate. *Genetical Research* **69**:45–48.

———. 2009. *Coalescent Theory*. Greenwood Village, CO: Roberts & Company.

Wakeley, J., and J. Hey. 1997. Estimating ancestral population parameters. *Genetics* **145**:847–855.

Wakeley, J., L. King, and P. R. Wilton. 2016. Effects of the population pedigree on genetic signatures of historical demographic events. *Proceedings of the National Academy of Sciences* **113**:7994–8001.

Wakeley, J., R. Nielsen, S. N. Liu-Cordero, and K. Ardlie. 2001. The discovery of single-nucleotide polymorphisms—and inferences about human demographic history. *American Journal of Human Genetics* **69**:1332–1347.

Wakeley, J., and O. Sargsyan. 2009. Extensions of the coalescent effective population size. *Genetics* **181**:341–345.

Wakeley, J., and T. Takahashi. 2003. Gene genealogies when the sample size exceeds the effective size of the population. *Molecular Biology and Evolution* **20**:208–213.

Wall, J. D. 1999. Recombination and the power of statistical tests of neutrality. *Genetical Research* **74**:65–79.

———. 2000a. A comparison of estimators of the population recombination rate. *Molecular Biology and Evolution* **17**:156–163.

———. 2000b. Detecting ancient admixture in humans using sequence polymorphism data. *Genetics* **154**:1271–1279.

———. 2003. Estimating ancestral population sizes and divergence times. *Genetics* **163**:395–404.

———. 2004. Estimating recombination rates using three-site likelihoods. *Genetics* **167**:1461–1473.

Wall, J. D., P. Andolfatto, and M. Przeworski. 2002. Testing models of selection and demography in *Drosophila simulans*. *Genetics* **162**:203–216.

Wall, J. D., and R. R. Hudson. 2001. Coalescent simulations and statistical tests of neutrality. *Molecular Biology and Evolution* **18**:1134–1135.

Wall, J. D., and M. Slatkin. 2012. Paleopopulation genetics. *Annual Review of Genetics* **46**:635–649.

Wang, D. G., J.-B. Fan, C.-J. Siao, et al. 1998. Large-scale identification, mapping, and genotyping of single-nucleotide polymorphisms in the human genome. *Science* **280**:1077–1082.

Wang, E. T., G. Kodama, P. Baldi, and R. K. Moyzis. 2006. Global landscape of recent inferred Darwinian selection for *Homo sapiens*. *Proceedings of the National Academy of Sciences* **103**:135–140.

Wang, G.-X., S. Ren, Y. Ren, H. Ai, and A. D. Cutter. 2010. Extremely high molecular diversity within the East Asian nematode *Caenorhabditis* sp. 5. *Molecular Ecology* **19**:5022–5029.

Wang, J. 2003. Maximum-likelihood estimation of admixture proportions from genetic data. *Genetics* **164**:747–765.

Wang, J., and M. C. Whitlock. 2003. Estimating effective population size and migration rates from genetic samples over space and time. *Genetics* **163**:429–446.

Wang, R. L., J. Wakeley, and J. Hey. 1997. Gene flow and natural selection in the origin of *Drosophila pseudoobscura* and close relatives. *Genetics* **147**:1091–1106.

Wang, Y., and J. Hey. 2010. Estimating divergence parameters with small samples from a large number of loci. *Genetics* **184**:363–379.

Waples, R. S. 1989. A generalized approach for estimating effective population size from temporal changes in allele frequency. *Genetics* **121**:379–391.

———. 2015. Testing for Hardy-Weinberg proportions: have we lost the plot? *Journal of Heredity* **106**:1–19.

Waples, R. S., and O. Gaggiotti. 2006. What is a population? An empirical evaluation of some genetic methods for identifying the number of gene pools and their degree of connectivity. *Molecular Ecology* **15**:1419–1439.

Wares, J. P. 2010. Natural distributions of mitochondrial sequence diversity support new null hypotheses. *Evolution* **64**:1136–1142.

Watterson, G. A. 1975. On the number of segregating sites in genetical models without recombination. *Theoretical Population Biology* **7**:256–275.

———. 1977. Heterosis or neutrality? *Genetics* **85**:789–814.

———. 1978. The homozygosity test of neutrality. *Genetics* **88**:405–417.

Watterson, G. A., and H. A. Guess. 1977. Is the most frequent allele the oldest? *Theoretical Population Biology* **11**:141–160.

Weckx, S., J. Del-Favero, R. Rademakers, L. Claes, M. Cruts, P. De Jonghe, C. Van Broeckhoven, and P. De Rijk. 2005. novoSNP, a novel computational tool for sequence variation discovery. *Genome Research* **15**:436–442.

Wegmann, D., D. E. Kessner, K. R. Veeramah, et al. 2011. Recombination rates in admixed individuals identified by ancestry-based inference. *Nature Genetics* **43**:847–853.

Wegmann, D., C. Leuenberger, S. Neuenschwander, and L. Excoffier. 2010. ABCtoolbox: a versatile toolkit for approximate Bayesian computations. *BMC Bioinformatics* **11**:116.

Weinberg, W. 1908. Über den Nachweis der Vererbung beim Menschen. *Jahreshefte Vereins vaterländische Naturkunde Württemberg* **1**:369–382.

Weir, B. S. 1979. Inferences about linkage disequilibrium. *Biometrics* **35**:235–254.

———. 1996. *Genetic Data Analysis II.* Sunderland, MA: Sinauer Associates.

Weir, B. S., L. R. Cardon, A. D. Anderson, D. M. Nielsen, and W. G. Hill. 2005. Measures of human population structure show heterogeneity among genomic regions. *Genome Research* **15**:1468–1476.

Weir, B. S., and C. C. Cockerham. 1984. Estimating *F*-statistics for the analysis of population structure. *Evolution* **38**:1358–1370.

Weir, B. S., and W. G. Hill. 1986. Nonuniform recombination within the human β-globin gene cluster. *American Journal of Human Genetics* **38**:776–781.

Weiss, G. H., and M. Kimura. 1965. A mathematical analysis of the stepping stone model of genetic correlation. *Journal of Applied Probability* **2**:129–149.

Weiss, G., and A. von Haeseler. 1998. Inference of population history using a likelihood approach. *Genetics* **149**:1539–1546.

Weissman, D. B., and N. H. Barton. 2012. Limits to the rate of adaptive substitution in sexual populations. *PLoS Genetics* 8:e1002740.

Welch, J. J. 2006. Estimating the genome-wide rate of adaptive protein evolution in *Drosophila*. *Genetics* **173**:821–837.

Wheeler, D. A., M. Srinivasan, M. Egholm, et al. 2008. The complete genome of an individual by massively parallel DNA sequencing. *Nature* **452**:872–876.

Whelan, S., and N. Goldman. 2004. Estimating the frequency of events that cause multiple-nucleotide changes. *Genetics* **167**:2027–2043.

Whitlock, M. C. 2011. G'_{ST} and D do not replace F_{ST}. *Molecular Ecology* **20**:1083–1091.

Whitlock, M. C., and K. E. Lotterhos. 2015. Reliable detection of loci responsible for local adaptation: inference of a null model through trimming the distribution of F_{ST}. *American Naturalist* **186**:S24–S36.

Whitlock, M. C., and D. E. McCauley. 1999. Indirect measures of gene flow and migration: $F_{ST} \neq 1/(4Nm + 1)$. *Heredity* **82**:117–125.

Wiehe, T., J. Mountain, P. Parham, and M. Slatkin. 2000. Distinguishing recombination and intragenic gene conversion by linkage disequilibrium patterns. *Genetical Research* **75**:61–73.

Wiehe, T. H. E., and W. Stephan. 1993. Analysis of a genetic hitchhiking model, and its application to DNA polymorphism data from *Drosophila melanogaster*. *Molecular Biology and Evolution* **10**:842–854.

Wilkins, J. F., and J. Wakeley. 2002. The coalescent in a continuous, finite, linear population. *Genetics* **161**:873–888.

Williamson, S. H., R. Hernandez, A. Fledel-Alon, L. Zhu, R. Nielsen, and C. D. Bustamante. 2005. Simultaneous inference of selection and population growth from patterns of variation in the human genome. *Proceedings of the National Academy of Sciences* **102**:7882–7887.

Williamson, S. H., M. J. Hubisz, A. G. Clark, B. A. Payseur, C. D. Bustamante, and R. Nielsen. 2007. Localizing recent adaptive evolution in the human genome. *PLoS Genetics* **3**:e90.

Williamson, S., and M. E. Orive. 2002. The genealogy of a sequence subject to purifying selection at multiple sites. *Molecular Biology and Evolution* **19**:1376–1384.

Williamson, E. G., and M. Slatkin. 1999. Using maximum likelihood to estimate population size from temporal changes in allele frequencies. *Genetics* **152**:755–761.

Wilson, G. A., and B. Rannala. 2003. Bayesian inference of recent migration rates using multilocus genotypes. *Genetics* **163**:1177–1191.

Wilson, I. J., and D. J. Balding. 1998. Genealogical inference from microsatellite data. *Genetics* **150**:499–510.

Wilson Sayres, M. A., K. E. Lohmueller, and R. Nielsen. 2014. Natural selection reduced diversity on human Y chromosomes. *PLoS Genetics* **10**:e1004064.

Wiuf, C., and J. Hein. 1999. Recombination as a point process along sequences. *Theoretical Population Biology* **55**:248–259.

———. 2000. The coalescent with gene conversion. *Genetics* **155**:451–462.

Wiuf, C., K. Zhao, H. Innan, and M. Nordborg. 2004. The probability and chromosomal extent of trans-specific polymorphism. *Genetics* **168**:2363–2372.

Wolf, J. B. W., A. Kunstner, K. Nam, M. Jakobsson, and H. Ellegren. 2009. Nonlinear dynamics of nonsynonymous (d_N) and synonymous (d_S) substitution rates affects inference of selection. *Genome Biology and Evolution* **1**:308–319.

Wolinsky, S. M., T. M. K. Bette, A. U. Neumann, et al. 1996. Adaptive evolution of human immunodeficiency virus-type 1 during the natural course of infection. *Science* **272**:537–542.

Won, Y. J., and J. Hey. 2005. Divergence population genetics of chimpanzees. *Molecular Biology and Evolution* **22**:297–307.

Wooding, S., and A. Rogers. 2002. The matrix coalescent and an application to human single-nucleotide polymorphisms. *Genetics* **161**:1641–1650.

Wray, G. A., M. W. Hahn, E. Abouheif, J. P. Balhoff, M. Pizer, M. V. Rockman, and L. A. Romano. 2003. The evolution of transcriptional regulation in eukaryotes. *Molecular Biology and Evolution* **20**:1377–1419.

Wright, S. 1922. Coefficients of inbreeding and relationship. *American Naturalist* **56**:330–338.

———. 1931. Evolution in Mendelian populations. *Genetics* **16**:97–159.

———. 1938. The distribution of gene frequencies under irreversible mutation. *Proceedings of the National Academy of Sciences* **24**:253–259.

———. 1940. Breeding structure of populations in relation to speciation. *American Naturalist* **74**:232–248.

———. 1943. Isolation by distance. *Genetics* **28**:114–138.

———. 1951. The genetical structure of populations. *Annals of Eugenics* **15**:323–354.

———. 1969. *The Theory of Gene Frequencies.* Chicago: University of Chicago Press.

———. 1978. *Variability Within and Among Natural Populations.* Chicago: University of Chicago Press.

Wright, S. I., and P. Andolfatto. 2008. The impact of natural selection on the genome: emerging patterns in *Drosophila* and *Arabidopsis. Annual Review of Ecology, Evolution, and Systematics* **39**:193–213.

Wright, S. I., I. V. Bi, S. G. Schroeder, M. Yamasaki, J. F. Doebley, M. D. McMullen, and B. S. Gaut. 2005. The effects of artificial selection on the maize genome. *Science* **308**:1310–1314.

Wright, S. I., and B. Charlesworth. 2004. The HKA test revisited. *Genetics* **168**:1071–1076.

Wu, C.-I. 1991. Inferences of species phylogeny in relation to segregation of ancient polymorphisms. *Genetics* **127**:429–435.

X

Xie, W., Q. Feng, H. Yu, X. Huang, Q. Zhao, Y. Xing, S. Yu, B. Han, and Q. Zhang. 2010. Parent-independent genotyping for constructing an ultrahigh-density linkage map based on population sequencing. *Proceedings of the National Academy of Sciences* **107**:10,578–10,583.

Xu, H., and Y.-X. Fu. 2004. Estimating effective population size of mutation rate with microsatellites. *Genetics* **166**:555–563.

Y

Yang, L., and B. S. Gaut. 2011. Factors that contribute to variation in evolutionary rate among *Arabidopsis* genes. *Molecular Biology and Evolution* **28**:2359–2369.

Yang, Z. 1997. On the estimation of ancestral population sizes of modern humans. *Genetical Research* **69**:111–116.

———. 1998. Likelihood ratio tests for detecting positive selection and application to primate lysozyme evolution. *Molecular Biology and Evolution* **15**:568–573.

———. 2002. Likelihood and Bayes estimation of ancestral population sizes in hominoids using data from multiple loci. *Genetics* **162**:1811–1823.

———. 2006. *Computational Molecular Evolution.* Oxford: Oxford University Press.

———. 2007. PAML 4: Phylogenetic analysis by maximum likelihood. *Molecular Biology and Evolution* **24**:1586–1591.

Yang, Z., and R. Nielsen. 2000. Estimating synonymous and nonsynonymous substitution rates under realistic evolutionary models. *Molecular Biology and Evolution* **17**:32–43.

Yi, X., Y. Liang, E. Huerta-Sanchez, et al. 2010. Sequencing of 50 human exomes reveals adaptation to high altitude. *Science* **329**:75–78.

Yoder, J. B., J. Stanton-Geddes, P. Zhou, R. Briskine, N. D. Young, and P. Tiffin. 2014. Genomic signature of adaptation to climate in *Medicago truncatula. Genetics* **196**:1263–1275.

Yu, Y., J. Dong, K. J. Liu, and L. Nakhleh. 2014. Maximum likelihood inference of reticulate evolutionary histories. *Proceedings of the National Academy of Sciences* **111**:16,448–16,453.

Z

Zanini, F., and R. A. Neher. 2012. FFPopSim: an efficient forward simulation package for the evolution of large populations. *Bioinformatics* **28**:3332–3333.

Zeng, K., Y.-X. Fu, S. Shi, and C.-I. Wu. 2006. Statistical tests for detecting positive selection by utilizing high-frequency variants. *Genetics* **174**:1431–1439.

Zeng, K., S. Shi, and C.-I. Wu. 2007. Compound tests for the detection of hitchhiking under positive selection. *Molecular Biology and Evolution* **24**:1898–1908.

Zhang, J., S. Kumar, and M. Nei. 1997. Small-sample tests of episodic adaptive evolution: a case study of primate lysozymes. *Molecular Biology and Evolution* **14**:1335–1338.

Zhang, J., R. Nielsen, and Z. Yang. 2005. Evaluation of an improved branch-site likelihood method for detecting positive selection at the molecular level. *Molecular Biology and Evolution* **22**:2472–2479.

Zhivotovsky, L. A., and M. W. Feldman. 1995. Microsatellite variability and genetic distances. *Proceedings of the National Academy of Sciences* **92**:11,549–11,552.

Zhu, Y., A. O. Bergland, J. González, and D. A. Petrov. 2012. Empirical validation of pooled whole genome population re-sequencing in *Drosophila melanogaster. PLoS ONE* **7**:e41901.

Zhu, Y. O., G. Sherlock, and D. A. Petrov. 2017. Extremely rare polymorphisms in *Saccharomyces cerevisiae* allow inference of the mutational spectrum. *PLoS Genetics* **13**:e1006455.

INDEX

Note: Material in figures, tables, or boxes is indicated by italic page numbers.

Printed in the USA/Agawam, MA
June 22, 2023

811901.009